Encyclopedia of Earth Sciences: Mining, Volcanology, Remote Sensing and Environmental Sciences

Volume II

Encyclopedia of Earth Sciences: Mining, Volcanology, Remote Sensing and Environmental Sciences
Volume II

Edited by **Joe Carry**

R CALLISTO
REFERENCE

New York

Published by Callisto Reference,
106 Park Avenue, Suite 200,
New York, NY 10016, USA
www.callistoreference.com

Encyclopedia of Earth Sciences: Mining, Volcanology,
Remote Sensing and Environmental Sciences
Volume II
Edited by Joe Carry

International Standard Book Number: 978-1-63239-233-6 (Hardback)

Printed in the United States of America.

Contents

Preface

The geographical and other environmental processes on Earth and the composition of the planet are of crucial significance in locating and utilizing its resources. This book is principally written for experts, geologists, civil engineers, mining engineers, and environmentalists. We are hopeful that the content will be used by students, and it will continue to be useful to them throughout their following professional and research careers. This does not mean that the book was printed solely keeping students in mind. Instead, from the point of view of experts in Environmental Science, it can be argued that this book contains more features than they will require in their primary studies or research. The book extensively covers topics related to mining, volcanology, remote sensing, and environmental sciences.

This book has been the outcome of endless efforts put in by authors and researchers on various issues and topics within the field. The book is a comprehensive collection of significant researches that are addressed in a variety of chapters. It will surely enhance the knowledge of the field among readers across the globe.

It is indeed an immense pleasure to thank our researchers and authors for their efforts to submit their piece of writing before the deadlines. Finally in the end, I would like to thank my family and colleagues who have been a great source of inspiration and support.

Editor

Part 1

Mining

Computer Aided Ore Body Modelling and Mine Valuation

Kaan Erarslan

Dumlupinar University, Mining Engineering Department, Kutahya
Turkey

1. Introduction

Mine valuation can be defined as the process of determining the worth of a specific mineral deposit and capability of making a return by a prospective investment (SME, 2005). Although the definition is very brief and compact, actually, it has a very wide content. Determination of worth of an underground asset requires a plenty of works to explain physical, structural and economical properties of it (Kennedy, 1990).

Underground assets are invisible bodies whose shapes, quality compositions and quantities are unknown. Geological explorations and investigations aim at determining all these unknowns (Sinclair and Blackwell, 2004). At the beginning of process, topographical and lithological data are gathered and a database is generated. Depth, thickness and grade changes, overburden structure, ore volume, shape and extensions, footwall and hanging wall properties are determined by various mathematical approaches using this database. All numerical estimations and visual supports help bringing out ore body model (Singer and Menzie, 2010).

The most concrete data to define shape, location, quality and quantity of an ore body is drill hole cores. GPS data is mostly used to draw topographical maps and surfaces. Additionally, underground maps such as thickness and grade contours are drawn as well. When topographical coordinates are combined with stratigraphical information, a three dimensional data set is handled. Eventually, after following several mathematical techniques, three-dimensional model of ore body can be obtained (Hustrulid and Kuchta, 2006). Beside physical ore model, quality composition should also be known. This is crucial because further engineering activities have an economical aspect. Mine design and production schedule is fairly related to both physical structure and quality composition of ore (Hartman, 1992).

Surveying data include three-dimensional components x, y, z (easting, northing, altitude/elevation) which enable surface modelling. Drill hole data including depth and layer information contribute to explain how geological structure is in the third dimension (Torries, 1998). Drill holes also carry the information of ore grade or calorific value. Geological interpretation of stratigraphical layers provides three-dimensional ore body model (Nieuwland, 2003).

Major instruments for computer aided mine valuation are;
- drill hole logs,
- contour maps,
- cross-sections,

- surfaces (topography, thickness, grade, etc.),
- surface sections,
- solid models (three dimensional models), volume and reserve estimation.

Some of outputs of mine valuation are visual and some of them are numerical. Visual outputs help researchers see drill hole sections, how ore body is, how it extents, how its shape looks like, how ore body and overburden relation is, how quality and thickness of deposit changes thru axes. Numerical outcomes are generally, area and volume reports, drill hole lengths, survey coordinate sets, composite calculations, economical assessments, etc (Torries, 1998).

In the last several decades, many approaches have been developed to clear up geological modelling problems (Agoston, 2005). The purpose is to estimate unknown values regarding limited data in hand. The methods such as geostatistics and neural networks have complicated mathematical and statistical bases and are utilised to model topography, ore body, and grade distribution (McKillup and Dyar, 2010). Iterative structure of computations makes computer use necessary for many of recent methods. There are various specialised and expert software to support geological and mining engineers. Ore body modelling with computer aid is faster and more reliable. New information addition is simple and updating is much quicker. Different scenarios can be studied and better decisions can be made. General work plan and flowchart of computer aided mine valuation is shown in Fig. 1.

Once ore deposit is visually and numerically modelled, next step is mine design and production scheduling. Engineering economics and optimisation concepts are considered at this stage. Optimisation and simulation methods such as graph theory, dynamic programming, linear and goal programming, mixed integer programming, moving cones, genetic algorithm and network analyses are main techniques that can be counted. Scope of optimisation and simulation is generally open pit limits and production (Erarslan and Celebi, 2001).

Purpose of geological database generation is to describe physical and quality properties of ore bodies and preparation economical of database. Limited data should be optimally utilised and unknowns should be cleared. Last decades have shown great improvements in ore body visualisation and figuring. Buried underground body with so many unknowns can be visualised with animations and virtual reality applications by the help of computer software. Ore body is represented as a three dimensional entity laying under topography.

Modelling approaches utilise topographical information and drill hole database. Frequently and representatively recorded GPS values and aerial views help to visualise photo realistic views of topography. Next step is to define ore shape in space. Drill holes values are the most crucial and critical data set for this purpose. Problem at this stage is to determine what happens between sample points. Answer of the question requires two-dimensional and three-dimensional representations of ore body. This may look easy for regular structures. However, in most cases the picture is just reverse. Geological interpretation is highly needed. Computer systems help interpreters at this point. Computer programs, in general trend, provide two alternatives to interpreters for 3D determinations; geological parallel cross-sections and block models.

In this chapter, the basic ideas and some mathematical approaches of computer aided mine valuation are given. How GPS and drill hole data are used to figure out an entire ore body model in visual and numerical aspects are explained.

```
                    ┌─────────────────────────┐
                    │   GPS & drill hole data  │
                    └─────────────────────────┘
                                 │
                                 ▼
                ┌───────────────────────────────────┐
                │  Data base generation & compositing │
                └───────────────────────────────────┘
                                 │
                                 ▼
            ┌─────────────────────────────────────────┐
            │  Gridding, triangular, rectangular mesh   │
            │   generation and assignment methods       │
            └─────────────────────────────────────────┘
```

2D visual & numerical processes	3D visual & numerical processes
Contour maps - topography - thickness - grade - ore top & bottom	**3D Surfaces** - topography - thickness - grade - ore top & bottom - ore cross-sections
- Triangular grid - Polygonal grid - Rectangular grid	**3D Models** - ore body blocks - 3D geo-sections
Volume & Reserve - isopach map - triangles - polygons	**Volume & Reserve** - ore body blocks - 3D geo-sections

```
                ┌───────────────────────────────┐
                │   Mine Design &               │
                │   Production Scheduling        │
                └───────────────────────────────┘
```

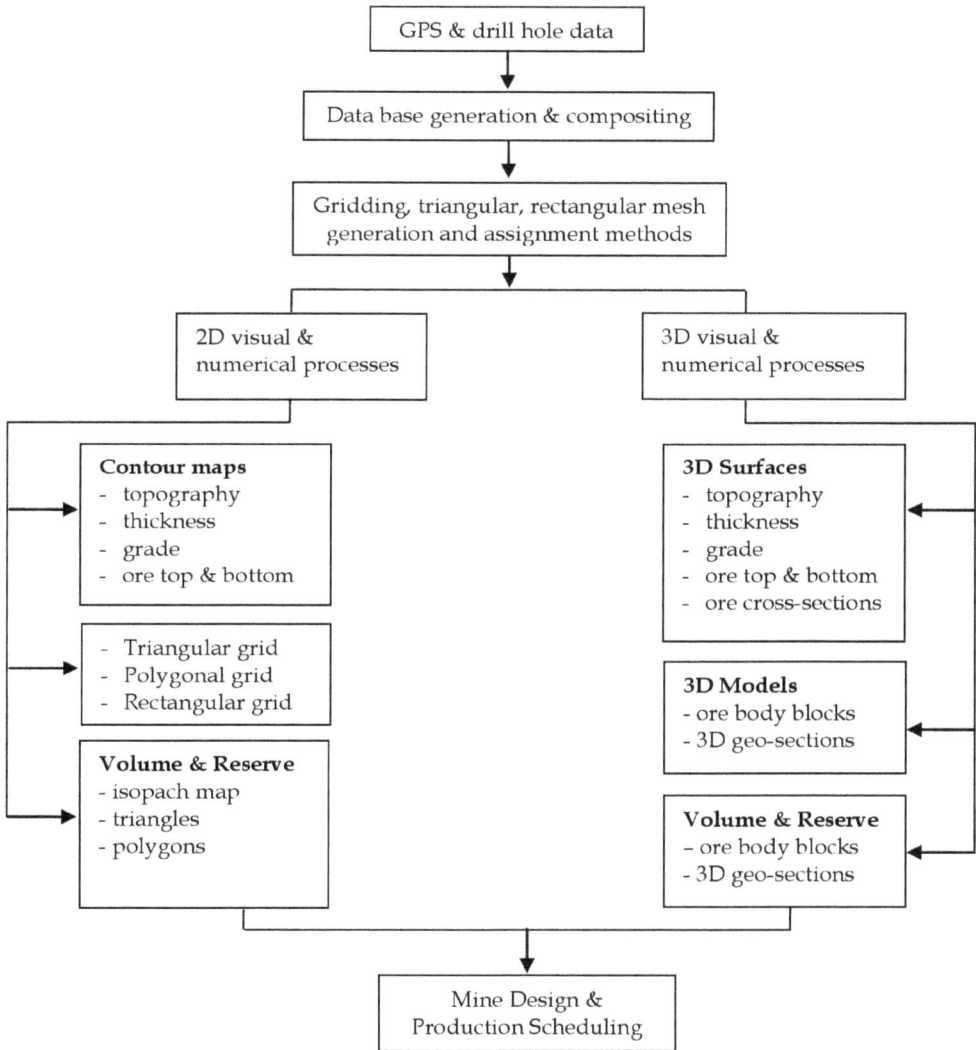

Fig. 1. Mine valuation work flow.

2. Database processes

Database is the base of every further study. Health of projects entirely depends on health of samples. GPS records are very crucial to model topography. However, the most concrete data that can be taken from field is drill hole cores. Mine valuation process starts with database building. It is composed of spatial coordinates of drill holes, geological formations that they intersect, depths of formations and their assay values. Surface determinations are, almost entirely related to GPS data.

2.1 Database structure

Input material for mine valuation systems is mainly topographical surveying data and drill hole cores. It is possible to categorise this database as; i- collar data, ii- survey data, iii- stratigraphical/lithological data, iv- assay values/grade analyses.

Collar data keeps (x,y,z) coordinates of drill ring. Survey database may include also physical coordinates of drill holes, depth information, azimuth and bearing angles. Stratigraphy database contains the information of geological formations through each hole. Assay database has the quality values of ore formations. Drill hole log/stamp is like an identity card of them (Fig. 2).

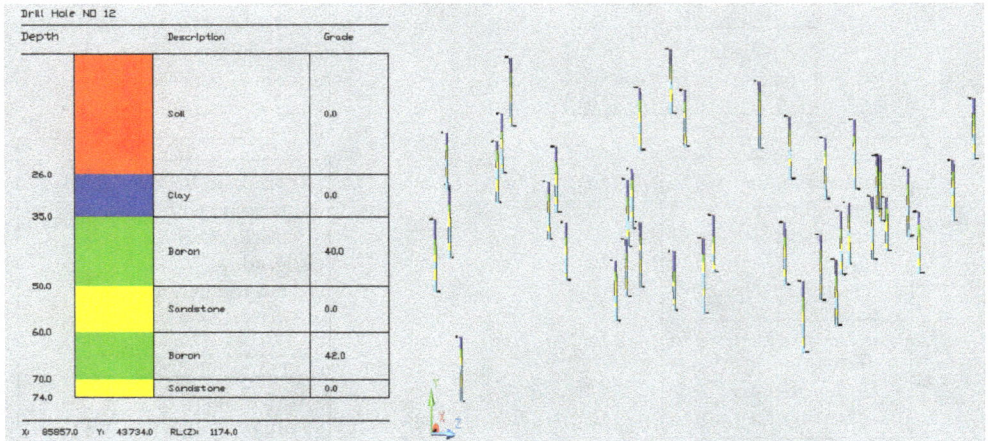

Fig. 2. A single drill hole log and drill holes in three dimensions (Erarslan, 2007).

Surveying data may also include three-dimensional components x, y, z (easting, northing, altitude/elevation) which enable surface modelling. However, GPS data that have been taken with frequent intervals enables photo realistic models. If data can take representatively, topography model looks like the field itself. Drill hole data including depth and layer information contribute to explain how its shape is in the third dimension. Drill holes also carry the information of ore grade or calorific value. Geological interpretation of stratigraphical layers provides three-dimensional ore body model.

2.2 Drill hole compositing

Drill holes cut intersect downwards successive layers. Mechanical properties of waste layers and quality values of ore layers are determined by applying several tests on hole cores. Generally, mineral formations are not monolithic and single piece bodies. Inter-burden layers may intersect mass or mineralisation may occur with waste layers in alternating forms. In other words, valuable mineral layers may exist in different thickness and quality amounts. During numerical calculations, a single thickness and grade value may be needed. Some classical reserve estimation techniques such as triangulation and polygon methods need composite values. In that case, what is the net thickness of valuable part and its quality? Here, compositing computation is applied to have total thickness and a unique grade (quality) value for each drill hole. Regarding a cut-off grade, ore thicknesses are summed up to give ore thickness. On the other hand, grade value is the thickness-weighted average (Hustrulid and Kuchta, 2006).

$$t_{total} = \sum_{i=1}^{n} t_i \tag{1}$$

$$g_{comp} = \frac{\sum_{i=1}^{n} g_i t_i}{\sum_{i=1}^{n} t_i} \tag{2}$$

where, t_{total} is total ore thickness, g_{comp} is composited grade, t_i is thickness and g_i is grade of ith core piece within n core pieces. After obtaining a single thickness and grade a value for each drill holes, volumes and reserve amounts in superimposed triangles or polygons can be calculated.

Another type of compositing is called as bench or level compositing. Ore field under investigation is divided into parallel horizontal levels and parts of drill holes corresponding to those levels are composited instead of compositing entire hole (Fig. 3).

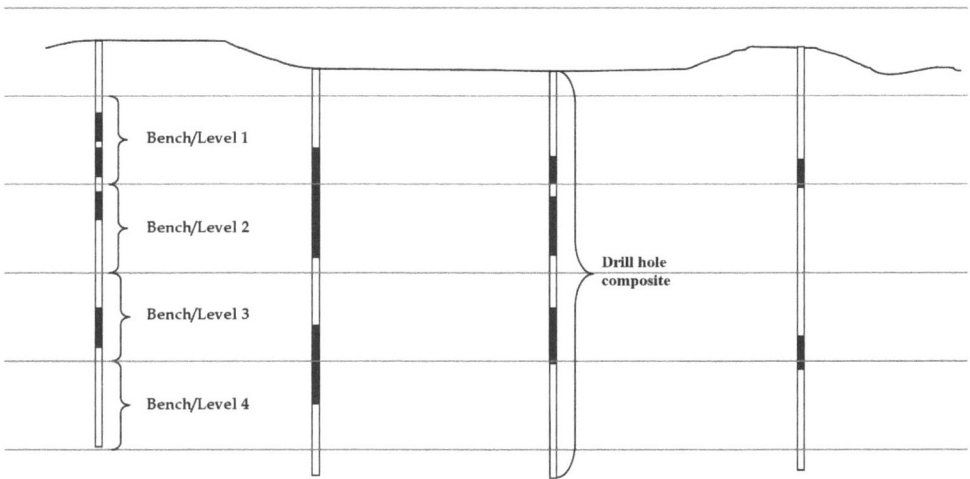

Fig. 3. Bench/Level Compositing

Level elevations may be same with bench elevations of prospective open pit. Such a compositing process gives ability and base to further block modeling and bench/level reserve estimation. Bench reserves help for production planning in further stages.

3. Data extension on a network

Drill hole data looks like spotting points in bird's eye view. Problem is to know how structure changes between these sample points. Sample points, where several parameters such as thickness and grade are already known, should be used to estimate these parameters at points where no sampling is available. This process can be named as data extension. Data can be extended thru two-dimensional planes or three-dimensional space. During extension process, square or rectangular grids, wireframes are imposed onto area. Triangulation is another type of artificial net, applied on field. Then, extension methods such as inverse distance square, geostatistics and artificial neural networks are applied.

3.1 Gridding

Although there are several methods to figure underground treasures such as aerial and seismic surveys, drill holes are still the most reliable and decisive data givers. In plan view, they look as irregularly distributed point data.

Grid is a regular network formed by triangle, square or rectangles generally. It is the result of a regular mesh need. Irregularly distributed sample points get a regular form if node points are assigned parameter values such as thickness and grade. In other words, thickness and grade values are estimated at node points, which results in a regular data structure (Fig. 4). Plenty of node points are generated by superimposing a wireframe grid onto a field (Knudsen, 1990, Parker, 1990).

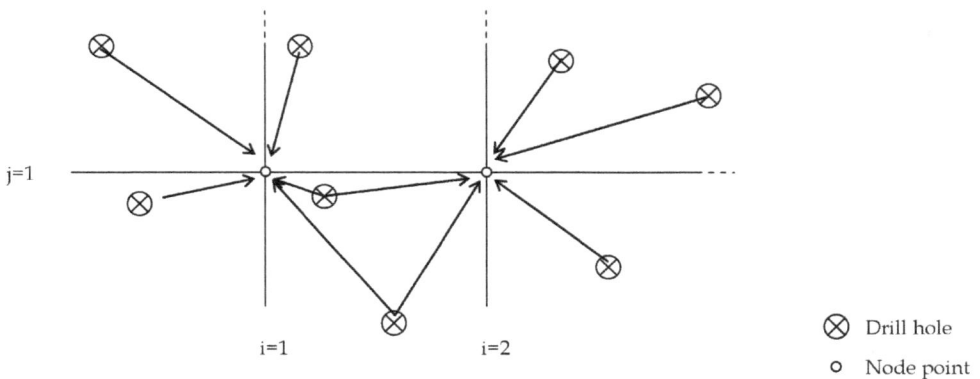

Fig. 4. Assignment to node points.

Irregularly distributed sample values (drill holes) are extended/distributed to field by several approaches;
a.   Classical methods (triangle, polygon)
b.   Inverse distance methods
c.   Geostatistical methods
d.   Artificial intelligence (neural networks)

These methods are generally employed to assign node values by using drill holes. Nodes are aligned through triangular or rectangular grids/networks (Fig. 5, Fig. 6).

Calculation of coordinates of grid nodes is a simple mathematical process. Let number of nodes in x direction is n and in y direction is m. Total number of nodes on grid is mxn. If each point is represented as $p(i,j \mid i=1,2,...,n\ j=1,2,...,m)$, then $p(i,j)$'s are function of x_i and y_i coordinates;

$$p(i,j) = f(x_i, y_j) \tag{3}$$

North (Y)

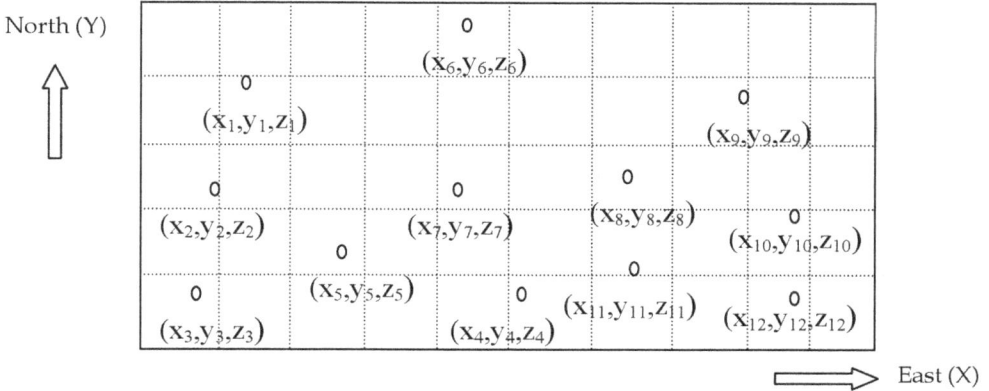

Fig. 5. Grid superimposed onto irregularly distributed drill holes.

Fig. 6. Polygon and triangle generation.

where, x_i is east coordinate of ith column and y_j is north coordinate of jth row. Then x_i and y_j can be calculated as;

$$x_i = s_x + i \cdot \Delta x \tag{4}$$

$$y_j = s_y + j \cdot \Delta y \tag{5}$$

where, Δx and Δy are spacing thru x and y lines and s_x and s_y are origin coordinates for east and north respectively. After computing x and y coordinates of all nodes, next step is calculating the third coordinate which may be thickness, grade, elevation or anything else. Third coordinate/parameter comes from surrounding sample points by carrying/extending drill hole parameters to these node points. For this purpose, there are several methods such as inverse distance square, geostatistics and neural networks.

Once a grid system is built, parametric values like thickness and grade are assigned. Then by means of rotation around an axis, third dimension can be sensed (Fig. 7).

3.2 Data extension in 2D

Data extension is a result of need to interpret how ore body behaves between present samples. Major two-dimensional extension applications of drill holes are triangular and polygonal area generation and of course, contour maps. The purpose is to carry pointwise data (drill hole) values into areas by certain mathematical methods and acceptations.

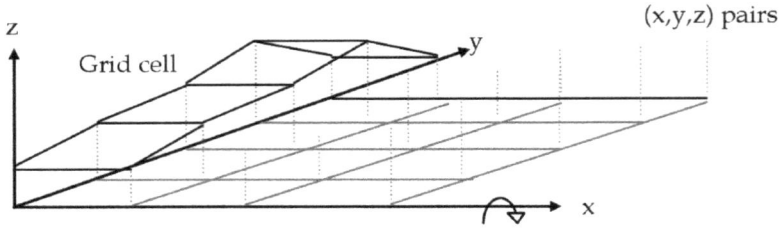

Fig. 7. Rotated grid system referring origin (s_x, s_y).

Additionally, in order to get a regular data structure, a grid is superimposed onto area. Each node point is assigned parametric values after computations. By this way, each grid node may act as a sample point.

3.2.1 Triangles and triangulation

Triangles are generated by joining drill holes. Area enclosed between three adjacent drill holes is calculated. Next step is to calculate volume $(V - m^3)$ by multiplying triangle area $(A - m^2)$ with average thickness $(t - m)$ of three drill hole composited thicknesses.

$$V = A \cdot t_{ave} \tag{6}$$

$$t_{ave} = \frac{t_1 + t_2 + t_3}{3} \tag{7}$$

It is also possible to find triangular reserve $(R - ton)$ if tonnage factor $(f - ton/m^3)$ and thickness weighted average grade $(g - \%)$ of three composited grades is known.

$$R = V \cdot f \cdot g_{ave} \tag{8}$$

$$g_{ave} = \frac{g_1 \cdot t_1 + g_2 \cdot t_2 + g_3 \cdot t_3}{t_1 + t_2 + t_3} \tag{9}$$

Triangle method is a classical and rough estimation one with certain acceptations and could be employed only for horizontally and regularly bedded fields. Sedimentary type fields such as coal may be applied. Each triangle corner is a drill hole and composite thickness and grade values are used for computations (Fig. 8).

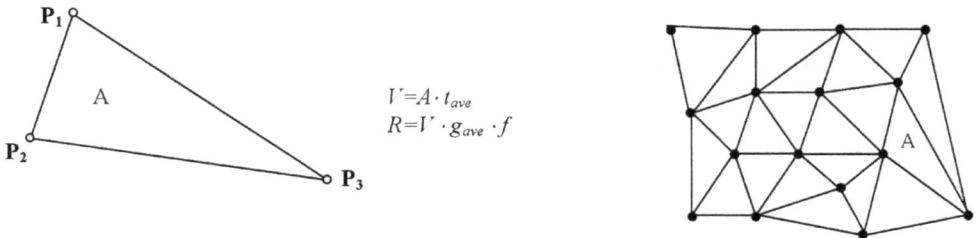

$$V = A \cdot t_{ave}$$
$$R = V \cdot g_{ave} \cdot f$$

Fig. 8. Triangular grid by drill holes and local reserve.

Triangulation method can also be applied to produce sub-triangles and find coordinates of sub-triangles' corners. DeLaunay, Voronoi and Thiessen approaches can be employed during triangulation (Sen, 2009). New triangular network is assigned parametric values by several methods such as inverse distance square, geostatistics and neural network. It can be employed not only for reserve estimation but also for surface drawing.

During node assignments, a question may arise; which sample points should be included in calculations? There should be a limitation for drill holes and answer is *radius of influence*. Samples within radius of influence area are used in calculation of weighted averages. Radius of influence may be result of trial-error attempts or may be geostatistical variogram *range*, which will be explained later.

3.2.2 Polygonal gridding

Polygon method is based on linear influence area concept. They are geometrically defined by the perpendicular bisectors of the lines between all points (Sen, 2009). Influence distances between drill holes are at their mid-points (Fig. 9).

Therefore, joint points of lines perpendicular to mid-points form a polygon and polygonal area can be calculated. Thickness and grade values within the area are assumed same with that drill hole's values. Volume of polygon is multiplication of area and thickness and reserve is product of volume, specific gravity and grade.

4. Assignment methods

Grid or mesh generation is followed by assignment stage. Nodes in 2D and blocks in 3D are assigned parametric values. This structure is the base for 2D and 3D models. 2D and 3D ore body modelling can be accomplished by several approaches. Joining cross-sections taken thru ore body and block modelling techniques are mostly used methods. The models do not only give the shape of ore body but also provide volume and reserve amount. Either triangular or rectangular mesh generation requires an assignment method. Here, commonly used mathematical approaches are;

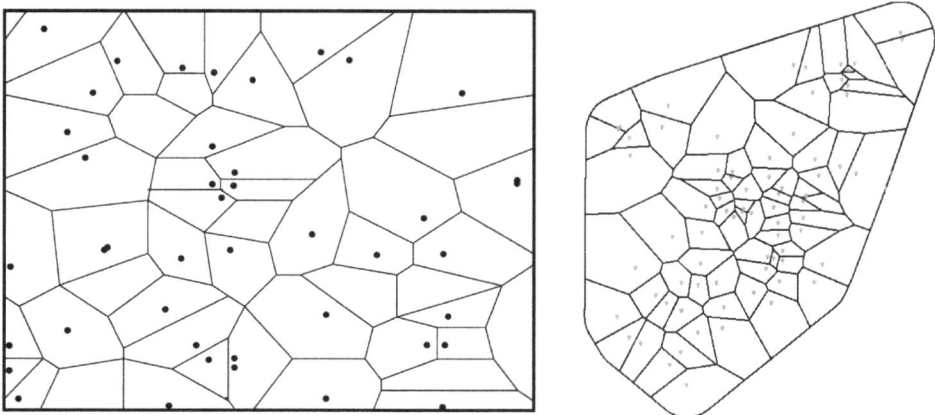

Fig. 9. Polygonalisation (Sen, 2009) and Voronoi polygons (BGIS, 2011).

i.   Inverse distance,
ii. Geostatistics,
iii. Artificial Neural Networks.

4.1 Inverse distance method

Inverse distance method is actually a weighted average method. However, weights are calculated inversely. As distance of a sample point (drill hole) to an assignment point (node) gets more distant, its contribution on the average result gets less. In other words, closer sample point means more effect on assigned value. The basic idea is providing effect of a drill hole to point on which assignment is to be performed, is inversely related with the distance between them. This is realised by taking weighted average of parametric values by distances as shown below:

$$z(i,j) = \frac{\sum_{k=1}^{K} \frac{z_k}{(\Delta_{(i,j)}^{k})^{\alpha}}}{\sum_{k=1}^{K} \frac{1}{(\Delta_{(i,j)}^{k})^{\alpha}}} \tag{10}$$

where,

$z(i,j)$	= assigned value (i.e. grade, thickness) at node point on i^{th} column and j^{th} row.
z_k	= parameter value carried by k^{th} drill hole (sample).
$\Delta^k_{(i,j)}$	= distance between k^{th} drill hole and (i,j) node.
α	= power of inverse distance process (α=2 in general and method is called as inverse distance square).

Distance between drill hole and node points can be calculated simply by;

$$\Delta_{(i,j)}^{k} = \sqrt{(\delta x(i,j) - D_x^n)^2 - (\delta y(i,j) - D_y^n)^2} \tag{11}$$

where,

$\delta x(i,j)$	= x co-ordinate of node (i,j).
D^n_x	= x co-ordinate of n'th drill hole.
$\delta y(i,j)$	= y co-ordinate of node (i,j).
D^n_y	= y co-ordinate of n'th drill hole.

This method has an acceptation that geological structure has a linear behaviour. In case of sudden thickness or grade changes, inverse distance method may fail in determining a healthy result. On the other hand, its application is not only limited to 2D calculations but it can also be employed in 3D computations.

4.2 Geostatistical methods

Basic concept of geostatistics is regional variability of parameters (Matheron, 1971, 1963; Krige, 1984). Not only deterministic and descriptive manner but also probability and statistics take place in calculations with geostatistics (Mallet, 2002). If there is mathematically explainable structure of an ore body within a limited region, it can be possible to state that behaviour by equations and use them for estimation (Sarma, 2009). When variability cannot be formulated by mathematical models then it is decided that sample values show random behaviour and probabilistic methods of statistics can be applied there.

Geostatistics do not only consider distances between sample points and assignment points but also their position and direction with each other (Webster and Oliver, 2007). Additionally, during estimation, not only variances between sample points and assignment points are taken into account but also variances of samples within themselves are regarded (David, 1977, Davis, 1973). This means, geostatistics regards not only distance relation between sample points and node point but also variation between them in relation with distance and direction (David, 1988). Main two stages of geostatistical process are variogram modelling and kriging.

4.2.1 Variogram modelling

Main tool of geostatistics is variogram model (Diggle and Ribeiro, 2007). It can be defined as the graph representing the relation between distance and parametric variance in a certain direction (Deutsch and Journel, 1998). In estimation stage, kriging interpolation, geostatistics utilises co-variances between sample points and point/area/volume on which assignment is to be performed, and co-variances between sample points affecting that point for extending a value (Kitanidis, 2003).

Initially, a variogram model representing the relation between distance and variance is defined. Variance is;
where,

$$\gamma(h) = \frac{\sum_{n=1}^{N} (z - z')^2}{2n} \tag{12}$$

$\gamma(h)$ = variance of sample pairs located h distance (lag) apart.
z = value at sample z.
z' = value at sample, h distance (lag) apart from z.
n = total number of pairs located h distance (lag) apart.

For each different h distance (lag), a variance $\gamma(h)$ is calculated (Fig. 10). According to distribution of h-$\gamma(h)$ pairs, a mathematical model is tried to fit. General trend is to use one of predefined variogram models such as, Spherical, Exponential, Gaussian, Linear, etc., (Davis, 1973; Krige, 1978).

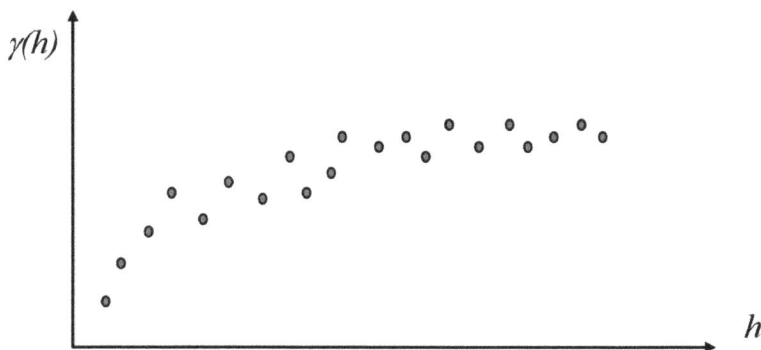

Fig. 10. Variance values marked for h distanced (lag) sample pairs.

For this purpose, sample pairs are found for a certain trigonometric direction $\theta°$ and lag distances h, $2h$, $3h$,..., ∂h. As it may not be possible to find pairs having direction and distance conditions absolutely, a tolerance distance $\pm\Delta h$ and direction (angle) $\pm\Delta\theta°$ is considered. Parameters, such as thickness and grade carried by sample points, are used in the formula and variance value is calculated. After marking variances on h-γ graph, *experimental variogram* is drawn (Leuangthong and Deutsch, 2004). Then, variogram is a graph explaining variance versus distance relation.

Experimental variogram points show a trend and needs interpretation for deciding which model best fits. Spotted points may have a mathematical determination in equation form, which is also called *variogram model*. It is similar to regression modelling. (Fig. 11).

In the graph C_o is called as *nugget effect*, C is *sill* and a is *range*. Nugget is the variance value at zero distance. Normally it has to be zero however due to sampling problems, tolerances in pairing and structural inconsistencies, nugget gets over zero. Best graph and its mathematical equation (model) for h-$\gamma(h)$ pairs are decided by visual interpretation regarding several predefined models such as spherical model (Matheron), exponential model, De-Wisjian, model, linear model.

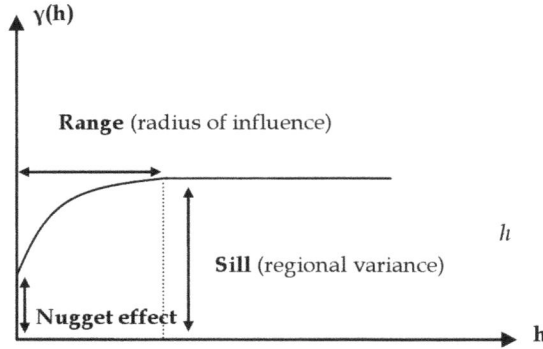

Fig. 11. Variogram model fitted for spotted points showing distance-variance; h-$\gamma(h)$ relation.

Models are classed as models with *sill* and *without sill*. Sill is defined as the platform where variance gets parallel to h axis. Distance where graph reaches at sill level is called as *range*. Range may also be considered as *radius of influence*. Models are given below;

Spherical Model

$$\gamma(h) = \begin{cases} c\left(\dfrac{3h}{2a} - \dfrac{h^3}{a^3}\right) + c_0 & h \le a \\ c + c_0 & h > a \end{cases} \tag{13}$$

Linear Model

$$\gamma(h) = a \cdot h \tag{14}$$

Exponential Model

$$\gamma(h) = c\left(1 - e^{-h/a}\right) \tag{15}$$

where,

c = Sill value (constant value where variances get parallel to lag axis).

c_0 = Nugget effect (variance where distance is zero).

a = Range (distance where variance reaches at sill value)

Another concept in geostatistics is direction and variability of variograms according to directions. If variogram models developed for different directions can be accepted tolerably identical to each other, then ore body is called *isotropic* and *anisotropic* vice versa. Same variogram model can be used during kriging estimations for all directions if existence of an isotropy is decided.

4.2.2 Kriging

In order to extend sample data and use it for the prediction of unknown values, kriging is applied. Kriging is an extension and estimation method where parameters are assigned to assignment points by means of surrounding sample points. Aim is to calculate a weight a_n for each sample point.

Aim of developing a variogram model is to state variance as a function of distance and determination of mathematical equation that will be used to compute covariance matrix terms that will be used in kriging stage. Kriging process, developed by Krige (1966), is an estimation method regarding covariances between assignment point-sample points and within sample points mutually as well. In equation and matrix form after derivations, respectively;

$$z^*(i,j) = \sum_{k=1}^{n} w_k \cdot z(x_k) \tag{16}$$

$$\sum_{k=1}^{n} w_k = 1 \tag{17}$$

where,

$z^*(i,j)$ = assigned value to (i,j) node point regarding surrounding n sample points (drill holes)

$z(x_k)$ = parametric value of x_k sample point

w_k = weight of sample point k.

Simple representation of matrices is below:

$$[\sigma_{xx}] \cdot [w_x] = [\sigma_{vx}] \tag{18}$$

where,

$[\sigma_{xx}]$ = matrix including variances between samples points mutually

$[\sigma_{vx}]$ = matrix including variances between assignment point and sample points.

$[w_x]$ = weight matrix

Weight matrix is computed and weights are used to estimate unknown parameter at assignment point. σ variance values in matrices are calculated by putting distance amounts between sample points mutually (Δ_{xx}) and sample-to-assignment points (Δ_{vx}) into h term in vaiogram model.

Geostatistics has a wide application field in geological modelling. Main problem of the method is visual evaluation and interpretation of experimental variogram during model

development, which may be controversial and relative. Number of samples should be an acceptable amount; may be twelve and mostly over thirty (Sen, 2007).
One of the advantages of geostatistics is error estimation ability;

$$e^2 = (Z^* - Z)^2 \tag{19}$$

where, e^2 is estimation error (variance), Z^* is estimated value, Z is actual value (Sinclair and Blackwell, 2004).

4.3 Artificial neural network method

One of most recent methods used to extend sample data is artificial neural networks (Rabuñal and Dorado, 2006; Haupt, et.al., 2008). Initially, the neural system is trained to learn the structure of samples. Each sample is assigned a random weight and total error is recorded. Iteration by iteration, error is distributed to approximate actual sample values. An acceptable error level is succeeded and weights of sample points are calculated (Fig 12).

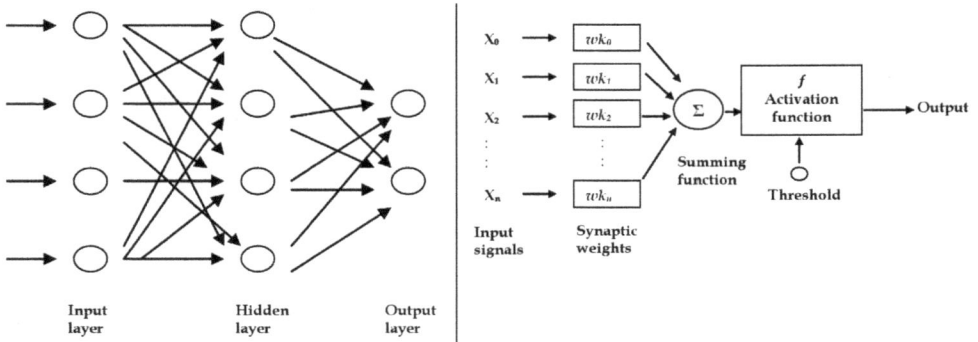

Fig. 12. Typical artificial neural network structures.

Then accordingly, a value is assigned to any desired location through area or space. In formulated form;

$$y = w_1 x_1 + w_2 x_2 + \ldots\ldots + w_n x_n = \sum_{j=1}^{n} w_j x_j \tag{20}$$

where, y is value to be assigned, w_j is the weight assigned to j'th sample and x_j is value of j'th sample. As explained earlier, calculation of w_j's requires an iterative process, which is called as $training$. A random weight assignment is followed by measurement of error between present input and desired output. An error above acceptable limit causes propagation of error through weights. During "squashing" the limit, an activation function, called as sigmoidal-function is employed;

$$f(x) = \frac{1}{1 + e^{-x}} \tag{21}$$

New weights are tested if they fulfil condition of error limit. When the system fits error level, weights can be used for assignment (Fausett, 1994).

Total error gets less and less at each iteration and process continues as expected error level is reached. In order to achieve minimum error, hundred thousands, even millions of iterations may be needed. At that stage, it is decided that system has learned database structure and ready for further interpretations and estimations. Any more, w_j coefficients are available for assignments (Freeman and Skapura, 1991; Dowla and Rogers, 1995).

5. Two-dimensional visual and numerical processes

After generation of triangular and rectangular nets, they can be used for different purposes such as volume and reserve estimation, surface representation and forming a contour database. Another very basic visual instrument to show behaviour of ore through area is contour map. Sample data is extended to area as isolines.

5.1 Contour maps

Contour maps are generally based on triangulated or gridded networks, which are superimposed onto mine area to represent topography in computerised environment. Node points on network wire are assigned several values such as topographical elevation, composited thickness and grade, ore seam upper or bottom surface elevations, etc. (Watson, 1992). Main idea and purpose is to estimate several parametric values at node points using sample values obtained by drill holes. Here, inverse distance square, geostatistics and artificial neural networks are applied to assign node values. Thereafter topographical, thickness and grade contour maps can be drawn (Fig. 8). Bézier curves, B-Splines and Cubic Splines are primarily applied mathematical techniques (Mortenson, 1999; Comnino, 2006; Foley et.al., 1990; Vince, 2005; Vince 2010). Points on grid nodes with (x,y) coordinates get third dimension coordinate as well after assigning a parametric value as z coordinate such as thickness, grade, etc. Then, point data set is ready for contour (curve) fitting.

Bézier curves employ *Bernstein polynomials* (Vince, 2010). Its general equation is given below:

$$B_i^n(t) = \binom{n}{i} t^i (1-t)^{n-1} \tag{22}$$

$$\binom{n}{i} = \frac{n!}{(n-i)!i!} \tag{23}$$

where $\binom{n}{i}$ is shorthand for the number of selections of i different items from n distinguishable items when the order of selection is ignored and the coordinates of any point on the circumference in terms of some parameter t. Bézier curves may also be in *quadratic* and *cubic Bernstein polynomials* form (Vince, 2010).

B-splines also use polynomials to form a curve segment. However, B-splines use a series of control points determining the curve's local geometry. This feature provides and ensures that only a small portion of the curve is changed with movement of a control point (Vince, 2010). Splines can be classed as *uniform* and *non-uniform* and also *rational* and *non-rational*.

Cubic splines enable continuity between segments, which puts them one-step away from simple quadratics. Cubic splines can also be considered as *piecewise polynomials* (Fig. 13).

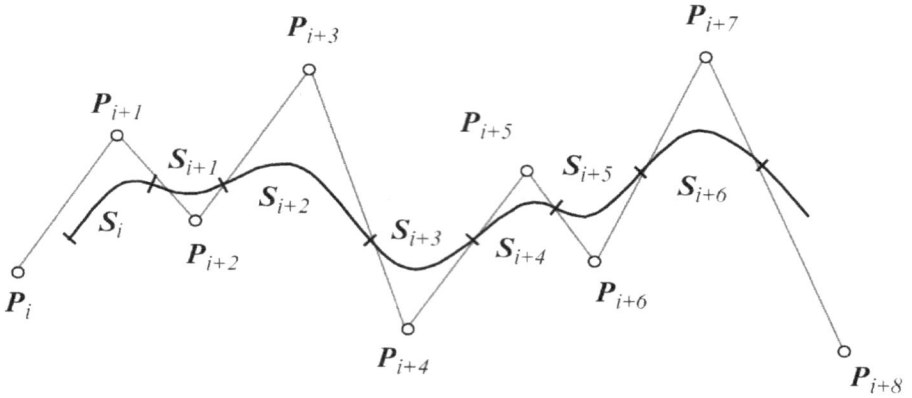

Fig. 13. Construction of a uniform non-rational B-Spline curve (Vince, 2010).

Here, curve segment S_i is under influence of points P_i, P_{i+1}, P_{i+2}, P_{i+3}, and curve segment S_{i+1} is related to points P_{i+1}, P_{i+2}, P_{i+3}, P_{i+4}. There exist $(m+1)$ control points and $(m-2)$ curve segments. Hence, a particular segment $S_i(t)$ of a B-spline curve is defined by

$$S_i(t) = \sum_{r=0}^{3} P_{i+r} B_r(t) \quad \text{for} \quad [0 \le t \le 1] \tag{24}$$

where,

$$B_0(t) = \frac{-t^3 + 3t^2 - 3t + 1}{6} = \frac{(1-t)^3}{6}$$

$$B_1(t) = \frac{3t^3 - 6t^2 + 4}{6}$$

$$B_2(t) = \frac{-3t^3 + 3t^2 + 3t + 1}{6} \tag{25}$$

$$B_3(t) = \frac{t^3}{6}.$$

These are the basic functions of cubic splines (Rogers and Adams, 1990). Finally, topography, ore seam upper and bottom contours, thickness (isopach), grade (isograde), etc., can be drawn (Fig. 14).
Curve fitting methods are modified for also surface fitting in three-dimensional environments. Eventual aim of computer systems is to display ore body in 3D space.

6. Three dimensional operations

3D ore body modelling can be accomplished by several approaches. Joining cross-sections taken thru ore body and block modelling techniques are mostly used methods. The models do not only give the shape of ore body but also provide volume and reserve amount.
Drill hole cores explain researchers definite coordinates of underground layers. Topographical information is combined with depths to yield three-dimensional coordinates.

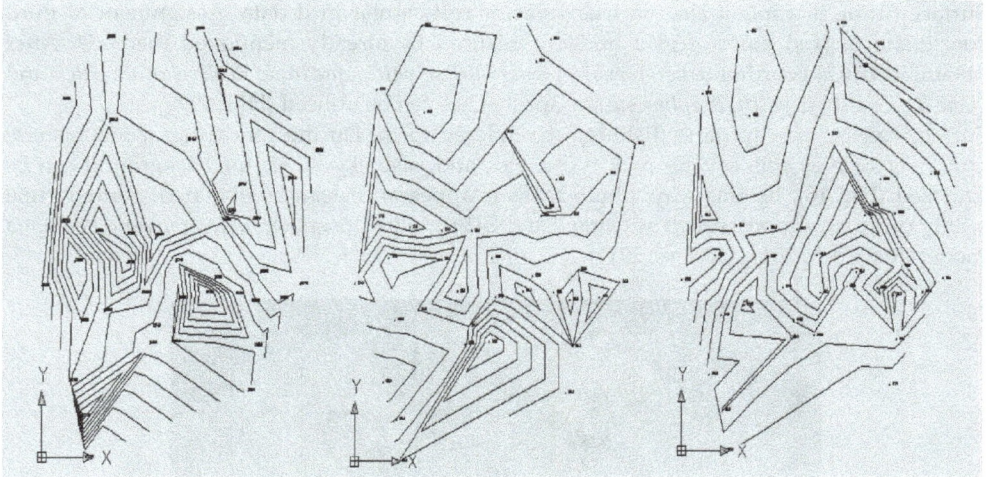

Fig. 14. Contour maps; topography, thickness and grade maps, respectively (Erarslan, 2007).

Major instruments for 3D computer aided mine valuation are;
- surfaces
- cross-sections,
- solid models (three-dimensional models).

6.1 Surfaces

Surfaces are important visual outputs of computer aided mine valuation systems. Surface can be prepared for any parameter assigned to grid system. (x,y) coordinate pairs of each node can be calculated easily. Fig. 15 shows how third dimension is sensed after rotation of 2D grid around y-axis.

Third coordinate may be topographical elevation, thickness, grade or anything else. Surface will be named according to its third parameter such as topographical surface, thickness surface, grade surface, etc. 3D visualisation of surfaces helps researcher imagine parametric changes.

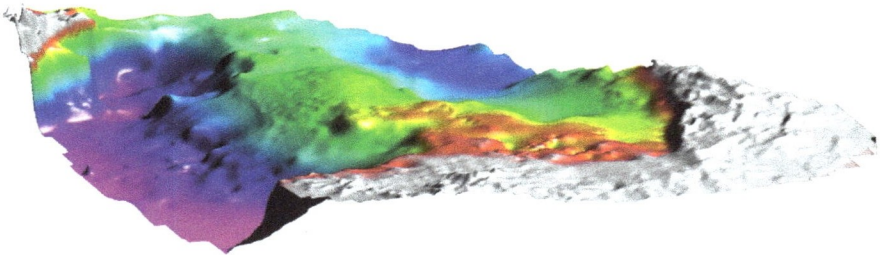

Fig. 15. Three-dimensional surface model (Pirsa, 2011).

Surface fitting is applied also on triangular or rectangular grid data. Assignment of third coordinate to (x,y) pairs of grid nodes is realised by already mentioned methods. After obtaining (x,y,z) coordinate set, *bicubic, planar surface patch, quadratic Bézier surface patch* and *cubic Bézier surface patch, B-Spline surface* approaches can be utilised (Hill, 1990).

Surface can be in wireframe (fishnet) or rendered form. During rendering action, several materials such as soil, several rock types, etc., and, also sky views for background can be imposed (Fig. 16). By this way, photo realistic appearances can be obtained. Some virtual reality program supports such as *OpenGl* and *GlView* allow users walk or fly over developed model (RealTech, 2011; GlView, 2011).

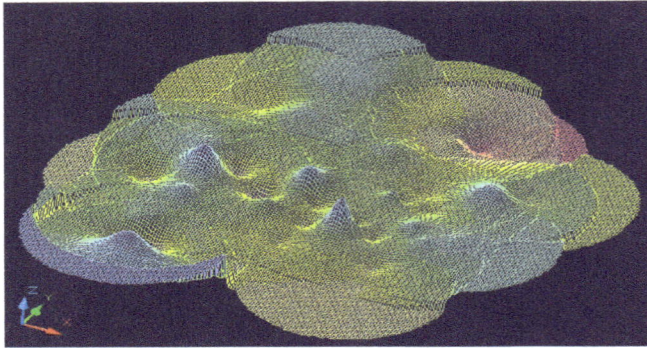

i)

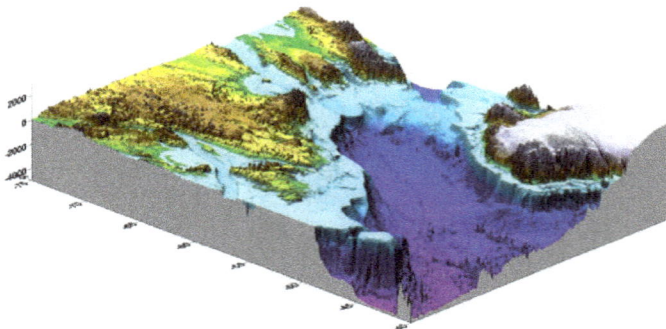

ii)

Fig. 16. 3D surface model; i- wireframe appearance (Erarslan, 2003), ii- rendered with texture (Golden, 2011).

6.2 Geological parallel cross-sections

Drill hole sections aligned thru or near a cross-section line are used to draw geological section view. Softwares show users hole lithologies and let them make interpretation on them. Closed polygons regarding stratigraphy determine ore body cross-section thru that section line (Figure 17).

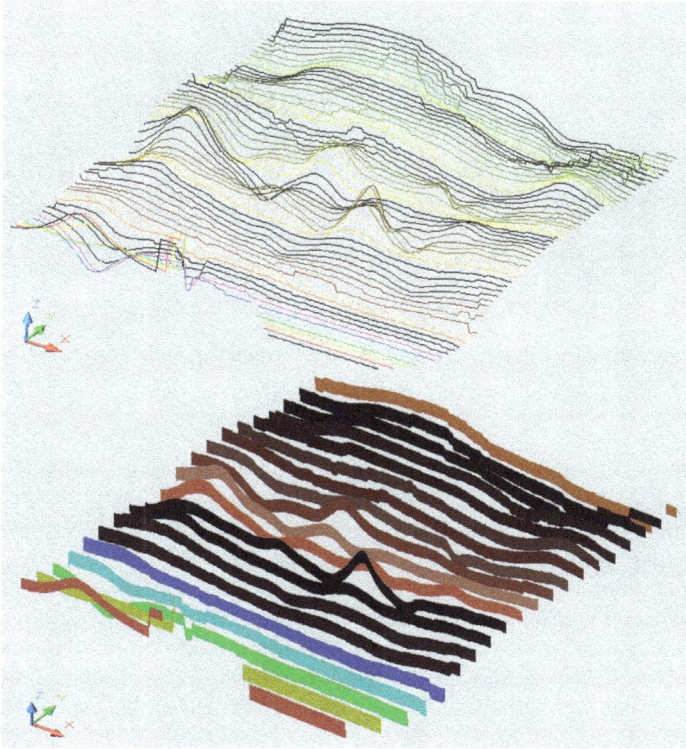

Fig. 17. Cross-section interpretation thru a section line (Erarslan, 2007).

6.3 Three-dimensional ore body models

Three dimensional ore body cross-sections do not only give an idea about structure of deposit but also prepare a base for three-dimensional ore model. Successive and parallel cross-sections are interpreted along ore body. Software can later on combine parallel cross-sections to form a 3D ore body model. Some software is also capable of combining cross-section, which are not parallel, and maybe intersecting each other. This type of sections are visualised on so-called fence diagrams (Fig. 18). Parallel cross-sections of ore body are fundamental tools for 3D modelling. They can be horizontal or vertical (Fig. 18).

Computer programs help researchers display sections and join them to build a 3D appearance. Each closed polygon enables calculation of section area and by means of average areas method, volume of ore body can be calculated. Gauss-Green area formulation can be employed to estimate sectional area:

$$A = \frac{\sum_{i=1}^{n-1}(y_{i+1} - y_{i-1})x_i}{2} \tag{26}$$

where, A is polygonal area, x_i and y_i terms are x and y coordinates of polygon nodes.

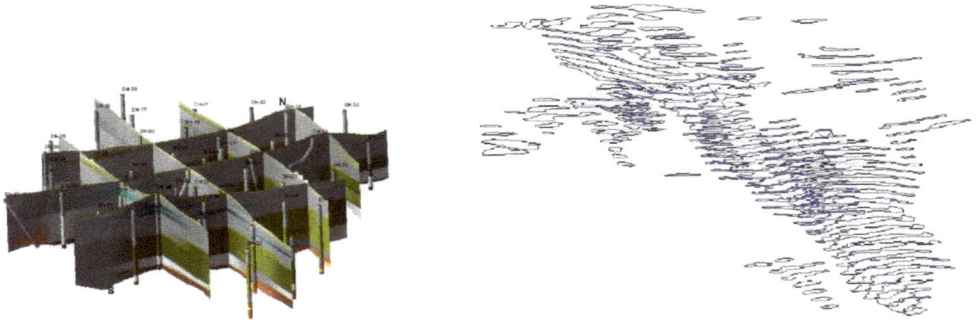

Fig. 18. Fence diagram (Rockworks, 2011) and wireframes through Zambujal Ore body (Lundin, 2007).

6.4 Block models

Another very frequently used ore body modelling method is block models. Block models could be thought as three-dimensional forms of 2D gridding. Field is divided into blocks, physical properties and quality composition are represented by this geometric form (Fig. 19). Centre and corner coordinates of each block can be calculated as (x,y,z) data sets. Reference/origin point is at top or bottom. In some cases, prospective open pit benches define height between levels (h). Regarding block width, length and height and referring to origin, coordinate computations are performed. Determination of physical position of each block is followed by thickness and grade assignments. Block assignments are performed by several mathematical approaches and *geological block model* is generated. Similar to calculations carried out on grid nodes, process is repeated for each elevation level. Geostatistics and neural networks are most advanced estimation models. Inverse distance square method can also be applied for more regular geological structures such as sedimentary beddings. According to assigned grade, blocks are coloured in software systems to enable observing quality tableau better.

6.5 Volume and reserve estimation

Volume and reserve of deposits is crucial subject of mine valuation as well as the visual aspect. There are various methods for numerical estimations. However, recent methods are computer dependent. Several approaches for ore volume and reserve estimations are given below (Hartman, 1992):

i.   The area enclosed by ore limits are multiplied by the average thickness. The volume is also multiplied with average tonnage factor to give inventory. Percent grade gives how much of this inventory is ore; that is reserve.

ii. Triangular net is used to calculate total reserve. Each triangular area is found and average thickness is multiplied by area to calculate volume of ore in triangle. Product of volume and weighted averages of tonnage factor and grade give triangular reserve. Summation of reserve of triangles yields total reserve.

iii. Field is divided into grids. Node points are assigned thickness, grade, etc. by inverse distance square, geostatistics and neural network methods. By this way, corner vertices of each node cell are assigned a value. Multiplication of cell area that is enclosed by grid

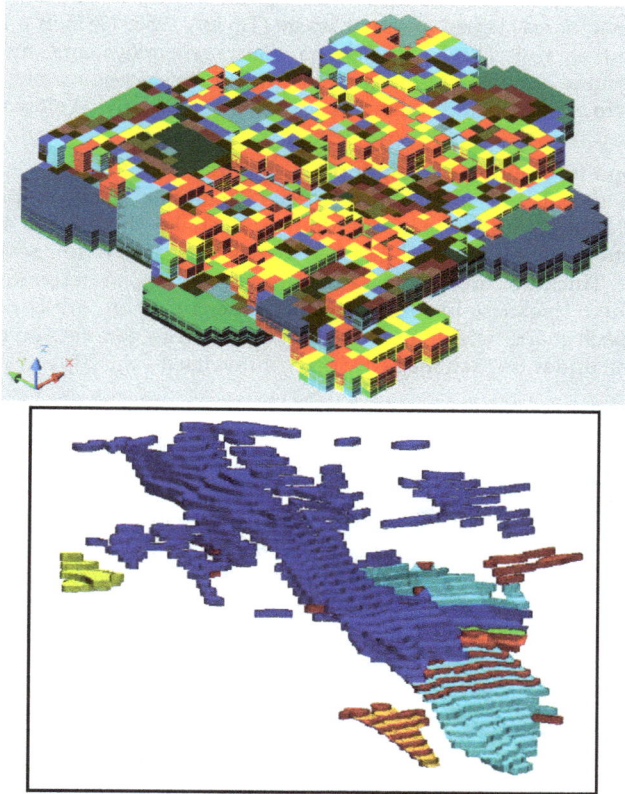

Fig. 19. 3D Ore body block model and Solid Model of Zambujal Ore body (Lundin, 2007).

nodes, thickness of ore in that cell and cell grade yields cell reserve. Total reserve is the total of all grid cells.

iv. Areas of parallel geological sections are calculated by Gauss-Green formula. Volumes between successive sections are estimated by average areas method.

v. Ore body block model gives also numerical results as well as visual. Each block reserve can be calculated. Total reserve is the reserve total of all blocks.

Block volume is simply multiplication of length, width and height of it. However, after thickness assignment to a block, thickness of ore should be considered rather than its geometrical height. Tonnage factor and ore grade give block reserve. Total reserve is the summation of all block reserves (Taylor, 1994, 1993).

$$Block\ reserve\ (t) = block\ volume\ (m^3) \cdot tonnage\ factor\ (t/m^3) \cdot grade\ (\%) \qquad (27)$$

Volume inside the open pit and ore volume bench by bench can later be estimated (Taylor, 1992). After determining open pit limits, border of each bench can be thought as a polygon and by means of Gauss-Green formula pit volume is computed. Stripping volume can also be calculated to find stripping ratio (Taylor, 1991).

An important factor in reserve estimation is *cut-off grade*, which can be described as the grade where excavated material is classed as ore or waste (Taylor, 1986, 1985). It may be considered as breakeven point as well (Taylor, 1972). So many researches and models have been developed to determine accurate cut-off grade. However, this economically crucial subject is special and unique to each field and it should be studied particularly (Wellmer et.al., 2008).

6.5.1 Isopach maps for volume calculation

Beside triangle and polygon methods, a special technique using isopach contours is employed for volume computation. Contour maps can be thought as two-dimensional extension and interpretation of drill holes. Isolines are drawn for several parameters provided by holes. Thickness and grade values are very crucial to determine ore body. It is possible to see where thickness increases and decreases or where ore is rich in grade and less valuable. Isopach maps, which are isolines for thickness, can also be used for volume calculation; volume under isopach maps are ore volume itself (Fig. 20).

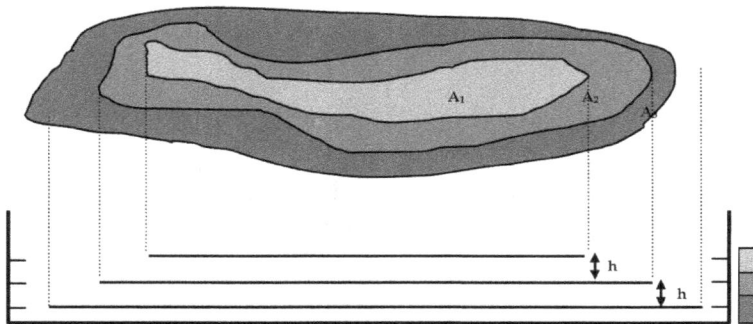

Fig. 20. Cross-section of isopach maps and volume calculation by using average areas rule.

Areas, at each thickness level are estimated and volume between each successive area pairs is calculated by using average areas rule:

$$V_j = \frac{A_j + A_{j+1}}{2} \cdot h \qquad (28)$$

where, V_j is volume at level j, A_j and A_{j+1} are successive areas of thickness level j and h is height (depth difference) between thickness isolines.

6.6 Mine design, production planning and mineral economics

Final stage of mine valuation is to decide if an ore deposit is worth making investment. Geological structure provides a database for economical assessment. Regarding physical and geological outputs of computerised systems, mine can be designed and production planning can be accomplished (Fig. 21).

Possible investment, annual costs and incomes are considered to estimate and foresee how that underground asset can be extracted optimally. Present worth, future worth, annual worth and rate of return of the net cash flows are calculated and reported. During calculations, straight line, double declining balance and sum of the years' digits methods are used for depreciation (Steiner, 1992). Taxes, salvage values, royalty costs, etc. are all taken into account.

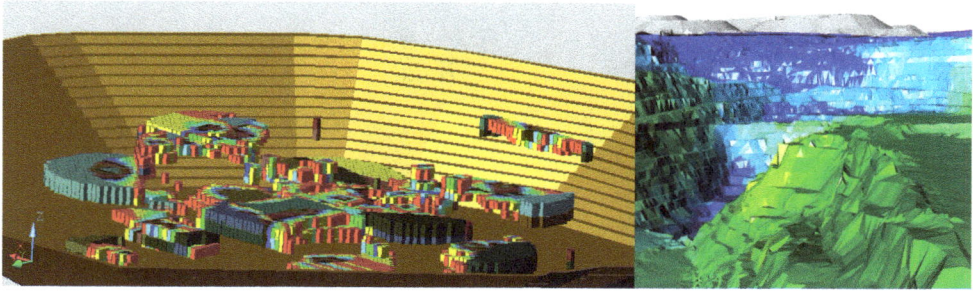

Fig. 21. Computer aided mine design (Beck, 2011).

Optimisation is another wide application branch of mine valuation. Geological block models are used to generate *economical block models* by using unit costs and income (Erarslan, 2001). As volume of a block, thickness and grade of ore at each particular block is known, then it becomes possible to convert this information to economical aspect. Multiplication of volume, tonnage factor and grade give block reserve. Unit production cost and expected income are considered and an economical value is assigned to each block. Economical block models have visual and numerical results. 3D appearances of them give an idea where ore body is rich and how quality changes. Mine design and optimisation applications such as optimum pit limit and optimum production planning are comprehensive numerical assessment methods (Erarslan, 2001).

7. Computer software for ore body modelling and mine valuation

Generally, size of the database may be too bulky to manage studies with hand effort. Hence, numerical algorithms and mathematical approaches necessitate computer applications to overcome huge computational time and processes. Today, many software and computer aided systems serve for geological modelling and mine valuation in this sense. The accuracy and speed of computers enable evaluation of various scenarios within reasonably short times.

Commercial softwares in general have robust database management capability. After building a healthy database structure, computer programs are ready for ore body modelling. Many mathematical models and approaches in literature take place in software packages to determine shape, location and quality composition of the entire body. Visual appearance of geological body is supported by numerical data such as ore reserve amount and quality composition, which are vital parameters for mine design and scheduling.

There are several integrated commercial packages for this purpose (MineSight, 2011; Gemcom, 2011; GDM, 2011; Techbase, 2011; Datamine 2011; Lynx Mining, 2011; RockWorks, 2011). They can successfully handle real cases, which may be fairly complex structures. Their processing and graphical capabilities and utilities have improved year and year (Fig. 22).

Computer systems make engineering designs, project preparation and management easier. Raw data such as GPS and drill hole cores are converted to contour maps, three-dimensional solid structures, volume and reserve reports, open and underground mine plans, economical assessments and production schedules. Various scenarios are examined quickly. Eventually, underground asset with so many unknowns becomes visual. Its properties thru space are clear. Million dollars are invested regarding this tableau. Thus, computer systems are vital partners of geological and mining engineers.

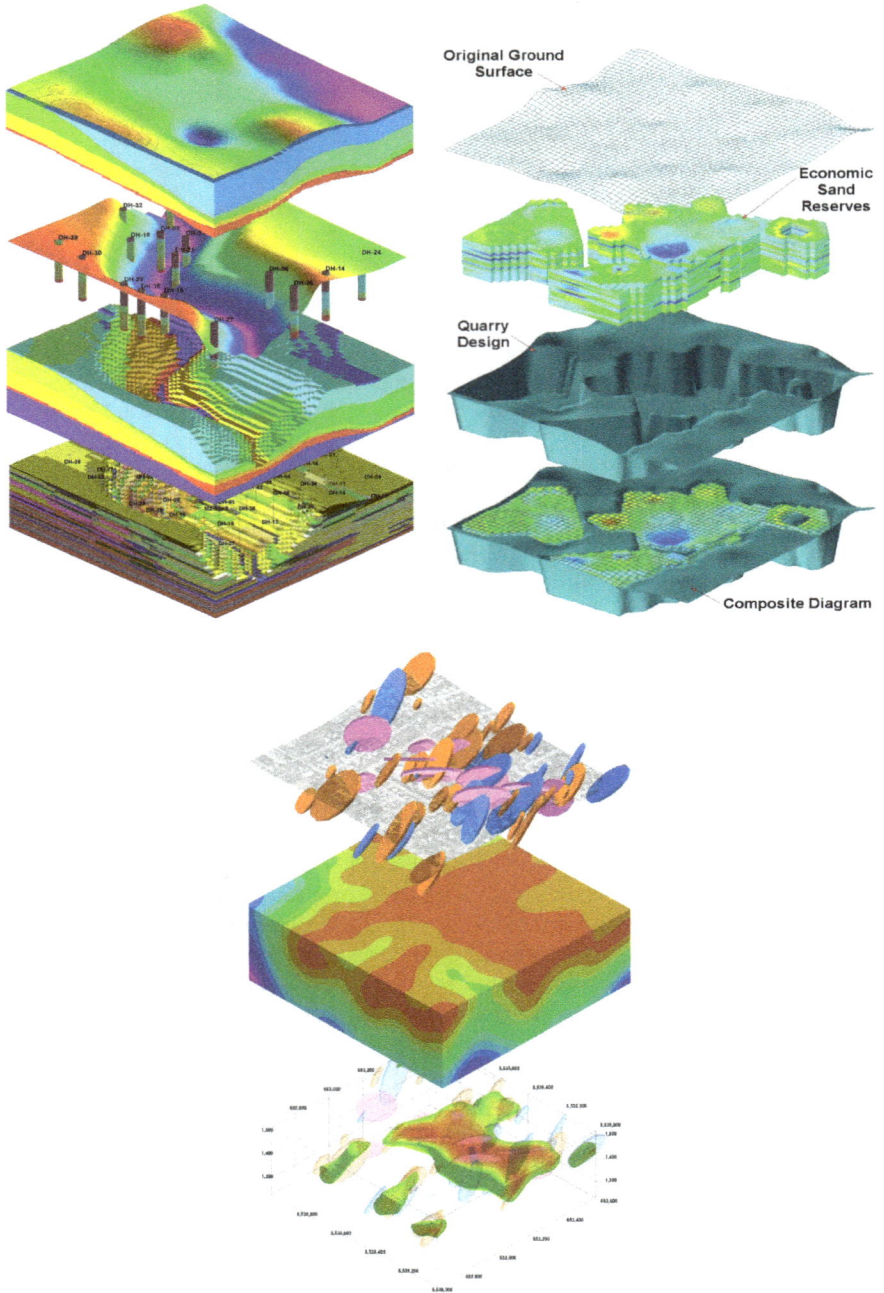

Fig. 22. Computer software functions for ore body modelling (RockWorks, 2011).

8. Conclusion

As mining is an industry requiring millions of dollars for investment and further operations, mine valuation is a crucial stage, which provides basic information for future stages. Before deciding on a mining investment, the preliminary process includes exploration, data gathering and valuation, determining geological structure, ore body modelling, mine design and planning.

In the last several decades, a number of mathematical and computational approaches have been developed to give the most accurate information related to ore bodies under investigation. In parallel, many computer programs have been developed to accomplish these complicated processes. This means that all investments are dependent on physical and economical characteristics of ore deposit. Similarly, profitability of the investment is strictly related to mine design and planning (Hartman, 1992). Eventually, data evaluation and ore body modelling is a very critical and basic process and mining operations are based on its results (Kennedy, 1990).

Drill hole database, geometrical and numerical analyses, contour maps, surfaces, sections, three dimensional ore body block models, volume and reserve calculations, economical assessment, classical valuation methods, triangulation, polygons, gridding, geostatistical approaches, neural network method, etc. are headings and instruments of mine valuation.

Computer aid has become inevitable and vital for contemporary ore body modelling and mine valuation applications. Long and complicated process starting with data base constitution and ending with mine design and production planning requires software support due to iterative mathematical structure of computations. Drill hole composites, triangulation, polygonal and grid nets, contouring, cross-sections, three-dimensional ore models, various volume and reserve estimations, geological and economical block models, mine design and production scheduling, optimisation and simulation works are highly computer dependent. At each stage of the processes, many developed mathematical approaches are utilised. One who deals with computer aided ore body modelling is faced with many and many studies and methods. However, geostatistics, neural networks, inverse distance methods are very basic subjects to be comprehended. Scientists work to improve present methods and develop new ones as well. Additionally, many commercial software are under a continuous development for a better service

9. References

Agoston, M.K. (2005). *Computer Graphics and Geometric Modeling*, Springer, ISBN 1852338172, USA.

Beck (2011). Beck Engineering Pty Ltd, Australia, http://www.beckengineering.com.au

BGIS (2011). Boston Geographical Information System, http://www.bostongis.com/

Comninos, P. (2006). *Mathematical and Computer Programming Techniques for Computer Graphics*, Springer, ISBN-10: 1-85233-902-0, USA.

Datamine (2011). Datamine Studio v. 2, CAE Mining, Montreal, Canada, http://www.cae.com

Davis, J.C. (1973). *Statistics and Data Analysis in Geology*, John Wiley and Sons, Inc., USA.

David, M. (1988). *Handbook of Applied Advanced Geostatistical Ore Reserve Estimation*, Elsevier, Amsterdam.

Deutsch C.V. & Journel A.G. (1998). *GSLIB Geostatistical Software Library and User's Guide*, 2nd Ed., Oxford University Press, ISBN 0-19-510015-8, New York.

Diggle P.J. & Ribeiro Jr., P.J. (2007). *Model-Based Geostatistics*, Springer, ISBN-10: 0-387-32907-2, New York.

Dowla F.U. & Rogers L.L. (1995). *Solving Problems in Environmental Engineering and Geosciences with Artificial Neural Networks*, The Massachusetts Institute of Technology, ISBN 0-262-04148-0, USA.Erarslan, K. (2007). Geology and Mining System (GMS) for Computer Aided-Mine Valuation, *Computer Applications in Engineering Education, Vol:15, No:1, 78-8*.

Erarslan, K. (2003). The Geology and Mining System (JMS) and Valuation of a Copper Field, *The Journal of Chamber of Mining Engineers of Turkey*, 42, (4), 3-14.

Erarslan K. & Çelebi, N. (2001), A Simulative Model for Optimum Open Pit Design, *Canadian Inst. Min. and Metall. CIM Bulletin*, 94, No. 1055, October, 59-68, Canada.

Fausett, L. (1994). *Fundamentals of Neural Networks*, Prentice Hall, ISBN 0133341860, USA.

Foley, J.D., van Dam, A., Feiner, S.K., Hughes, J.F. (1990). *Computer Graphics Principles and Practice*, 2nd Ed., Addison-Wesley, ISBN 0-201-12110-7, USA.

Freeman J.A. & Skapura, D.M. (1991). *Neural Networks Algorithms, Applications, and Programming Techniques*, Addison-Wesley Publishing Company, ISBN 0-201-51376-5, USA.

GDM (2003). BRGM International Division, Orléans Cedex, France. www.brgm.fr

GemCom (2011). GEMCOM, Inc., Canada. www.gemcomsoftware.com

GlView (2011). CAE Systems GMBH, Witten, Germany, http://www.glview.de

Golden Software (2011). Golden Software Inc., Colorado, http://www.goldensoftware.com/products/surfer/surfer-3Dsurface.shtml

Haupt S.E.; Antonello Pasini A. & Marzban C., Eds. (2008). *Artificial Intelligence Methods in the Environmental Sciences*, Springer, ISBN 978-1-4020-9117-9, USA.

Hartman, H.L. (1992). *Mining Engineering Handbook*, 2nd Edition, SME, AIME, Littleton.

Hill, Jr, F.S. (1990). *Computer Graphics*, McMillan Pub. Co., ISBN 0-02-354860-6, New York.

Hustrulid W. & Kuchta M. (2006). *Open Pit Mine Planning and Design*, Taylor & Francis, ISBN 9780415407410, USA.

Kennedy, B.A. (1990). *Surface Mining*, 2dn Edn, SME, AIME, Littleton.

Kitanidis, P. K. (2003). *Introduction to Geostatistics: Applications to Hydrogeology*, Cambridge University Press, ISBN 0 521 58312 8, UK.

Krige, D. G. (1984). Geostatistics and the definition of uncertainty; *Trans. Inst. Min. Metall.* Sect. A, v. 93, pp. A41–A47.

Krige, D. G. (1978). Lognormal–deWijsian geostatistics for ore evaluation; *South African Inst. Min. Metall.*, Monograph Series, Johannesburg, South Africa, 50 pp.

Leuangthong, O. and Deutsch, C.V. (2004). *Quantitative Geology and Geostatistics - Geostatistics Banff 2004*, Springer, ISBN-10 1-4020-3515-2, Netherlands.

Lynx Mining (2011). Lynx Geo Systems, S.A. Pty. Ltd., South Africa, http://www.lynxgeo.com/

Mallet J.L. (2002). *Geomodeling*, Oxford University Press, ISBN 0-19-514460-0, USA.

Matheron, G. (1971) The theory of regionalized variables and its applications; *Les cahiers du Centre de Morphologie Mathematique, Fontainebleau*, No. 5, 211 pp.

Matheron, G. (1963). Principles of geostatistics; *Economic Geololgy*, v. 58, pp. 1246–1266.

McKillup, S. & Dyar, E.D. (2010). *Geostatistics Explained- An Introductory Guide for Earth Scientists*, Cambridge University Press, ISBN-13 978-0-521-76322-6, New York.

MineSight (2011). MinTech, Inc., Tucson, Arizona, USA. www.mintec.com; www.minesight.com

Mortenson, M.E. (1999). *Mathematics for computer graphics applications*, 2nd Ed., Industrial Pres Inc., ISBN 0-8311-3111-X, USA.

Nieuwland D. A., Ed. (2003). *New Insights into Structural Interpretation and Modelling*, The Geological Society of London, ISBN 1-86239-133-5, London.

Pirsa (2011). Pirsa Minerals, Adelaide, Australia, http://www.pir.sa.gov.au/minerals/geology/3d_geological_models

Rabuñal, J.R. & Dorado J. (2006). *Artificial Neural Networks in Real Life Applications*, Idea Group Publishing, ISBN 1-59140-902-0, USA

RealTech (2011). RealTech-Virtual Reality Co., USA. http://www.realtech-vr.com

RockWorks (2011). RockWare, Inc., Golden, Colorado, USA, http://www.rockware.com/product/gallery.php?id=165

Rogers, D.F. & Adams, J.A. (1990). Mathematical Elements for Computer Graphics, McGraw Hill, Intl. Ed.2, ISBN 0-07-053529-9, USA.

Sarma D.D. (2009). *Geostatistics with Applications In Earth Sciences*, 2nd Ed., Springer, ISBN 978-1-4020-9 379-1, India.

Sen, Z. (2009). *Spatial Modeling Principles in Earth Sciences*, Springer, ISBN 978-1-4020-9671-6, New York.

Sinclair A.J. & Blackwell G.H. (2004). *Applied Mineral Inventory Estimation*, Cambrigr University Press, ISBN 0-521-79103-0, UK.

Singer D.A. & Menzie W.D. (2010). *Quantitative Mineral Resource Assessments an Integrated Approach*, Oxfrod University Press, ISBN 978-0-19-539959-2, New York.

SME (2005). *Exploring Opportunities: Careers in the Mineral Industry*, Society of Mining Engineers, Career Guidance Booklet, Littleton.

Steiner, H.M. (1992). *Engineering Economic Principles*, McGraw Hill International Editions, Singapore, 559p.

Taylor, H. K. (1994). Ore reserves, mining and profit; *Can. Inst. Min. Metall. Bull.*, v. 87, no. 983, pp. 38–46.

Taylor,H. K. (1993). Reserve inventory practice-reply; *Can. Inst. Min. Metall. Bull.*, v. 86, no. 968, pp. 146–147.

Taylor, H. K. (1992). The guide to the evaluation of gold deposits; integrating deposit evaluation and reserve inventory practices – discussion; *Can Inst. Min. Metall. Bull.*, v. 85, no. 964, pp. 76–77.

Taylor,H. K. (1991). Ore reserves – the mining aspects; *Trans. Inst. Min. Metall.*, v. 100, pp. 146–158.

Taylor, H. K. (1986). Ore Reserve Estimation, Methods, Models and Reality, *Can. Inst. Min. Metall.*, Montreal, May 10–11, pp. 32–46.

Taylor, H. K. (1985). Cutoff grades – some further reflections; *Trans. Inst. Min. Metall.*, v. 94, sect. A, pp. A204–A216.

Taylor, H. K. (1972). General background theory of cutoff grades; *Trans. Inst. Min. Metall.*, v. 81, sect. A, pp. A160–A179.

Techbase (2003). Techbase International, Lakewood, Colorado, USA. www.techbase.com

Torries T.F. (1998). *Evaluating Mineral Projects*, Society for Mining, Metallurgy, and Exploration, Inc., ISBN 0-87335-159-2, USA.

TUHH (2011). E-Learning Platform for IFM, Tech. Univ. Of Hamburg-Harburg, http://daad.wb.tu-harburg.de/?id=279

Vince, J. (2010). *Mathematics for Computer Graphics*, Springer, 3rd Ed., ISBN: 978-1-84996-022-9, New York.

Vince, J. (2005). *Geometry for Computer Graphics Formulae, Examples and Proofs*, Springer, ISBN 1-85233-834-2, USA.

Watson, D.F. (1982). Acord: Automatic Countouring of Raw Data, *Computers and Geosciences*, 8, (1), 97-101.

Wellmer F.W.; Dalheimer M. & Wagner M. (2008). *Economic Evaluations in Exploration*, Springer, ISBN 978-3-540-73557-1, New York.

Webster, R. & Oliver, M.A. (2007). *Geostatistics for Environmental Scientists*, 2nd Ed., John Wiley & Sons Inc., ISBN-13: 978-0-470-02858-2, England.

Part 2

Volcanology

Mud Volcano and Its Evolution

Bambang P. Istadi[1,*], Handoko T. Wibowo[2], Edy Sunardi[3],
Soffian Hadi[4] and Nurrochmat Sawolo[1]
[1]Energi Mega Persada
[2]Independent geologist
[3]Universitas Padjajaran
[4]Sidoarjo Mudflow Mitigation Agency
Indonesia

1. Introduction

The term mud volcano refers to topographical expressions of naturally occurring volcano-shaped cone formations created by geologically excreted liquefied sediments and clay-sized fragments, liquids and gases. Ejected materials often are a mud slurry of fine solids suspended in liquids which may include water and hydrocarbon fluids. The bulk of released gases are methane, with some carbon dioxide and nitrogen. Mud volcanoes may be formed by a pressurized mud diapir which breaches the Earth's surface or ocean bottom. Flowing temperatures at the ocean bottom may be as low as freezing point and are associated with the formation of hydrocarbon hydrate deposits. Flowing temperatures can also be hot if associated with volcanic gases and heat escaping from deep magma which can turn groundwater into a hot acidic mixture that chemically changes rock into mud and clay-sized fragments. These mud volcanoes are built by a mixture of hot water and fine sediment that either pours gently from a vent in the ground like a fluid lava flow; or is violently ejected into the air as a lava fountain of escaping mud, volcanic gas, stream and boiling water.

Mud volcanoes are most abundant in areas with rapid sedimentation rates, active compressional tectonics, and the generation of hydrocarbons at depth. Typically they are also found in tectonic subduction zones, accretionary wedges, passive margins within deltaic systems and in active hydrothermal areas, collisional tectonic areas, convergent orogenic belts and active fault systems, fault-related folds, and anticline axes. These structures act as preferential pathways for deep formation fluids to reach the surface. (see Pitt and Hutchinson, 1982, Higgins and Saunders, 1974; Guliyiev and Feizullayev, 1998; Milkov, 2000; Dimitrov, 2002; Kopf, 2002, Mazzini, 2009).

The existence of mud volcanoes are controlled by tectonic activity where fluid escapes from areas undergoing complex crustal deformation as a result of transpressional and transtensional tectonics. Collisional plate interactions create abnormal pressure condition and consequently overpressured buildup of deep sedimentary sediment which in turn result in formation of diapirs. Over pressured zones typically are under-compacted sedimentary layers which have lower density than the overlying rock units, and hence have an ability to

---

flow. They are the product of rapid deposition where the connate water is trapped, unable to escape as the surrounding rock compacts under the lithostatic pressure caused by overlying sedimentary layers. In thick, rapidly deposited shale dominant sedimentary sequence, the low and reduced porosity and permeability due to compaction inhibit the expulsion of water out of the shale. As burial continues, fluid pressure increases in response to the increasing weight of the overburden. This Non-equilibrium compaction is believed to be the dominant mechanism in formation of overpressured sediments. Over pressure however can also result from maturing organic rich predominantly clay sediments which are generating methane and other heavier gases that are still trapped within the sedimentary sequences. The above geological elements that result in diapirsm and mud volcano is often known as "elisional" basin mainly characterized by rapid deposition of thick young sediments, presence of abnormally high formation pressure or overpressures fluids, under-compacted sediments, petroleum generation, compressional setting, high seismicity and occurrence of faults (see Milkov, 2000, and Kholodov, 1983).

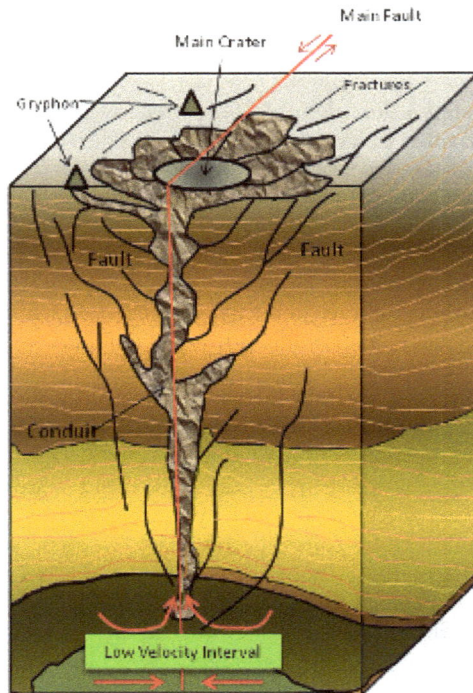

Fig. 1. Basic structure and anatomy of a conical mud volcano. The mud volcano is formed by the escaping natural gas that rises to the surface when it finds a conduit (strike slip fault) and carries mud which has a lower density (and typically found as low velocity interval) than the surrounding sedimentary succession. Fluid, gas, and surface water are ejected in a cone shape like a mountain and forms craters, mud pools (salses) and cones (gryphons). Tectonic movement is very influential, as well as rapidly deposited sediments and burial of organic rich sediments. Strike-slip faults in active tectonic regions are the most ideal place for the formation of mud volcanoes.

Overpressure buildup mechanisms contribute to the brecciation of the deep sedimentary units include for example the dewatering of thick clay-rich sedimentary units, and geochemical reactions in sedimentary units with high temperature gradients are the mechanism for the eruption (Mazzini, 2009). He further suggest that when the subsurface overpressure reaches a threshold depth where the overburden weight is exceeded, fracturing and breaching of the uppermost units occur, sometimes facilitated by external factors such as earthquakes. The upward movement of the mud to surface is due to buoyancy and differential pressure.

Geological structures like faults and anticlines where mud volcanoes are commonly found are easily perturbed by earthquakes as they represent weak regions for the seismic wave's propagation. This mechanism is well described by Miller et al. (2004) where earthquakes initiating local fluid movements cause fractures that propagate to the surface manifesting with a time delay from the main earthquake. Miller et al. (2004) propose a link between earthquakes, aftershocks, crust/mantle degassing and earthquake-triggered large-scale fluid flow where trapped, high-pressure fluids are released through propagation of coseismic events in the damaged zones created by the mainshock. The resulting disturbance of the gravitational instability triggers the beginning of flow, while the pressure drops and the lower cohesion media is easily fluidized and ultimately vacuumed to the surface through piercement structures which provide the conduits for high pressure mud/fluid and gas release.

The geometry of mud volcanoes is variable. They can be up to a few kilometers in diameter and several hundred meters in height. The main morphological elements of a mud volcano are the crater(s), hummocky periphery mud flows, irregularly shaped terrains, gryphons, and mud lakes or salses. A classification of mud volcanic edifices morphology was proposed by Kholodov, 2002 (in Akhmanov and Mazzini, 2007), these are: (1) "classic" conic volcanic edifice with main crater and mud flow stratification reflecting periodical eruption; (2) Stiff mud neck protrusion, typically due to its high viscosity and hence able to form steep hills; (3) swamp-like area; contrary to no (2), due to its low viscosity the mud spreads over a large area; (4) "collapsed synclinal" depression; and (5) crater muddy lake, is the most abundant type in various mud volcanic areas. It is often that mud volcano morphology shows a combination of the common types described above depending on the viscosity of the mud and the stage of its development.

Mud volcanoes show different cyclic phases of activity, including catastrophic events and periods of relative quiescence characterized by moderate activity. It appears that each eruptive mud volcano has its own period of catastrophic activity, and this period is variable from one volcano to another. The frequency of the eruptions seems essentially controlled by local pressure regime within the sedimentary sequences, while the eruptive mechanism and evolution seem strongly dependent on the state of consolidation and gas content of the fine-grained sediments. This is shown in the compilation of historical data onshore Trinidad as described by Deville and Guerlais (2009).

Approximately 1,100 mud volcanoes have been identified on land, in shallow as well as deep waters. It has been estimated that well over 10,000 may exist on continental slopes and abyssal plains. The largest known structures are 10 km in diameter and reach 700 m in height. Occurrences of mud volcanoes on the seafloor have been documented more frequently since the intensive use of side scan sonar began in the late 1960's. Mud volcanoes have been found in many parts of the world, and have been documented in Rumania, Italy, Iran, Iraq, New Zealand, India, the Myanmar, Malaysia, Gulf of Mexico, Trinidad,

Venezuela, Colombia and the USSR. The largest number of mud volcanoes is found in the Azerbaijan trend which continues into the Southern Caspian area. In the Indonesian region, mud volcanoes are found on the Islands of Sumatera, Nias, Pagai, Sipora, East Java, East Kalimantan '(Borneo), Rote, Barbar, Aru, Timor, Tanimbar, Yamdena and Papua. They are found in high rate subsidence basins such as Madura-East Java Basin, Kutai Basin, in high seismicity areas such as islands in the Banda Sea and in tectonically complex areas such as Timor and Papua (Sukarna, 2007).

In Papua, Indonesia, mud volcanoes are found along a zone of disruption 400 km long and nearly 100 km wide, occupying hilly terrain with low relief scarred by landslip. They are aligned along structural trends of up to 50 km long and 25 km wide. Individual mud volcanoes range from 3 m to 2.5 km in diameter and reach a maximum height of about 110 m. The ejected mud consist of mud stone containing various shapes and sizes of clasts of older rock assumed as exotic block, which is believed as part of mélange diapirsm (Sukarna, 2007).

Information on mud volcanoes can be used to study the subsurface condition and used as pathfinders of the conditions indicative of subsurface hydrocarbon accumulations in unexplored areas. Gas geochemical data from mud volcanoes can be examined for possible presence of source rocks and their maturity levels. Mud volcanoes are often related to active petroleum systems, especially if the released gas shows a deep thermogenic character. A global data-set of more than 140 onshore mud volcanoes from 12 countries shows that in 76% of cases the gas is thermogenic, with 20% mixed and only 4% purely microbial (Etiope et al., 2009). The thermogenic nature of most of mud volcanoes is related to the relatively high thermal maturity of gas-generating organic-rich rocks. On the other hand, mud volcanoes which release large amounts of CO_2, such as those related to magmatic activity, may not indicate the presence of significant hydrocarbon reservoirs (e.g., Milkov, 2005). Many large onshore hydrocarbon fields were discovered after drilling around mud volcanoes in Europe, the Caspian Basin, Asia and the Caribbean (see Etiope et al, 2009, Link, 1952; Guliyev and Feyzullayev, 1997). Gas origin, composition and secondary post-genetic processes such as secondary methanogenesis which follows anaerobic biodegradation of petroleum or heavy hydrocarbons however, are fundamental factors for determining depth and quality of the related petroleum system, especially in frontier or unexplored areas (see Etiope et al., 2009).

Apart from providing information and evidence of hydrocarbon potential and a working petroleum system, mud volcanoes also provide useful data about the sedimentary section which can be determined by examination of ejected rock fragments incorporated in mud volcano sediments (breccia).

Mud volcanoes, depending on their size and activity, can pose ecological hazards and disaster to the environment as well as to the population of the surrounding area. Mud volcanoes typically ejected breccia and/or mud flows and/or flame in temporal association with earthquakes. Active mud volcanoes can vent a large amount of carbon dioxide and flammable methane, and may influence global climate. Large eruptions are known to have occurred in the Black Sea and in areas around the Caspian Sea where gas exploded in a flame several hundred meters high that burns vegetation within the vicinity of the mud volcano. Mud volcanoes may also pose a geohazard for drilling and platform constructions due to the potentially violent release of large amounts of hydrocarbons and mud breccia. When the viscosity of the mud breccias is low, it may flood large area and inundate villages, homes, roads, rice fields, and factories and displace people from their homes.

2. Evolution of LUSI Mud volcano

Kusumadinata, 1980 described mud volcanoes as any extrusion on the earth's surface in the form of clay or mud in which morphology forms a cone in which there is a lake and coupled with the discharge of water and is driven by strong gas flow. Often the release of gas is followed by an explosion and burns, thus the extrusion appearance greatly resembles a magmatic volcano. Apart from this description, Sangiran dome (e.g. Watanabe and Kadar, 1985) and 1936 Dutch maps (see Duyfjes, 1936), very little has been published in journals and scientific papers on mud volcanoes in Western Indonesia, unlike the ones found in Eastern Indonesia such as in Timor (e.g. Barber et al., 1986). In fact, mud volcanism in Indonesia, particularly in East Java is poorly understood. This lack of understanding changed when the LUSI mud volcano (Lumpur "mud"- Sidoarjo), was born as it offered a unique opportunity to study the dynamic development of a mud volcano from its birth on 29th May 2006. In contrast, studies on mud volcanism are typically conducted during the dormant periods between eruptions of already existing mud volcanoes.

The subsequent eruption of LUSI mud volcano has been closely observed and analyzed by the geological community (see Mazzini et al., 2007, 2009; Sunardi et al, 2007; Kadar et al., 2007; Sudarman and Hendrasto, 2007; Kumai and Yamamoto, 2007; Hutasoit, L, 2007; Sumintadireja et al., 2007; Deguchi et al, 2007; Abidin et al. 2007, 2008; Satyana, 2007, 2008; Satyana and Asnidar, 2008; Fukushima, 2009; Mori and Kano, 2009; Hochstein and Sudarman, 2010; Sutaningsih, 2010 etc). Its birth occurred just 2 days after a devastating Yogyakarta earthquake in May 2006. The trigger of LUSI mud volcano is controversial, and has been the center of debate among geoscientists and drilling engineers as it is located near the Banjarpanji-1 oil and gas exploration well drilling that was being drilled. (i.e. Mazzini et al., 2007; Davies et al., 2007, 2008; Manga, 2007; Manga et al., 2009; Tingay et al., 2008, 2010; Sawolo et al., 2008, 2009, 2010; Istadi et al., 2008, 2009). Mazzini proposed that LUSI was caused by fracturing following the May 27th earthquake and accompanied depressurization of > 100 °C pore fluids from > 1700 m depth. This resulted in the formation of a quasi-hydrothermal system with a geyser-like surface expression and with an activity influenced by the regional seismicity. Davies, on the other hand, suggested that an underground blowout in the Banjarpanji-1 well breached to the surface, thus creating a conduit for the high pressured fluid to escape and created a mud volcano.

The LUSI mud volcano eruption has continued for over five years, and potentially will continue for many years to come, impacting an ever larger area. The mud eruptions occurred in at least five separate locations forming a NNE–SSW lineament about 200 m away from the Banjarpanji-1 (BJP-1) exploration well in Sidoarjo, approximately 30 km south of Surabaya, East Java, Indonesia (Fig. 2 & 8). The approximately 700 m lineament is contiguous with the Watukosek fault zone (Istadi et al., 2009).

Hot mud erupted at 5,000 m^3 a day at the beginning, increased to 50,000 m^3/day in the initial months, then escalated to 125,000 m^3/day and reached a high rate of 156,000 m^3/day by October 2006. The high flow rates coincide with a series of earthquakes. In less than one year after its birth, LUSI displaced some 24,000 people, inundated toll road, several villages, housing estates, paddy fields and farm land, factories, schools, mosques, shops, offices, destroyed a gas pipeline, killing 13 workers and has covered about 700 Ha of land. The extent of the damage caused by LUSI is substantial and has caused subsidence of up to 5.53 cm/day in the area next to the main eruption. Simulation shows that it will affect an area of 3 km radius, disrupt the subsurface condition and subsidence of up to 60 meters. In a highly populated area of Sidoarjo, this could potentially disrupt more than 10,000 families; a major issue to the people, infrastructures and environment. The eruption rate however, has decreased to less than 10,000 m³/day at the time of writing in July 2011; perhaps LUSI is now entering a new phase, from an eruptive one to a mature and quiescence phase.

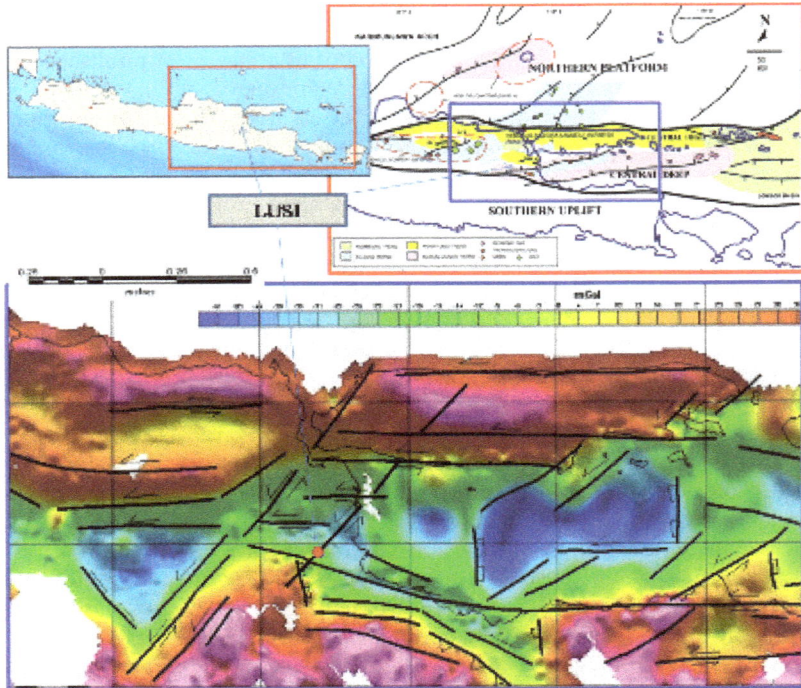

Fig. 2. LUSI is located in East Java, about 30 km South of Surabaya (top left). Regional tectonic framework of East Java (top right) shows major NE-SW and E-W major fault trends. Bouguer Gravity map (processed with 2.65 g/cc density) of East Java (bottom) showing East Java Basin's depositional centers (blue) which are controlled by the major faults in the area.

2.1 Stratigraphic setting of the LUSI Mud volcano

The LUSI Mud Volcano is located about 10 km northeast of Penanggungan Mountain, in Reno Kenongo village, Porong District, Sidoarjo Regency, East Java. Its location is in the Southern part of the hydrocarbon prolific East Java inverted back-arc Basin which was formed during the Oligocene- Early Miocene (Sribudiyani et al, 2003), on the Eastern tip of the Kendeng Zone (De Genevraye and Samuel,1972). The Geology of the area is characterized by the rapid deposition of thick organic rich sediment as part of the Brantas delta, influenced by the extensional tectonic regime (Willumsen and Schiller, 1994, Schiller et al, 1994). Due to the rapid deposition, shales in the area are undercompacted and overpressured (Mazzini et al., 2007). The geological condition is similar to other areas where mud volcanoes are found such as the Caspian and the Black Sea (Planke et al, 2004, Mazzini et al., 2007; Tingay et al., 2008)

Java Island, located at the southern part of the Sundaland, was formed by rock assemblages associated with an active margin of plate convergence. The island has recorded plate convergence between the Australian plate and the Sundaland continental fragment since Late Cretaceous. Therefore, the island is made up of complex of plutonic-volcanic arcs, accretionary prisms, subduction zones, and related sedimentary rocks (Satyana and Armandita, 2004). The structural history is divided into two phases: a Middle Eocene to Oligocene extensional phase, and a Neogene compressional or inversion phase. Grabens and half-graben structures were developed during the extensional phase, which was followed in the Neogene by compressional deformation with some wrenching. The most recent sedimentation in the East Java Basin occurred during the Late Pliocene to Holocene (3.6–0 Ma), during which time the southern part of the basin (Kendeng depression zone) was affected by north verging thrusts and uplift. The depression developed as a response to the isostatic compensation of the uplift of the southern Oligo-Miocene volcanic arcs. The uplift was accompanied by an influx of volcaniclastic rocks from the southern volcanic arc provenance and were deposited into the depression and causing the depression to subside.

Other provenance of the Kendeng Depression is the northern uplifted area that filled the basin with shallow-marine carbonates and marine muds from the Oligocene to the Holocene. Very thick sediments from the two provenances were deposited rapidly into the Kendeng Depression mostly as turbiditic deposits. The East Java geosyncline has thick Tertiary sediments of more than 6000 m (Koesoemadinata, 1980) with an estimated sedimentation rate of 2480 m/ma in the vicinity of LUSI (Kadar et al.,1997). The high sedimentation rates followed by rapid subsidence caused non-equilibrium compaction, and along with the maturation of organic materials resulted in the overpressured sediments within the Kendeng zone (see Willumsen and Schiller, 1994; Schiller et al., 1994). The overpressured sediments were later compressed, become mud diapirs and pierced the overlying sediments in many parts of East Java as mud volcanoes.

Outcrops of sedimentary rocks are very rare as they are covered by alluvial sediements and weathering. Therefore fresh rock outcrops at the rock mine in Karanggandang Village, 28 km to the northwest of the LUSI (Kadar et al, 2007) is important to complete the stratigraphic column of LUSI and the Banjarpanji-1 well.

The stratigraphy at LUSI (figure 3) consists of (1) alluvial sediments, (2) Pleistocene alternating sandstone and shale of the Pucangan Formation (to about 500 m depth), (3) Pleistocene clay of the Pucangan Formation (to about 1000 m depth), (4) Pleistocene bluish gray clay of the Upper Kalibeng Formation (to 1871 m depth), and (5) Late Pliocene volcaniclastic sand of at least 962 m thickness. The stratigraphy below the Late Pliocene sand is not well known; however, when the Banjarpanji-1 well reached 2834 m depth, cuttings from the bottom of the well did not contain limestone fragments indicating that drilling of Banjarpanji-1 well had not reached the carbonate reservoir target. Davies et al. (2007) suggested that these porous rocks were Kujung Formation limestone and that this formation is the source of the fluids erupting at LUSI. Seismic correlations from the Porong-1 well, 6.5 km to the northeast of LUSI, indicate that the rocks underlying the Late Pliocene volcaniclastic sands are carbonates which contain coralline red algae fossils, corals and foraminifera fragments. The strontium isotope (Sr) of the carbonates shows the absolute age of 16-18 ma Early Miocene, and therefore the carbonates are correlated with the Tuban Formation outcrops found extensively in the western part of East Java basin which show age range from 15.2 ma to 20.8 ma based on analysis of strontium isotopes (Sharaf et al., 2005).

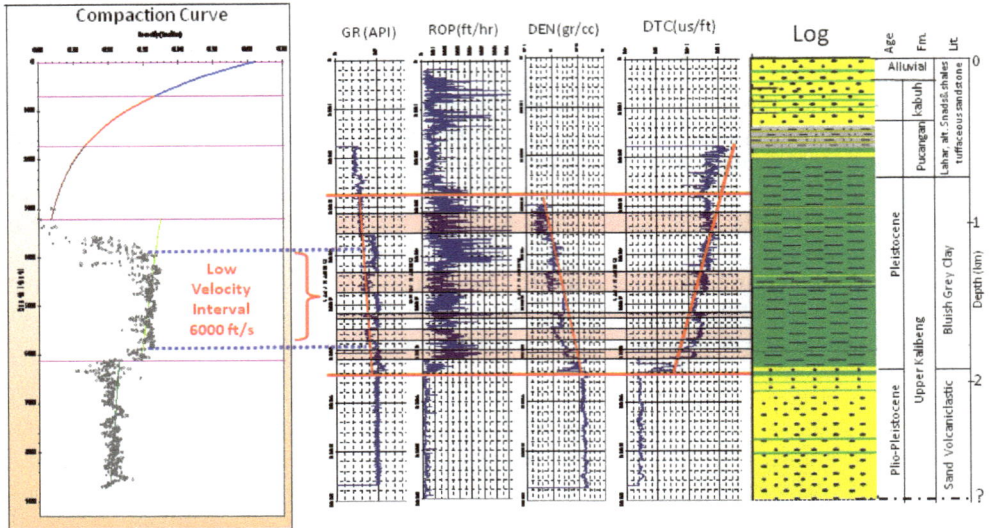

Fig. 3. Stratigraphy column in BJP #1 well nearby LUSI MV. Density, GR, ROP and DT suggest the presence of overpressured zones. These are probably **highly plastic,** undercompacted shale, controlled by rapid sedimentation that trapped water and resulted in an overpressured condition.

The seismic cross-section of this trajectory (figure 4) has a path through three wells, and track from SW to ENE. The wells are Banjarpanji-1 (BJP-1), Tanggulangin-1 (TGA-1) and Porong-1 (PRG-1). At the bottom of the well of Banjarpanji-1 well is a mass of shale that appears mounded and has a large structural dimension. Faulting in the form of positive flower structure between the wrench faults that continuously cut the low velocity intervals, is interpreted as a mud diapir (a depth> 9292 ft or 2834 m).

The structural feature that resembles a flower structure suggests the presence of a wrench fault, thus horizontally sliding components and oblique movements are suggested. The presence of overpressured zones probably consists of highly plastic, undercompacted shale. On seismic it is correlated with a Low velocity zone, and it is characterized by a chaotic discontinues pattern.

The collapse structure adjacent to the well Porong-1, located approximately 7 km from LUSI forms a depression around the crater. This structure likely represents an extinct mud volcano that, once it terminated its activity, gradually collapsed around a vertical feeder channel. The multiple contemporaneous faults indicated in the seismic section is due to gravitational slumps and intrusive structure suggesting piercement from the upward moving over-pressured sediments of the underlying shale diapir are evident (see Istadi, 2009). By the same mechanism, the multiple faults at LUSI may have been reactivated and served as conduit for the mud eruptions along the fault planes.

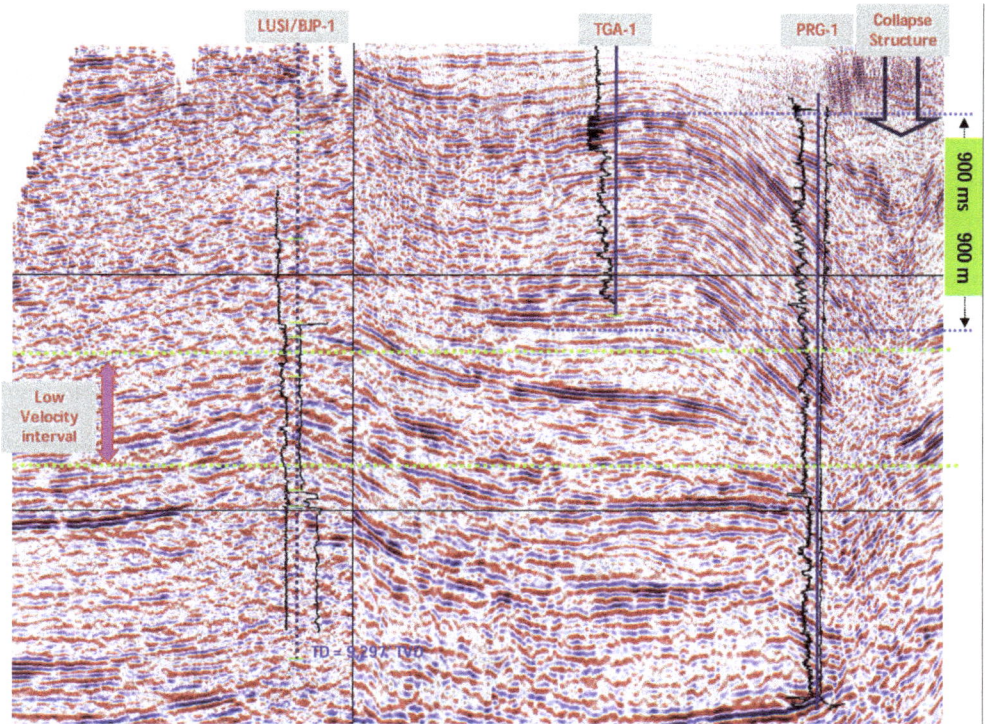

Fig. 4. Seismic section of LUSI – Banjarpanji-1 – Tanggulangin-1 – Porong-1 – Porong collapse structure. The Porong collapse structure located approximately 7 km from LUSI is a paleo mud volcano where subsidence is evident and the multiple faults present likely served as conduits for the mudflow. Similarly, the multiple faults near the BJP-1 well (200 m from LUSI) may have been reactivated and served as conduit for the mud eruptions and escaping gas, hence the appearance of gas bubbles along fault lines.

2.2 Controversy on the trigger

The trigger of the LUSI mud volcano is controversial because of its location and time of birth. The insinuating factors include:

1. The surface location of LUSI is approximately 200 meters away from an exploration well, the Banjarpanji well (Davies et al., 2007, 2008 and 2009). This suggests a possible connection between the drilling of the well and LUSI.

2. A major 6.4 Richter scale magnitude Yogyakarta earthquake rocked Java Island two days earlier (Mazzini et al., 2007 and 2009). This may have reactivated local fault and allowed an overpressured mud to escape to the surface.

3. The exploration well suffered two major drilling problems after the earthquake; two mud losses following the earthquake and a kick during the drill string removal (Sawolo et al., 2009 and 2010). This suggests a possible link between the earthquake and the change in the subsurface condition of the region.

4. A number of mud volcanoes in the vicinity of Watukosek fault were reported to suddenly become active at the time of the earthquake (Mazzini et al., 2009)
5. Three mud eruptions that appeared to be in line with the direction of Watukosek Fault (Sawolo et al., 2009)
6. The limited amount of well data available in the public domain especially on the events prior and following the initial eruptions as well as discrepancies over the interpretation of drilling data from the Banjarpanji-1 well (Tingay, M., 2010).

The process of LUSI mud volcano creation is generally believed to be a natural process of an over pressured shale eruption through seismically reactivated faults as conduits (Mazzini et al., 2009). The trigger of the eruption, however, is still a hotly debated issue within the engineering community. Was the creation triggered by a reactivation of nearby Watukosek fault or an underground blowout in Banjarpanji well? A well blowout generally means an uncontrolled flow of formation fluid into a wellbore that travelled up the well to the wellhead. An underground blowout, on the other hand, is when the flow of reservoir fluid does not exit at the wellhead but flows to a low pressured formation in the well and eventually breaches the surface.

Banjarpanji-1 is an exploration well that was drilled in March 2006 in a densely populated part of East Java, in the Sidoarjo region. In May 2006, the well was drilling toward its target, the Kujung limestone. There was little subsurface drilling problems during the drilling process as the well was drilled using a Synthetic Oil Based mud that eliminated much of the problems in the thick overpressured shale section. The schematic of the well is shown in figure 5.

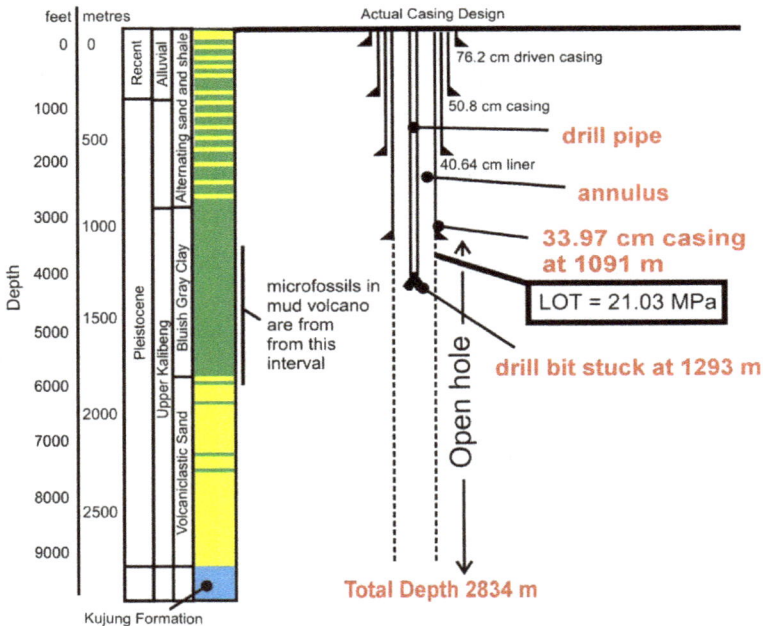

Fig. 5. Summary of the stratigraphy drilled by the Banjarpanji-1 well and the casing setting depths (Davies et al.,2008).

The problem free drilling condition changed in May 27th 2006 when a 6.4 magnitude earthquake strike near Yogyakarta about 250 km away. Within ten minutes of the earthquake, a 20 bbls of mud loss was observed in the well. Another mud loss occurred following two major aftershocks of 4.8 and 4.6 magnitude where a 130 bbls of drilling mud was lost to the open hole. Drilling mud loss is a serious problem in drilling as the mud hydrostatic pressure is the only mechanism to balance the reservoir pore pressure. Such major mud losses typically occur when drilling mud flows into a newly drilled cavity or a fractured zone or newly formed fractures.

2.3 The underground blowout hypothesis

An underground blowout is not an uncommon problem during well drilling, especially in exploration wells where the geology and reservoir pore pressure is uncertain. If the effective mud weight used to drill the well falls below the pore pressure, a kick (an influx of formation fluid into the well bore) can happen. This kick displaces the heavy drilling mud which reduces the hydrostatic pressure of the mud column and causes an even larger influx of formation fluid. This kick must be killed correctly and a proper hydrostatic pressure restored in order to retain the control of pressure in the well. If the kick is not controlled properly and the well is subjected to a pressure above its critical fracture pressure, then the weakest formation, generally immediately below the casing shoe, may fracture. When there is sufficient force behind the kick, this fracture can propagate upward into shallower formations, or even breach to the surface. This is the mechanism of an underground blowout during drilling of a well.

The possibility that the mud eruption was caused by an underground blowout in the Banjarpanji-1 well was initially suggested by the media immediately after the eruption. There was very little official information of its cause and very little well data was available in the public domain at the time. A similar charge was later iterated in the Davies et al., 2007 and 2008 papers with a pressure analysis that suggested a fracture at the casing shoe. Davies proposed that the well was subjected to a pressure above its critical fracture pressure that caused its weakest formation to fracture. The hydro-fracturing of the formation was propagated and finally it breached to the surface and caused LUSI.

In his papers, Davies proposed a number of operational sequences during the drilling of the Banjarpanji-1 well that caused a kick that leads to the underground blowout. Originally he proposed that the over pressured Kujung limestone was drilled, and this provided the high pressure fluid that caused a kick in the well (Davies et al., 2007). In his later paper (Davies et al., 2008), Davies refined his proposition to the kick was caused by a pressure under-balance in the well since 46 stands of drill pipe were removed without replacing the lost volume due to the removal of the drill pipes. And in his latest discussion paper (Davies et al., 2009) he proposes that the kick was due to a 'swabbing' effect. Swabbing means a negative pressure due to an upward movement of drill string.

Davies propositions of Underground Blowout hypothesis is rejected as pressure inside the well is too low to break the formation at the casing shoe of the Banjarpanji-1 well (Nawangsidi, D., 2007). The assumption used by Davies to develop his hypothesis is incorrect as the well is not full of mud but contains water as well as gas; this in effect lower the specific gravity and the hydrostatic pressure in the well. With a lower specific gravity and the published annulus pressure cited by Davies, it is impossible to fracture the casing shoe.

Davies's postulations on what happened, are not supported by well data and the actual condition of Banjarpanji-1 well (Sawolo et al., 2009). In his paper, Sawolo showed that (i) the

Kujung carbonate was not penetrated by the well, as the calcimetry data during drilling remained low - suggesting that no carbonate formation is drilled, and the well experienced a loss of mud (the opposite of a kick). (ii) The charge that mud was not pumped to compensate for the pulling of drill string out is incorrect. Automatic data recorder showed that mud was pumped appropriately to compensate for the pulling of the drill string. (iii) Analysis on speed when pulling out of the hole and the condition of the wellbore preclude the possibility of swabbing. And finally, (iv) pressure analysis based on the automatic data recorder (Real Time Data-RTD) showed that the pressure exerted on the casing shoe is too low to be able to fracture the formation. It is therefore concluded that the well was intact, no underground blowout occurred in the well and the Banjarpanji well could not have been the trigger of LUSI.

The geologists and drillers in charge of drilling the well have reviewed all the data and cannot support the underground blowout theory. They charge that it is an over simplification to the actual condition (Sawolo et al., 2010). Their approach in analyzing the trigger of LUSI was broader than that of Davies. Instead of relying on limited data and assumptions to fill the missing gap, they collected all available drilling data, logs and cross checked it with other reports to come up with an enriched credible set of data. The various data were assembled to form a mosaic that clearly showed that the well remained intact and an underground blowout in the Banjarpanji-1 well did not occur.

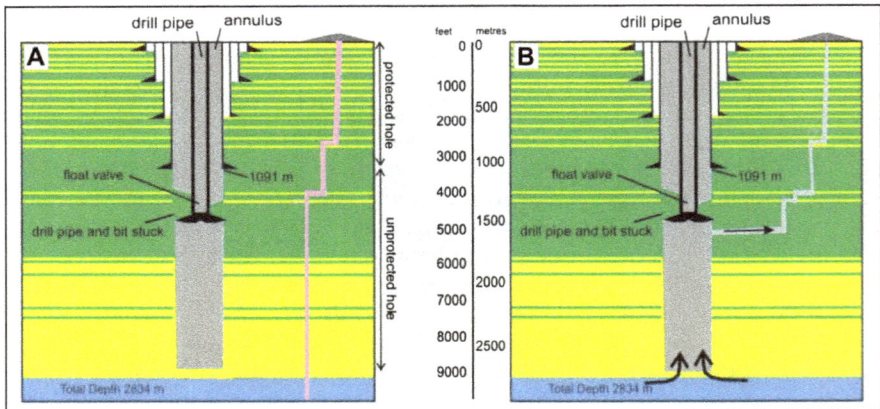

Fig. 6. The controversy on LUSI's trigger; Sawolo's view (A) – where the well is intact, the mud flow did not pass through the well. Davies (B), contends that the mud passed through the well, fracturing the casing shoe and triggered the LUSI mud volcano (Sawolo et al., 2010).

The geologists believe that LUSI is more likely to be related to seismic activity. The loss of drilling mud that coincided with the time of the earthquake suggests that the earthquake may have altered the subsurface condition of the region.

The two opposing views of the LUSI trigger as a seismic related event (Sawolo et al) and an underground blowout (Davies et al.,) are shown in figure 6. The scenario proposed by Sawolo is shown in A, where the well is intact and the mud flow did not pass through the well. Davies view is shown in B, where the mud flow passed through the well and fractured the casing shoe.

The root cause of the difference of opinion between Davies and Sawolo is believed to be access to well information; especially the quality and quantity of data used in the pressure analysis. Analysis based on limited data can result in a misleading conclusion. It should be understood that the well information on the public domain at the initial stages of the eruption was limited as majority of data was kept confidential by the oil and gas company. It was not until 2009 that Sawolo et al. published for the first time a wide range of well data and observations previously not available in the public domain; this includes important data on formation strength such as the Leak Off Test, Bottom Hole Pressure, minute by minute well pressures and fluid density in the well. The intent of opening the data to the scientific community was that future research on LUSI could be based on actual and credible data in order to minimize assumptions.

The engineering debate on the trigger of LUSI continued until 2010, when Tingay, 2010 published a paper outlining the pros and cons of the competing theories. The paper presented the first balanced overview of the LUSI mud volcano by identifying critical uncertainties of the plumbing system, events prior and following the initial eruptions as well as discrepancies over interpretation of petroleum engineering data from the Banjarpanji-1 well. These uncertainties caused much of the trigger controversy. It is obvious that more studies are needed before LUSI's actual trigger can be identified.

2.4 The fault reactivation hypothesis

Mud volcanoes are spatially associated with both major and secondary faults within the regional stress field. Surface and subsurface geology shows that faults exist in the region that crosses LUSI areas with the NE-SW and NW-SE direction trends. Faulting in the NE-SW direction is known as the Watukosek fault, an oblique strike-slip fault, whereas the NW-SE trend dextral strike-slip fault pattern is the Siring fault. These faults are partly buried by alluvial sediments.

Eruption of mud volcanoes along the fault line is one of several possible subsurface hydrological responses to earthquakes. Manga, however, dismissed the possibility that LUSI was triggered by the Yogyakarta earthquake that occurred two days earlier because of its distance and magnitude (Manga, M. 2007). An empirical plot from past earthquakes in the world shows a distinct relation between the earthquake's magnitudes to the distances from their epicentres (figure 9).

When the data-point from the Yogyakarta 27th May 2006 earthquake was entered, it shows that the Yogyakarta earthquake was unlikely to have caused the LUSI mud volcano. It is interesting to note that two past earthquakes that were larger and closer to the area failed to cause a mud volcano in the region.

Mellors has a similar view on the relationship between large earthquakes and triggering effect on mud volcanoes (Mellors et al., 2006). This is because large earthquakes produce strong static and dynamic stress changes near its centre that could trigger mud volcanoes. His research is based mainly on a 191 years record of mud volcano eruption and large earthquake in the Caspian Sea. It suggests that the triggering effect is strongest where the shock at the mud volcano has the intensity of approximately Mercalli 6 and above and at a distance of less than 100 km. However, he also found that even when the intensities exceed the apparent threshold, only a fraction of active volcanoes erupt. This indicates that other factors also play an important role.

However, there are too many uncertainties in predicting the triggering effect of earthquakes on mud volcanoes (Mori and Kano, 2009). Mori pointed out that until recently, seismologists would not believe that triggering due to earthquakes over hundreds of kilometres is

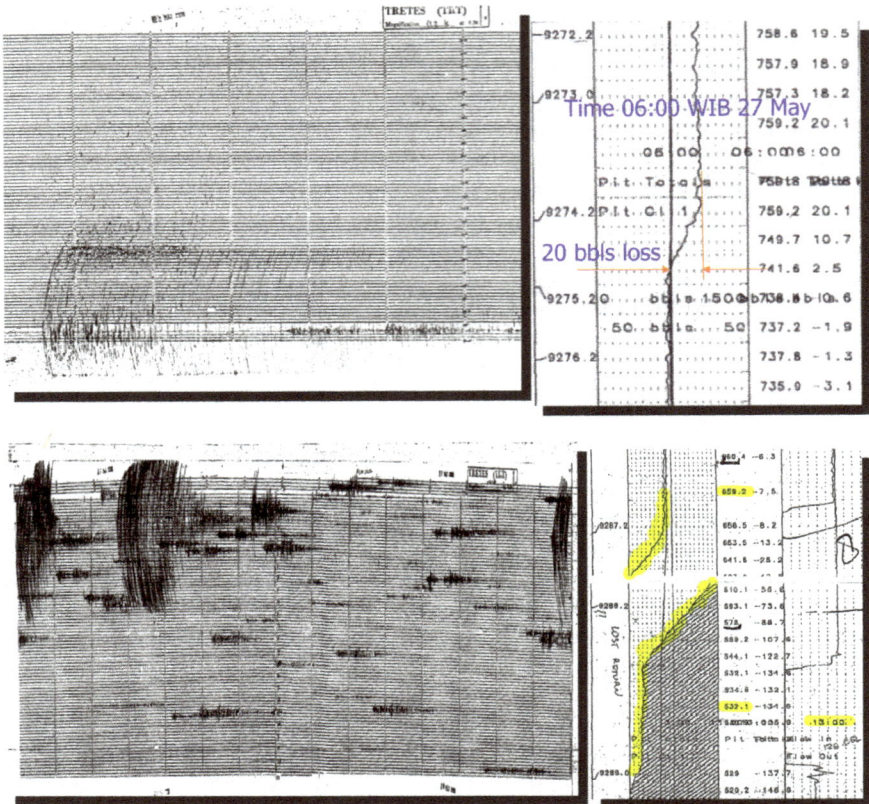

Fig. 7. Losses of mud after the Yogyakarta earthquake. The top left picture shows the seismograph reading at Tretes BMG station about 15 km away and the right picture shows the 20 bbls loss seven minutes after the main earthquake. The bottom left showed the aftershocks and the right shows the 130 bbls complete loss of circulation from the wellbore that happened two hours after two aftershocks (Sawolo et al., 2009).

possible. This view has changed especially in the hydrothermal areas. When Mori entered the Yogyakarta earthquake and LUSI data in Fisher's Dynamic Stress vs. Frequency plot (figure 10), it is located within the value that has triggered mud volcanoes in other regions. Mazzini, one of the very few earth-scientists who conducts field work and data measurement right after the mud eruption, also disagrees with Manga's conclusion (Mazzini et al., 2007). He cited that eruptions can be affected by earthquakes several thousands of kilometres away and that a delay of few days between the time of the earthquakes and the eruption is not uncommon.

His field work showed that a regional fault, the strike-slip Watukosek fault, crosses The LUSI area. A number of extinct mud volcanoes are found aligned with this fault from Java to Madura island. In addition, seismic profiles acquired prior to the eruption shows evidence of a vertical piercement structure with upwards dipping strata around the LUSI conduit zone. He argued that this could be interpreted as evidence for a long history of

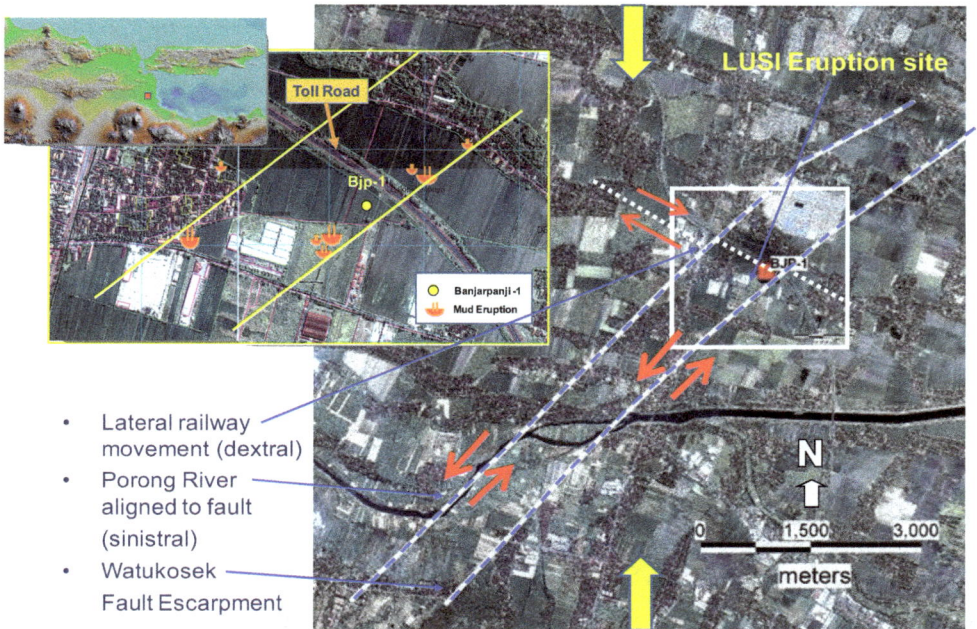

- Lateral railway movement (dextral)
- Porong River aligned to fault (sinistral)
- Watukosek Fault Escarpment

Fig. 8. Watukosek fault, consisting of 2 parallel faults where the Porong River is aligned along the fault line, while the Watukosek fault escarpment represents the up thrown fault block. LUSI eruption sites are along the Watukosek fault line. The Watukosek fault, striking from the Arjuno volcanic complex, crosses the LUSI mud volcano and extends towards the northeast of Java island.

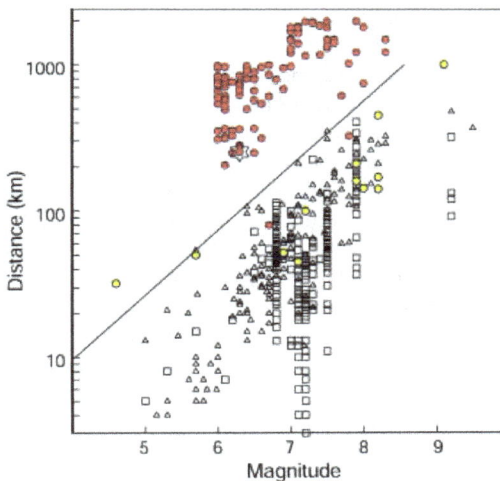

Fig. 9. Distance between the earthquake epicenter and hydrologic response as a function of earthquake magnitude (Manga, M., 2007).

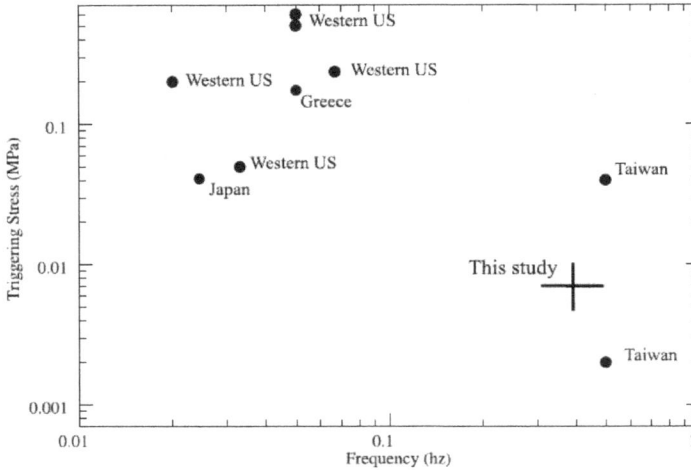

Fig. 10. Values for dynamic stress and frequency of seismic waves that have triggered small seismic events, compiled by Fisher et al. (2008). The cross shows the estimate for the Yogyakarta earthquake at LUSI. Source: Mori and Kano, 2009

active vertical movements of mud underneath LUSI, possibly with former eruptions or as a disturbed signal due to the fault that crosses this area. He suggested that the Yogyakarta earthquake ultimately triggered the eruption through the already overpressured subsurface piercement structure. This is supported by a partial loss of well fluid recorded in the Banjarpanji well nearby 10 minutes after the earthquake, and a major loss of well fluid after two major aftershocks (see previous chapter on The Underground Blowout Hypothesis – figure 7). These mud losses, he argued, could be the result of movements along the fault that was reactivated, lost its sealing capacity and become the passageways for overpressured subsurface fluid to escape. These fluids ultimately reached the surface at several locations aligned NE–SW in the Watukosek fault zone direction.

Davies disagreed with Mazzini's conclusion that the Yogyakarta earthquake reactivated the Watukosek fault and triggered LUSI mud volcano (Davies et al., 2007). He argued that the earthquake was too small and too distant to trigger an eruption when in the recent past, two bigger and closer earthquakes failed to trigger an eruption. He considered the static and dynamic stresses caused by the magnitude 6.4 earthquake too small to trigger LUSI.

Mazzini backed his hypothesis by presenting further field data that support his hypothesis that a strike-slip faulting was the trigger mechanism that released overpressure fluids through already present piercement structures (Mazzini et al., 2009). He presented several observations on the fault reactivation evidence, among others:

- Residents close to the Gunung Anyar, Pulungan, and the Kalang Anyar mud volcanoes, located along the Watukosek fault almost 40 km NE of LUSI (Fig. 1), reported increased venting activity of the mud volcanoes after the Yogyakarta seismic event. Simultaneously, boiling mud suddenly started to erupt in Sidoarjo, later forming the LUSI mud volcano.
- A 1200 m long alignment of several erupting craters formed during the early stages of the LUSI eruption. The direction of these aligned craters coincides with the Watukosek

fault. The craters were formed during May-early June 2006, but were later covered by the main LUSI mud flows.

- Large fractures several tens of centimetres wide and hundreds of meters long were observed in the proximity of the BJP-1 exploration well with identical NE-SW orientation. However no fluids were observed rising through these fractures, which suggests a shear movement rather than a deformation from focussed fluid flow.

The intersection of the fault with the nearby railway clearly indicates lateral movement. The observed lateral movement recorded at the railway during the first four months was 40– 50 cm. The lateral movement recorded at the neighbouring GPS stations during the same time interval reveals at total displacement of 22 cm (2 cm in July, 10 cm in August, 10 cm in September) (see figure 11). This later displacement was possibly related to the gradual collapse of the LUSI structure. In any case, the difference between these two records shows that an initial 15–20 cm of displacement that must have occurred during the early stages (i.e. end of May–June) related to the Watukosek fault shearing. Since 27th May earthquake, the rails have had to be repaired four times. Two of these repairs were done within the first three months after the earthquake to remove the bending due to the continuous shearing.

Fig. 11. Shear stress have damaged nearby infrastructures such as the dextral movements of a railway, bursting of a gas pipeline and numerous breakages of water pipelines at the same location further supports displacements along faults. (A)The railway bent to the west of main vent on September 2006. Offsets that occurred approximately 40 cm with orientation direction NW - SE. (B) At the same location, the railway was bent again in October 2009, with an offset of approximately 45 cm. The bending of the railway line is due to fault reactivation that often has differential movements which created shear stress.

- A water pipeline experienced significant bending and ruptures at the intersection with the fault (Fig. 5A–B). Since the May 2006 earthquake occurred, the pipeline has been repaired sixteen times. Note that neither the rails nor the water pipeline had kink problems before the earthquake.

He also found seismic sections taken in the 1980s that showed a dome-shaped piercement structure; the most spectacular is the collapse structure in the nearby Porong 1 well (Istadi et al., 2009) (see figure 4). This structure is likely to represent an extinct mud volcano that gradually collapsed around its own vertical feeder channel.

Mazzini further showed shear-induced fluidization mechanism through experiment that a relatively small displacement resembling a fault movement can turn a pressurized sand box model from once sealing layers, to become non-sealing. He demonstrated that the critical fluid pressure required to induce sediment deformation and fluidization is dramatically reduced when strike-slip faulting is active. (see Mazzini et al., 2009).

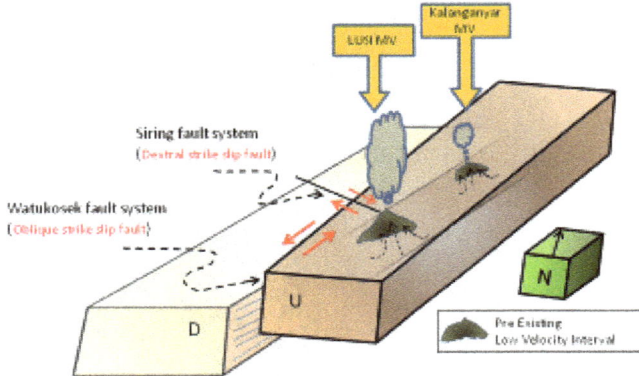

Fig. 12. Schematic cartoon (not to scale) of a mud volcano appearing along strike-slip faults. The shear zone along the Watukosek fault system and Siring fault that crosses LUSI where a low velocity interval existed before the eruption. Reactivation of the strike-slip fault after the earthquake caused the draining of fluids from the low density units towards the fault zone as the preferential pathway.

2.5 Response to earthquake

Due to its tectonic position at the front of the subducting Australian plate under the Sunda plate to the south, Java has been seismically active (see figure 13A). The compressional stresses, either due to subduction or its secondary effect that compresses the Sunda plate in a N-S direction, puts strain on local faults, especially those trending NE-SW. The latter caused a rupture on the NE-SW Opak fault, and had resulted in the magnitude 6.4 Yogyakarta earthquake, on 27 May 2006. This earthquake led to a new understanding of its effect on the volcanic plumbing system of Java Island. At the time of the earthquake, two Javanese volcanoes - Merapi and Semeru, were active; the distance of these volcanoes from the epicenter are around 50 km and 260 km respectively (see figure 13). It was observed that while there was no new volcanic eruption, the eruptive response of the heat and volume flux of these two volcanoes changed considerably by a factor of two-to-three starting on the third day after the earthquake (Harris and Ripepe, 2007). Their work revealed immediate eruptive response through processing of thermal data for volcanic hot spots detected by the Moderate Resolution Imaging Spectrometer (MODIS), (http://hotspot.higp.hawaii.edu). This implies that the earthquake triggered enhanced simultaneous output and identical trends in heat and volume flux at both volcanoes.

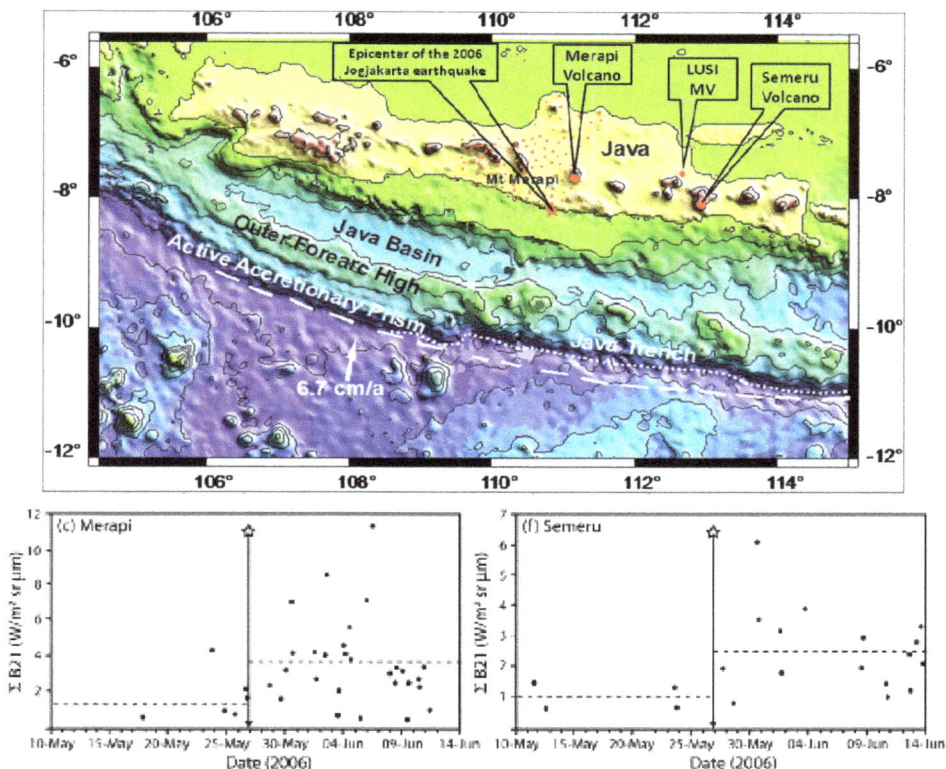

Fig. 13. Map of Java, showing the location of the Merapi and Semeru volcanoes. Increases in heat and volume flux occured 3 days after the Yogyakarta earthquake in the Merapi and Semeru Volcanoes. Thermally anomalous pixels detected by MODVOLC showing all band 21 pixel radiance. Source: Harris and Ripepe, 2007.

It was also reported that the magma extrusion rate and the number of pyroclastic flows from the volcano suddenly tripled [Walter et al., 2008]. This change did not last long, and everything was back to normal again after 12 days. This observation suggests that while this magnitude 6.4 earthquake may not able to trigger a new eruption, it is able to change the intensity of an erupting volcano at a long distance (260 km).

The May 2006 earthquake was one of the deadliest earthquakes in Java in historical times. Although it was as a magnitude 6.4, the scale of destruction was unprecedented in the region. The large scale destruction was concentrated in a 10 – 20 km distance along the Opak River Fault where the subsurface lithology consists mainly of soft volcaniclastic lahar deposit (Walters et al., 2007). Walters study suggests that such deposits have the property to amplify the ground motion such that even a relatively small magnitude earthquake could result in large scale destruction.

The two works of Harris and Ripepe, and Walters suggest the complex interdependency of the causes and effects in a seismically and volcanically active environment. The 27th May

2006 earthquake changed the static and/or dynamic stresses of the area. Their studies suggest a link between earthquake, changes in subsurface condition and its effect on the volcanic activity.

To monitor and record seismic waves around LUSI seismograph installation was carried out at several stations between April and July 2008 (see Figure 14). Seismic waves can be generated by the existing fault activity or by new cracks in the rock layers that had lost their cohesive strength as a result of subsidence around the main eruption vent of LUSI. The microseismic or seismic waves and energy released during crack formation in the rocks is relatively small compared to the energy released by earthquakes.

Microseismic activity recorded by the seismograph network installed around LUSI consists of 6 sensor units, of short period type and broadband seismographs. Each seismograph was

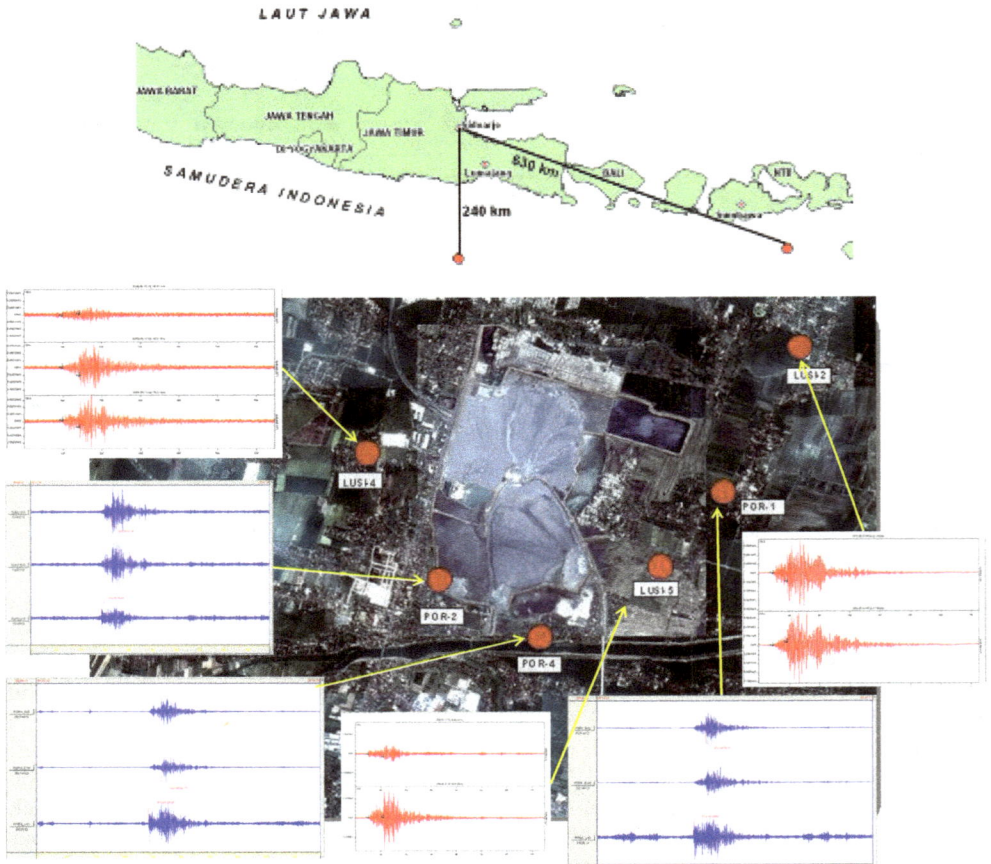

Fig. 14. (A) Epicenter locations of June 1st and 12th 2008 earthquakes located about 240 km and 630 km respectively from LUSI. (B) LUSI Microseismic monitoring network located around the center of the main crater. Seismographs show the June 12th 2008 earthquake with an epicenter located about 240 km South of LUSI.

equipped with a digital recorder system that records continuously for 24 hours, and GPS was used as timing marks on the seismic wave data. Data was processed by analyzing the arrival time of the P wave and S wave. The results of "picking" or "reading arrival rate" was analyzed with appropriate software, to determine the source of vibration.

To determine the location of the vibration source or microseismic hypocenter requires seismic wave velocity data at LUSI location. Wave velocity data was obtained from seismic surveys and wells logging data during drilling. Processed results in the form of coordinates of the location of the source of the wave system are plotted in three dimensions, so that the pattern of its occurence can be seen clearly. To facilitate processing, field data which is a mixture of different frequencies and microseismic noise are filtered, so as to identify microseismic events, arrival time, P wave and S wave, maximum amplitude and duration. All data was processed to determine the parameters of microseismic, namely: the timing, location coordinates, depth and magnitude. The results of the data processing are classified into two types of earthquakes, namely: the earthquake which occurred outside LUSI, and those that occurred around LUSI. In this case we will focus on earthquake data that occurred outside the LUSI area to determine earthquake response to changes in temperature, gas flux and behaviour that occur in the main vent.

The ability to detect an earthquake depends on the magnitude of the earthquake, the sensitivity of the sensors (seismometers), and the distance between the hypocenter and the location of the sensors. In general, earthquakes in Indonesia with magnitudes above 5.0 on the Richter scale, will be recorded by almost all seismograph networks in Indonesia. Like the two above mentioned tectonic earthquakes, wave energy can propagate from the source to the sensor around LUSI, with greater strength than the noise level around the sensors.

No.	Stations	Coordinates		Periods	Agency
		Latitude	Longitude		
1	POR 1	-7.53084	112.73086	29 April – 5 July 2008	BMKG
2	POR 2	-7.54043	112.70377	29 April – 5 July 2008	BMKG
3	POR 4	-7.54414	112.71470	29 April – 5 July 2008	BMKG
4	LUSI 2	-7.51485	112.74049	29 April – 5 July 2008	BMKG
5	LUSI 4	-7.52660	112.69772	29 April – 5 July 2008	BMKG
6	LUSI 5	-7.53700	112.72535	29 April – 5 July 2008	BMKG

Table 1. Coordinates of microseismic network stations in the area LUSI.

During the monitoring period two tectonic earthquake occurred outside LUSI. These are:

1. June 1, 2008, Time 15:59:50.2 GMT, the epicenter was located at latitude 9.53° South - longitude 118.04° East, at a depth of 90 km with a magnitude of 5.5 SR, about 630 km from LUSI

2. June 12, 2008 At 05:19:55 GMT, the epicenter was located at latitude 9.68° South - longitude 112.67° East, at a depth of 15 km and magnitude of 5.4 SR, about 240 km from LUSI.

In addition to microseismic monitoring, temperature, LEL (low explosive limit- in air where 20% LEL corresponds to 10000 ppm), and H_2S concentration monitoring was continuously performed using portable monitoring equipment by BPLS (Sidorajo Mud Mitigation Agency) officers in the field. Measurements from 1 to 20 June 2008 showed a fluctuation LEL, H2S, and temperature at the center of eruption. The peak value of the measurement period occurred on June 12 and 13, 2008, in which all measurement parameters rose sharply, particularly temperature and the concentration of H_2S (see figure 15).

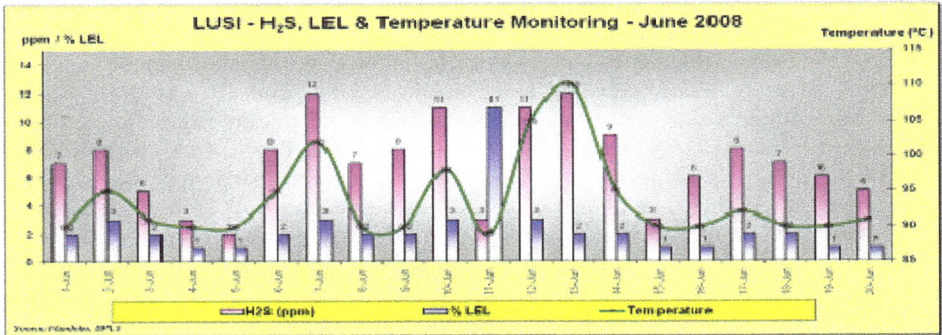

Fig. 15. Correlation between LUSI mud volcano activity and earthquakes. Increasing gas expulsion, temperature and mud eruption rates after earthquake are shown in the above graph after the 12th of June 2008 5.5 Mw earthquake. The epicenter was located some 240km South of LUSI.

The increase in temperature positively correlates with data from the installed seismograph network around LUSI which showed an earthquake occurred approximately 240 km south of LUSI on June 12, 2008. In The case of LUSI, the earthquakes have affected the rheology of fluid in term of permeability, changing the viscosity and the rate of mud eruption, consequently the increased concentration of expelled gases and temperature.

2.6 Horizontal displacement

Geodetic measurements were conducted at the LUSI site to quantify the ongoing deformation processes. The primary data sources were the GPS surveys periodically conducted at monitoring stations to measure vertical and horizontal movements relative to a more stable reference station. Seven GPS survey campaigns were conducted between June 2006 and April 2007. The GPS measurements were conducted at 33 locations using dual-frequency geodetic type receivers over various time intervals. Each measurement lasted from 5 to 7 h. (Istadi et al., 2009).

Areas within a 2–3 km radius of LUSI's main mud eruption vent are experiencing ongoing horizontal and vertical movement aligned to major faults. The horizontal displacements have spatial and temporal variations in magnitude and direction, but generally follows the two major trends, namely in the direction of NE - SW and NW – SE (see figure 16). Rates of horizontal displacement are about 0.5–2 cm/day, while vertical displacements are about 1–4 cm/day, with rate increasing towards the extrusion centre (Abidin et al, 2008).

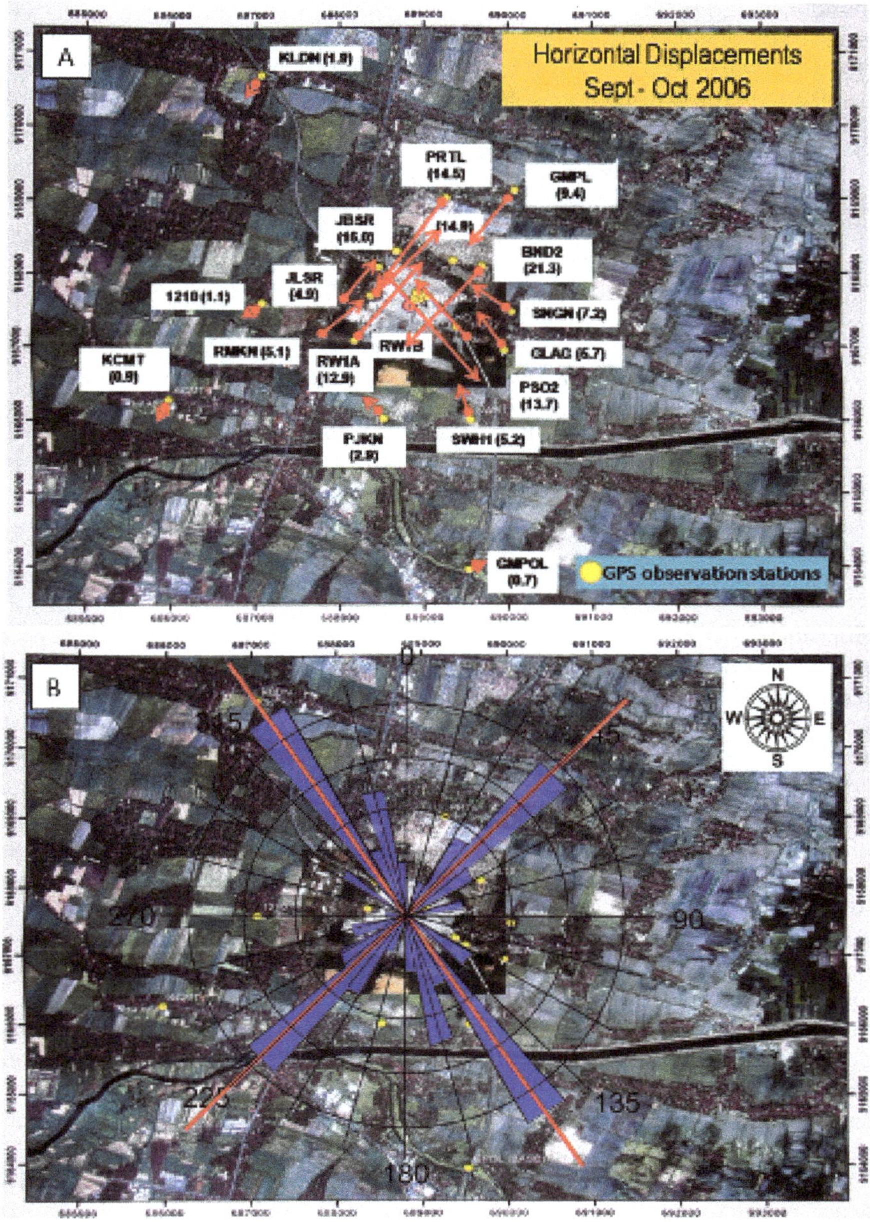

Fig. 16. (A). Horizontal displacement measurements in September - October 2006. Directions of the red arrows show the direction and magnitude of movement. (B). Measurements from June 2006 - March 2007 indicate the major trends are NW-SE and NE-SW as seen in the rose diagram.

2.7 Subsidence and uplift

Five years after the mud eruption, the area near LUSI has subsided at a considerable rate. Buildings and houses near the eruption site have completely disappeared under layers of mud. However, in the east and northeast uplift is occurring. To measure both the subsidence and uplift, four survey campaigns were conducted (Table 2):

Start	End	Points	Method
July 2006	March 2007	25	GPS
Dec. 2007	April 2009	30	Total Station
Dec. 2008	Feb. 2011	15	GPS
Dec. 2008	Feb. 2009	5	Level

Table 2. Four survey methods to measure elevation near LUSI MV

Data from these four surveys was used to show the changes in elevation, subsidence and uplift, as well as horizontal movement over time. Subsidence contour maps were created using GIS software by interpolating the measurement data. The results showed an almost concentric pattern shown in Figure 17.

The subsidence started as a crack in the ground that continued to grow and decrease its elevation. The existence of subsidence was evidenced by, among other things, the pattern of ground cracks, tilting of houses, cracking of flyover and bridges, as well as collapsing of buildings. The direction of the cracks varies depending on its location. In the Renokenongo area, southeast of LUSI, the cracks direction is NE- SW, whereas in West Siring area, west of LUSI, the cracks are North-South.

Subsidence and horizontal movements indicate the dynamic geological changes in the area. These movements have caused reactivation of pre-existing faults or newly formed faults. The continued movements along faults would likely result in the emergence of more fractures and gas bubbles (see figures 17 and 18).

Subsidence continues as the mud eruptions progress. The subsidence might result from any combination of ground relaxation due to mudflows, loading due to the weight of mud causing the area to compact, land settlement, geological structural transformation and tectonic activity (Abidin et al., 2007).

Based of field measurements, areas up to 3 km from the main eruption vent are experiencing subsidence to some degree. Presently however, due to much reduced volumes of mud eruption, the measured rate of subsidence on the West side of main eruption vent indicate a decrease from the original 25 cm/month when LUSI was very active in the first year, to less than 5 cm/month. If the decreasing trend continues, the affected subsidence area will likely decrease from earlier prediction of more than 3-4 km.

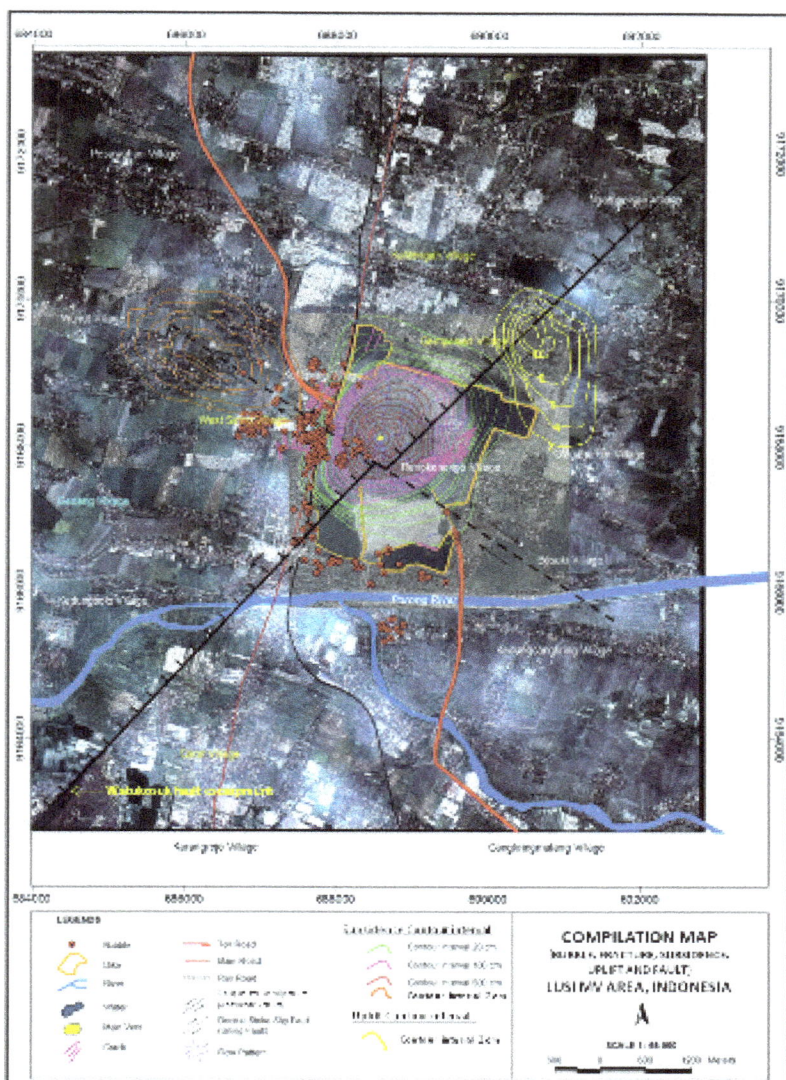

Fig. 17. LUSI post eruption map. The subsidence contour is status as of January 2010, constructed by interpolating the measurement data, and was created using GIS software. The contour showed an almost concentric pattern. The area West of the main vent was subsiding faster than other areas.

The map also shows fractures distribution around LUSI. East of the main vent, fractures trend NE - SW, whereas West of the main vent the fracture trend is North-South.

The Gas bubble distribution around LUSI status in May 2011 where more than 220 gas bubble locations have been recorded since the start of LUSI eruption in May 2006. Presently only a few are still active.

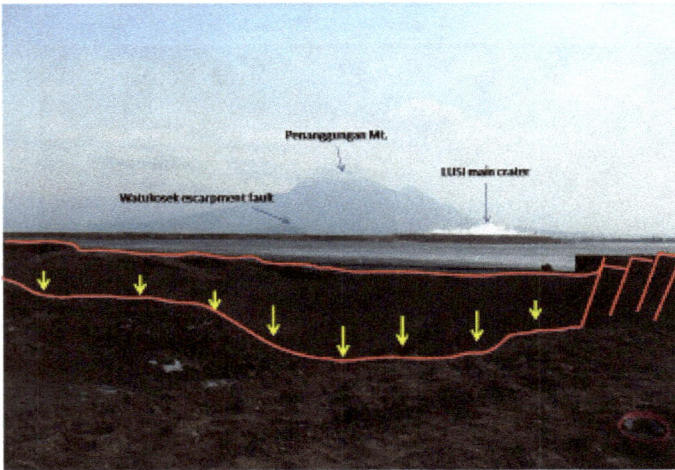

Fig. 18. Photo showing subsidence and collapse of the retaining mud dyke northeast of the LUSI main vent that occurred on 21 May 2008. In some parts, where slumping and subsidence occurred, local small scale faulting at the edge of subsiding wall occured. The continued subsidence proves very difficult to maintain the dyke.

2.8 InSAR data

InSAR (Interferometric Synthetic Aperture Radar) is a technique to map ground displacement with a high resolution of up to centimeter-level precision (e.g. Massonnet and Feigl, 1998; Hanssen, 2001). InSAR is effective tool to measure the amount of ground deformation caused by earthquake, volcanic activity has been useful for studying land subsidence associated with ground water movements (e.g. Amelung et al., 1999; Gourmelen et al., 2007), mining (e.g. Carnec and Delacourt, 2000; Deguchi et al., 2007a), and geothermal as well as oil exploitation (e.g. Massonnet et al., 1997; Fielding et al., 1998). The amount and pattern of deformation are shown by a range of colors in the spectrum from red to violet. The computed interferograms are interpreted using an inversion method that combines a boundary element method with a Monte-Carlo inversion algorithm (Fukushima et al., 2005). In LUSI, this technique was used to determine the surface deformation due to the mudflow starting from 19 June 2006 (three weeks after the mud eruption) to 19 February 2007. The measurement was done using PALSAR (Phased-Array L-band SAR) onboard the Japanese Earth observation satellite ALOS. Measurement of land subsidence is possible as the L-band microwave is less affected by vegetation (Deguchi et al., 2007a).

Deguchi et al. (2007a, 2007b) and Abidin et al. (2008) performed a study and measured the ground subsidence temporal changes of deformation obtained by applying time-series analysis to the deformation results extracted by InSAR.

- From 19 June 2006 to 4 July 2006 the subsidence showed an elliptical pattern, suggesting subsidence around the main vent and west of the main vent.
- From 4 July 2006 to 19 February 2007, the scale of subsidence and uplift became more significant. Both subsidence and uplift East of the main vent became more pronounced. In contrast to the high rate of mud eruption however, the InSAR results clearly showed that the ground deformation associated with mud eruption decreased after November 2006.

The results from the use of InSAR indicate subsidence has occurred in this area. Four different areas of deformation is suggested, these include areas centered around the main eruption vent; areas to the west-northwest of the main vent; areas to the northeast of main vent; and to the southwest of the main vent. Apart from the areas to the west-northwest which is associated with the deformation due to gas production in Wunut gas field, the other 3 deformation areas follow the regional fault pattern, contiguous to the Watukosek NE-SW fault trend.

The results also demonstrate the progressive subsidence evolution from time to time during the period of measurement. Subsidence in the main eruption area showed the most rapid subsidence rates. The 8-months measurements period showed ellipsoidal subsidence pattern covering an area of approximately 2 x 3 km2 with a long axis trending NE-SW.

Another area to the west-northwest of the main eruption area is also experiencing subsidence. This particular area is within the Wunut gas field which covers approximately 2 X 2.5 km2 with long axis trending NW-SE. This trend corresponds to the regional Siring NW-SE fault trend.

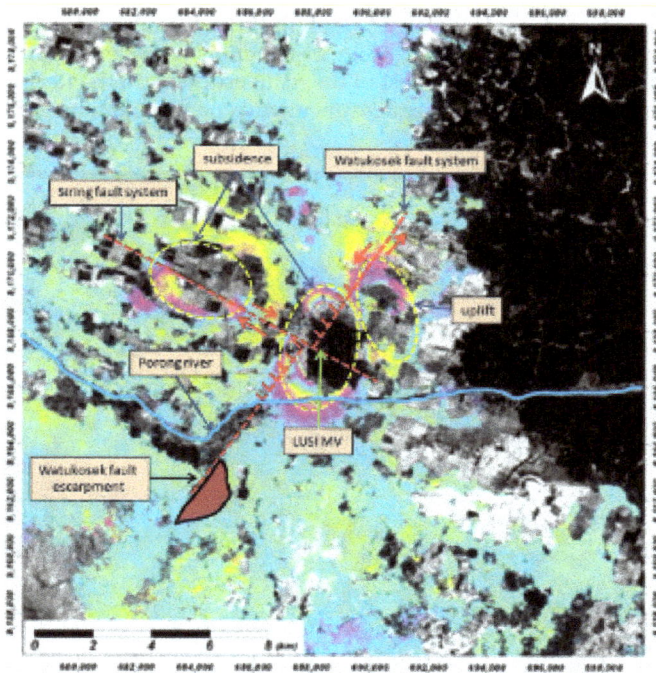

Fig. 19. The interpreted results of InSAR satellite imagery in February 2007 suggest an elliptical subsidence along the NW - SE long axis with a distance of 1-2 km from the main eruption vent, namely in the area around West Siring and Pamotan. In the vicinity of the main mudflow and the eastern regions about 2.5 km northeast of the main eruption, the subsidence occurred elliptical on the N-S long axis.
(figures modified from Deguchi et al, 2007)

Fault reactivation resulted in horizontal and vertical movement, which later manifested in the formation of uplift and subsidence or vertical and horizontal offset. An overlay of the ellipsoidal InSAR measurements with regional faults in these areas indicate a correlation between the two. Elipsoidal uplift suggest the long axis trending NNW - SSE is a restraining stepover to offset oblique strike slip fault of the reactivated Watukosek fault.

It is interesting to note that the InSAR measurements found that the deformation diminished after November 2006, only 6 months after the start of the eruption. Interpretation of interferogram for each periodic cycle for the period of May to July 2006 (beginning of eruption) showed more temporal change of deformation compared to the period of November 2006. In contrast, during the period of October - November 2006 field observations indicate increasing intensity of subsidence in the western side of the main vent, particularly in the village of Siring Barat. The main eruption vent and surrounding central area were experiencing most rapid rate of subsidence and continual collapse of the mud retaining dykes. Areas to the E-NE of the main vent were experiencing increases in uplift. The indication of contrasting InSAR measurements could be interpreted as lesser or diminishing effect of initial fault reactivation that triggered LUSI.

Interpretation of interferogram by Deguchi suggesting psudo anomaly in an area to the northeast of the main vent and does not indicate uplift based on conversion to rectangular coordinates (see Deguchi et al, 2007b). Field observation however suggest an uplift has occurred in areas to the east and northeast of the main eruption, in the Renokenongo village and surrounding areas. The uplifted area covers an area of approximately 1 X 1.5 km2 with a long axis trending NNW – SSE.

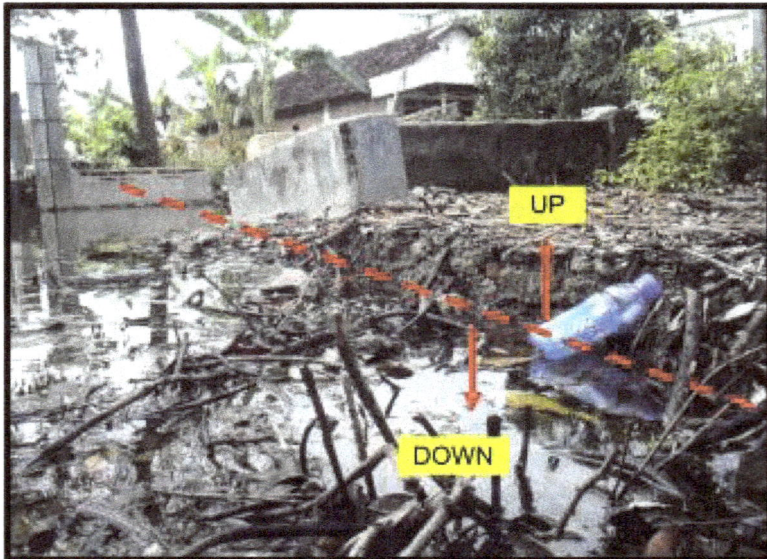

Fig. 20. The pattern of fractures trending NE -SW in the Village Renokenongo. A section of land on the right hand side of the picture is uplifted (east side) while the left is the downthrown block (west side). Note: The mineral water bottle is used as a comparison to indicate the amount of displacement (~20cm). In contrast to Degushi et al., 2007b psudo anomaly interpretation, the above photo taken 2 months after the eruption suggests displacement due to fault movement. Movement due to subsidence was unlikely as it was minimal at the early stages of the eruption.

2.9 Fracture orientation

Fractures appeared around LUSI area as a result of loss of cohesion due to ground movement, both vertical and horizontal movements. These fractures were concentrated mainly to the East of the main eruption (Renokenongo village), around the main vent and to the West (Siring Barat village), with displacements of varying degree and magnitude. The fractures follow the sinistral Watukosek NE – SW trend. Juxtaposed with the Watukosek fault reactivation, is the Siring fault movement that trends NW – SE which has dextral strike slip movement. These fractures were caused by reactivation of faults but their orientation pattern are often not apparent due to thick alluvial cover.

Fig. 21. (A). On June 2, 2008 the dyke on the East side of the main vent broke with an orientation NE-SW. Then on June 8, 2008 the 40 m long dyke collapsed as deep as 6 meters. (B). Fractures on the West Siring village west of the main vent showed an orientation trending North – South. (C)&(D) an active fault is located west of the main vent and trends North – South.

2.10 Gas bubbles

Gas bubbles of various sizes and pressures started to appear two days after the mud eruption. Those that appear from water wells generally have a higher pressure and high methane concentration than bubbles from surface fractures (see figure 22). The ejected materials from these gas bubbles typically had some water, mud with minor sand. A total of over 220 gas bubble locations have been identified since the start of the eruption, however the number that are still active continually decrease. Presently less than 20 gas bubbles are still active, suggesting LUSI is entering a more stable and less active phase.

Gas bubbles are not continuous; they may burst for several weeks or months then stop and reappear elsewhere. Some gas bubbles appear in straight lines that are contiguous with the fault trends. These gas bubbles are mainly concentrated on the West and South of the main eruption which reflect the existence of subsurface gas accumulation breached by deep fractures. The gas accumulation is believed to be a part of the Wunut gas field flanks with its sealing capacity breached by the reactivated Watukosek faults or newly formed fractures as a result of rapid subsidence in the area.

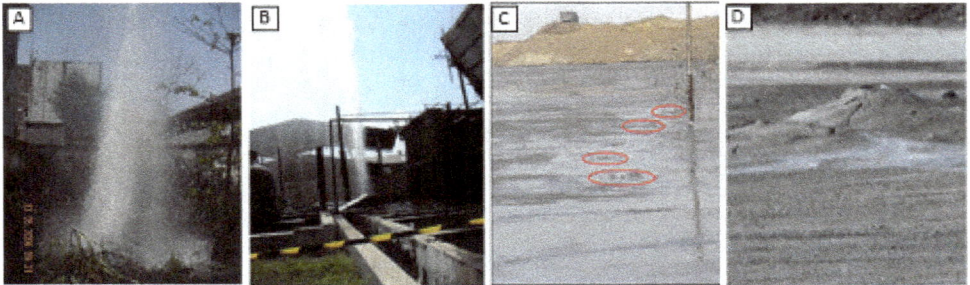

Fig. 22. (A) & (B) The gas bubble originating from water wells, with tremendous pressure and high content of methane gas. Besides removing water,the bubbles also ejected sand, shell fossils and a bit of mud from the swamp sediments. (C) Gas bubbles along the fracture to the west of the main vent, low pressure and in clusters. (D) a Gryphon located approximately 400 m west of the main vent.

Gas bubbles around the mud volcano have formed gryphons of around 30 cm in diameter and height of around 40 cm (see figure 22D). The ejected material was mainly methane gas and some water (see figure 22 A and B).

2.11 Source of mud, water, gas and heat

Mud material ejected from the mud volcanoes is believed to have originated from shale layers known as 'Bluish Gray' clay of the Upper Kalibeng Formation of Plio-Pleistocene in age. The similarity between the mud and the cutting samples from the nearby well Banjarpanji-1 from a depth of 1220 – 1828 meters is based on the following:

1. The similarity of foraminifera and nanno fossil collection, as well as index fossils containing Globorotalia truncatulinoides and Gephyrocapsa spp. that are Pleistocene in age. Benthos Foram collection shows that the sediment was deposited in the marine environment in the inner to middle neritic zones, ranging from shoreline to a depth of 100 meters.
2. Kerogen composition correlates with the side wall core from Banjarpanji -1 at a depth of 1707 m.
3. Thermal maturity based Vitrinite reflectance (Ro) correlates with cuttings and side wall core samples from Banjarpanji-1 at a depth of 1554-1920 meters.
4. Clay mineral composition has similarities with samples from the side wall core from Banjarpanji -1 at a depth of 1615-1828 meters where the illite content in illite-smectite mixture reached 65%.

LUSI muds contain various types of clay including smectite, kaolinite, illite and minor chlorite. It is known that illite minerals form at temperatures between 220 to 320 °C, smectite forms at surface temperatures of up to 180 °C and altered minerals chlorite forms at temperatures between 140 - 340 °C. The XRD analyses carried out on core samples from Banjarpanji-1 imply an intensive progressive smectite-illite transformation with the depth. This suggests that the intersected Upper Kalibeng Formation was exposed to a minimum temperature of 220 °C.

In the initial stages the volume of water in LUSI was very large reaching up to 70% of the total volume of mud with an average salinity of 14,151 ppm NaCl. The lower salinity than sea water suggests dilution. At the time of writing, the liquid composition made up 30% of the total volume. The source of water has been debated by various researchers. Davies et al (2007) states that the water originates from the carbonate Kujung formation, while Mazzini et al (2007) based on geochemical data concluded that the high-pressure water is derived from clay diagenetic dehydration of Upper Kalibeng Formation.

Indonesian Geological Agency, Ministry of Energy and Mineral Resources in 2008 conducted water analysis by using the Oxygen isotope method (ο18O) and Deuterium (ο D) to determine the magmatic origin of LUSI. Results showed deuterium concentration (οD) from -2.7‰ to -13.8 ‰ and Oxygen-18 (ο18O) from +7.59‰ up to +10.11‰, and high Chloride content of 12,000 – 17,000 ppm. Based on the above they concluded that the water source of LUSI is associated with igneous rock sourced from magma (Sutaningsih et al., 2010).

Perhaps it is quite impossible to determine the source of water as it could be a mixture of different sources, ie. clay diagenetic dehydration, carbonates, deeper source linked to geothermal, trapped water due to disequilibrium compaction and mixing with shallow meteoric waters. The importance of determining the source is for hydro-geological purposes, in handling the impact of the mud flow, its effect to the environment and contamination to the ground water. If the fluid is old (Tertiary), it is trapped water, whereas, if the water is young (Quaternary), it is likely to be recharged upslope. For the former, naturally, the eruption will stop after a certain time, whereas, for the later condition, the eruption will never stop (Hutasoit, 2007). However, Sunardi et al., 2007 suggested that LUSI will likely stop when hydrostatic pressure equilibrium is reached.

Groundwater samples from LUSI and its surrounding gas bubbles near the main vent have been chemically analyzed for major anions (Cl-, HCO_3-, and SO_4^{2-}) and cations (Na+, K+, Ca^{2+}, and Mg^{2+}). The result shows that there is a significant difference in water chemistry between the main vent and the bubble. The concentration of Cl-, Na+, Ca^{2+}, and Mg^{2+} in the main vent water are much higher. This suggests that the water may be from different sources, or both are from the same source, but the gas bubble water has been diluted by shallow groundwater. The second case implies that the pressure in the gas bubble areas may be depleting so that shallow groundwater is mixed with deeper sourced water. If the pressure is still high, then flow from the eruption area will contaminate the shallow groundwater. In either case, the ongoing subsidence is also caused by the decreasing pore pressure as the water is discharged to the surface.

The composition of the erupted gas sampled in July in the proximity of the crater showed CO_2 contents between 9.9% and 11.3%, CH_4 between 83% and 85.4%, and traces of heavier hydrocarbons. In September, the steam collected from the crater showed a CO_2 content up to 74.3% in addition to CH_4. Simultaneously, the gas sampled from a 30.8 °C seep 500 m away from the crater had a lower CO_2 content (18.7%). The four gas samples collected during the September campaign were analysed for $\delta^{13}C$ in CO_2 and CH_4. The $\delta^{13}C$ values for CO_2 and CH_4 vary from $-14.3‰$ to $-18.4‰$ and from $-48.6‰$ to $-51.8‰$, respectively (Mazzini et al., 2007). The relatively low $\delta^{13}C_{CH4}$ ($-51.8‰$) indicates input from biogenic gas mixed with a thermogenic contribution. The biogenic gas was derived from immature shale layers, probably from the overpressured shale at a depth of 1323-1871 meters, whilst the thermogenic gas was derived from shale layers that are more mature, probably of Eocene age. The CO_2 is postulated to come from the dissolved CO_2 in the water of the shale layer, at temperatures above 100 °C and low pressure. The constant presence of H_2S since the beginning of the eruption could also suggest a contribution of deep gas or, most likely, H_2S previously formed at shallow depth in layers rich in SO_4 and/or methane or organic matter. The rapidly varying composition of the erupted gas indicates a complex system of sources and reactions before and during the eruption (Mazzini, 2007).

Temperatures measured from a mud flow within 20 m of the LUSI crater revealed values as high as 97 °C (Mazzini, 2007, 2009). Given the visible water vapor and steam this suggests temperatures above 100 °C. The heat source of the erupted mud is believed to be from a formation at a depth of over 1.7 km where the temperature is over 100 °C. Geothermal gradients of c. 42 and 39 °C/km have been reported in the area. With such a high temperature gradient, LUSI can be viewed as a geo-pressured low temperature geothermal system that discharged hot liquid mud close to its boiling point the first four years of its life (Hochstein and Sudarman, 2010). Hochstein believed that the high temperature gradients are likely due to the low thermal conductivity of the highly porous, liquid saturated reservoir rocks. Mazzini, on the other hand, believed that the high geothermal gradient is due to the close proximity to Mount Arjuno-Welirang (about 40 km), which is part of the Java volcanic arc that formed since the Plio-Pleistocene (Mazzini, 2007, 2009).

Two shallow ground temperature surveys carried out in 2008 showed anomalously low temperatures at 1 m depth (possibly due to a Joule-Thompson effect of rising gases) and liquid mud temperature that varied between 88 and 110 °C with the highest temperatures occurring after a large, distant earthquake. The mud temperature of mud volcanoes is controlled by the gas flux (endothermic gas depressurizing induces a cooling effect), and by the mud flux (mud is a vector for convective heat transfer) Deville and Guerlais (2009).

2.12 Geomorphology of the area

In general, the geomorphology in Porong and the surrounding area is divided into 5 units: Under the volcanic slopes unit, Foot volcanic plateau unit, Cuesta unit, Alluvial plains unit, and Mud volcano unit. The geomorphological units division is based on morphology, the height difference and slope (Desaunettes, 1977).

2.12.1 Under the volcanic slopes unit

The unit is located at the northern foot of the Penanggungan mountain or in the Proximal facies. This unit is distributed mainly in the southern area of LUSI, adjacent to the mountain range. Lithologic constituents of the unit are generally in the form of volcanic breccia, tuff, lava, tuffaceous breccias, lava and agglomerates and the presence of shallow andesite intrusions in small dimensions. The dominant process in this unit is volcanism. Volcanism processes of Penanggungan Mountain produce volcanic cone morphology. The pattern of distribution in this area is a radial pattern.

2.12.2 Foot volcanic plateau unit

This unit has the morphology of the plains at the foot of Penanggungan mountain or in the medial facies. The unit was formed from the deposition of material surrounding the volcano eruption as laharic. Laharic deposits are found in the form of loose sand and gravel to boulder-sized fragments as products of volcanic eruptions. There is a wide variety of bedding igneous rock fragments to the level of weathering, colors and dimensions.

Lithologic constituents of this unit are fine tuff, sandy tuff, tuff and tuffaceous breccia. The dominant processes in this unit are erosion and sedimentation. The pattern of distribution in this area is a radial pattern.

2.12.3 Cuesta unit

The Cuesta unit is primarily distributed in the southern area of LUSI. The highest point is at an elevation of 150 m at the top of the Watukosek hill. The lowest point is at an elevation of 20 m on the valley of Watukosek. The dominant process in this unit is a tectonic process of faulting, which resulted in shear faults and the down thrown block to the West to form a steep escarpment in the area of Watukosek. This escarpment is known as the Watukosek Escarpment. Lithologic constituents are of andesite breccia, sandstone and tuff. Morphology in the region reflects the existence of Watukosek fault as indicated by the presence of steep slopes on the western escarpment while relatively gentle on the eastern slopes. The pattern of distribution in this unit is trellis pattern.

2.12.4 Alluvial plain unit

Alluvial plains unit make up most of the area and are widely distributed near LUSI. Geomorphological slope is approximately 0 -5%. Lithologic constituents are loose sand deposits, clay, sandy clay.

Fig. 23. LUSI area showing the division of volcanic facies. The central facies is located at the top Penanggungan mountain , proximal facies on the upper slopes and medial facies on the foot slope below the mountain. LUSI overlies the alluvial plains which are approximately10 km from Penanggugan mountain.

This geomorphological unit is controlled by alluvial rivers. Geologic processes that act on this unit are erosion, transport and deposition. Lateral erosion took place due to slopes of the mountains to the South causing lateral erosion to be more effective than vertical erosion. In this area there are large rivers namely Porong River which is flowing from West to East that ends up in the Madura Strait. Structural control is clearly visible on the morphology in this area evidenced by the abrupt deflections in the Porong River that follows the fault pattern.

2.12.5 Mud volcano unit

The unit was formed due to discharge of mud from formations below the surface. The morphology is like a low relief hill. The mud volcano Unit is limited by the retaining dykes so that the mud does not spill over into surrounding areas. This unit includes the Village of East Siring, Jatirejo, Tanggulangin Glagaharum, Ketapang and surrounding areas. Lithologically this unit is predominantly the mud itself that contain some fossils.

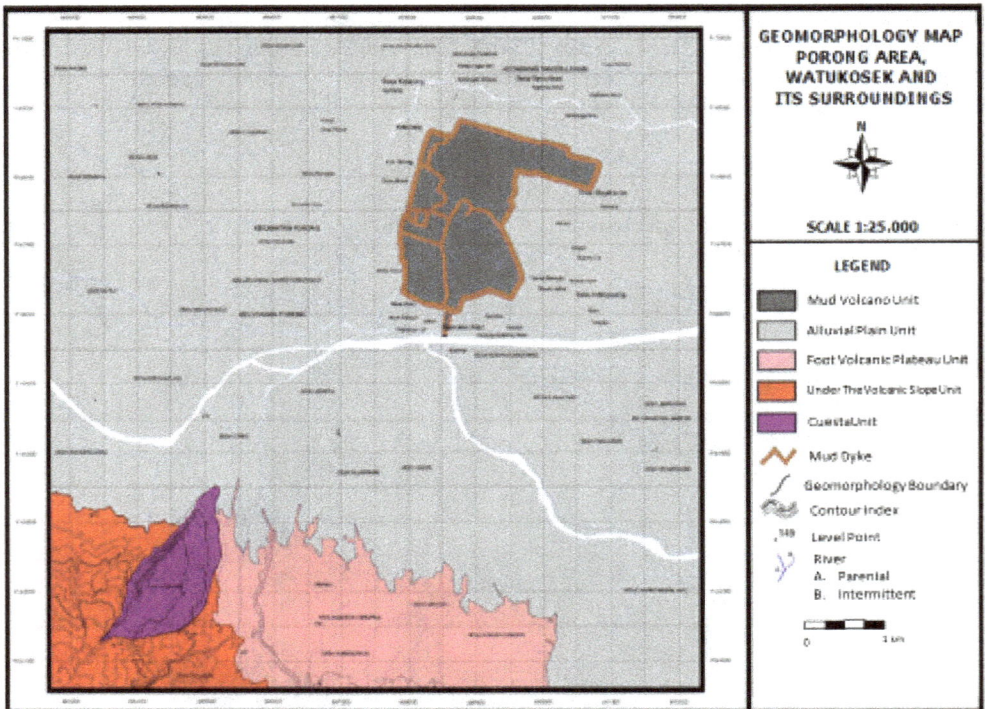

Fig. 24. Geomorphology map of the Watukosek area.

The morphological shape of LUSI is a semi-conical buildup with a peak around the main eruption vent. It is similar with the mud volcano models developed by Kholodov (1983) and Kopf (2002) where LUSI is classified as a swampy mud volcano type. The peak is not high due to the low viscosity of the extruding mud.

2.13 Forming of the Crater

The series of photographs in figure 25 below represent the changes through time at the main vent of LUSI Mud Volcano. In a time span of 5 years, LUSI has evolved from a small eruption of steam, hot water and mud to a destructive high rate mud flow, engulfing houses, schools, factories, neighboring villages and caused a large-scale ground deformation, damaged the highways, railroad, pipelines, electrical power lines and others; to presently a much more calm low rate ejection of mud and fluid and occasional intermittent stopping of steam eruption. LUSI evolved through time from a localized kilometre-scale fault zone in 2006 and expanded through pre-existing NE-SW Watukosek fault zone pathways in 2010.

May 2006
LUSI mud volcano on its first day, May 29th, 2006. The mud eruption is approximately 200 m from the Banjarpanji-1 well location. Initial eruption in the form of mud and hot water and clouds of steam with a discharge rate of less than 5,000 m³/day.

June 2006
In June, the crater had swelled and the discharge has reached approximately 50,000 m3/day, with water temperatures as high as 97 °C.

August 2006
A semi conical structure is starting to form. The volume of water in LUSI is very large reaching up to 70% of the total volume of mud, shown in the picture as water reflection. The low viscosity of the mud results in mud spreads across, extending to large areas instead of building up vertically.

May 2007

In May 2007, retaining walls/dykes were built to prevent the mud from spilling over to the villages and major roads. The height of the dykes encircling the center of eruption reached approximately 15 m. The average mud flow rate at the time was around 100,000 m3/day.

May 2008

Ring levees/dyke were rebuilt and raised to prevent overflow of mud into the closest villages. Discharge rate was still around 100,000 m_3/day with surface water temperatures remain constant at 97 °C.

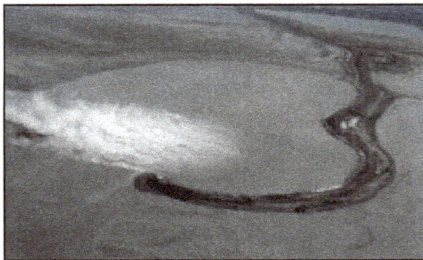

February 2009

By February 2009, the ring dyke on the south and north sides are rapidly sinking due to subsidence and are difficult to maintain despite efforts to continuously pile with soil and gravel.

July 2009

The ring dyke around the main vent collapsed and sank in July 2009. The eruption discharge rate at this time is reduced by 60% to approximately 40,000 m^3/day.

January 2010
The diameter of the main crater at this time is approximately 120 m and flowing continuously 30-50,000 m3/day. At times instead of a single crater, it changes to two or three points aligned in the direction of the Watukosek fault. The flow is mainly liquid and hot steam. A gently sloping cone is starting to form. The mud covered area is mostly wet, covering 80% of the total area.

January 2011
At this time the mud flow rates and the scale of steam clouds are reduced. The eruption rate has decreased to less than 10,000 m3/day. LUSI is now entering a new phase, from an eruptive one to a mature and quiescence phase. The mud around the main vent is solidifying forming a dome.

May 2011
The mud volcano viewed from the west side. Note the reduced scale of the clouds of steam.

May 2011
The mud volcano viewed from the north side with the Watukosek escarpment hills and Mt. Penanggungan in the background.

Fig. 25. Changes from time to time at the main vent of LUSI Mud Volcano

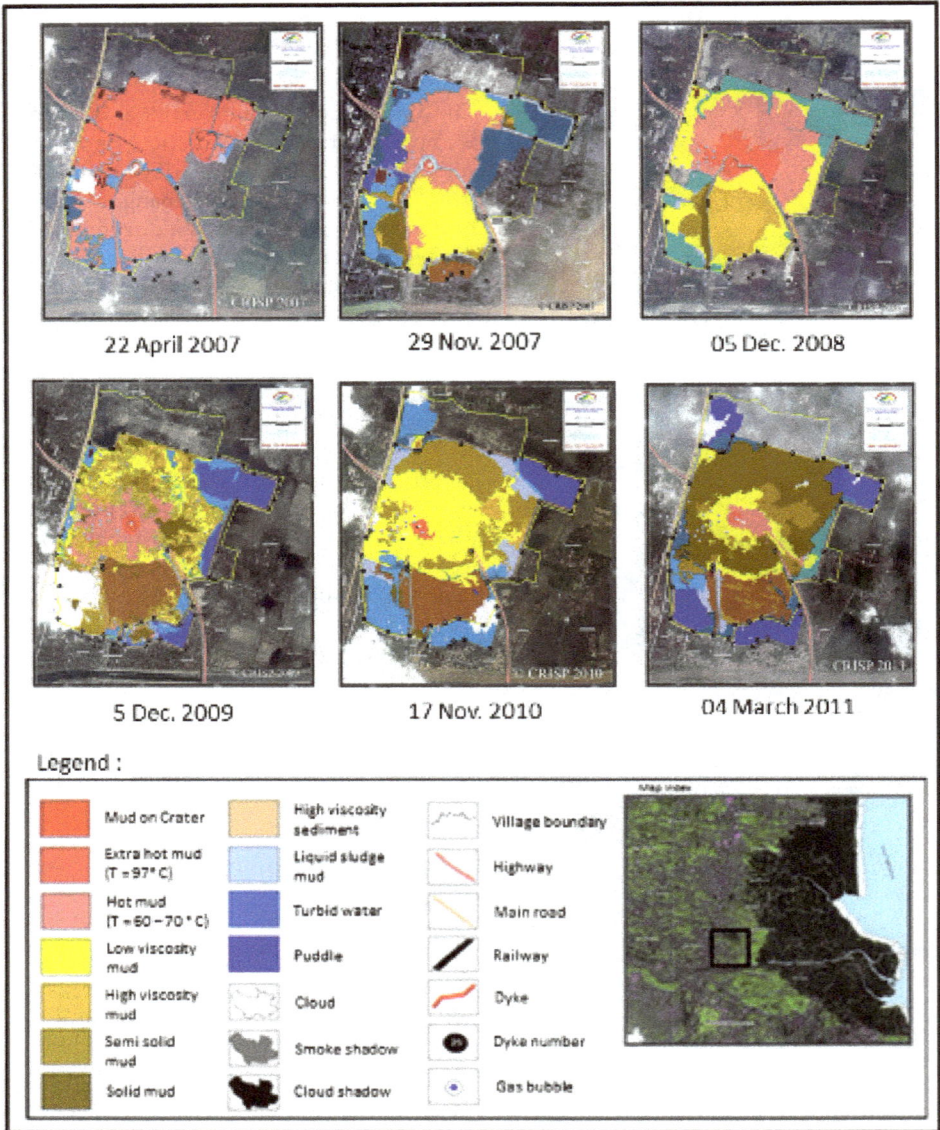

Fig. 26. Map interpretation of the results around the center of LUSI from IKONOS imagery using ERMAPPER Software from 2007 to 2011. The interpretation shows a decrease in the volume of hot mud around the main vent. In 2007, almost all the fluid inside the dyke is above 60°C. By November 2007, the rate of hot mud has begun to diminish. A year later, in December 2008, the LUSI morphology dome has begun to form. The next phase was the reduced production of hot mud in the main vent and the heightened dome coupled with the formation of the patterns of mud and water flow in the vicinity. (source: BPLS 2011)

2.14 Morphological changes

Spatial Aerial Photo Analysis was performed utilizing the CRISP satellite map regularly obtained from the Centre for Remote Imaging, Sensing and Processing, at the National University of Singapore (http://www.crisp.nus.edu.sg). Changes at the center of the eruption and the adjacent slopes can be observed.

The area of observation was between -7°27'04" / 112°40'27" and -7°35'52" / 112°49'35" an area of about 3.7 km x 4.0 km or 14.9 km^2, with the focus of the coverage area on the mud ponds.

The data processing stages were as follows:

1. Processed multitemporal IKONOS image data obtained from the CRISP in 2007 until 2010. ERMAPPER software was used for image enhancement, image correction, interpretation and image classification.
2. Spatial or geographical information analysis, information visualization, organization of information, combining thematic information using GIS software.
3. Classification of information based on the detection and identification of objects on the surface of the earth from a satellite image, to the next as the primary identifier elements and limits of an object is done by coloring.

3. Geohazard

The mud eruption in Sidoarjo has buried houses, villages, schools, factories, and displaced thousands of people and continues to pose a geohazard risks in a densely populated area with many activities and infrastructures. Studies of other mud volcanoes in East Java were used in the geohazard assessment.

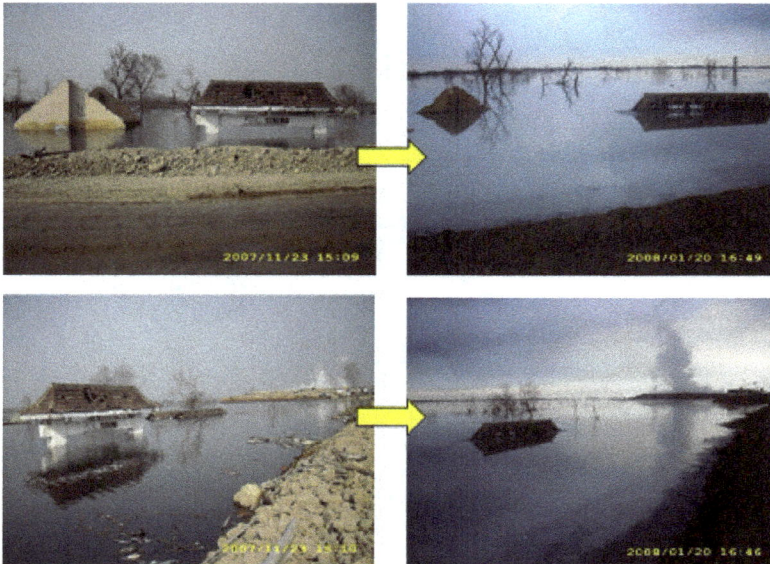

Fig. 27. Dramatical sinking of a village gate of Siring Timur, located to the west of the main vent. The gate and the rubble was half buried but still visible in the 23 November 2007 photograph. Two months later just a part of the tile roof and walls remain visible on 20 January 2008.

LUSI initially had five mud eruption vents, but only one remains active. There is a possibility that inactive mud eruption vents may reactivate or new ones will emerge in other locations. The study suggests a possibility of mud erupting at one or more of the known gas bubble locations or at a new location along zones of weakness on reactivated pre-existing faults or on new fault zones. Methane gas bubbles have been identified in more than 220 locations, and are generally associated with fractures. Some are more active than others while some have died. In most cases, the methane is non-flammable because it is in such low concentrations from rapid dispersion in the air. However where bubbles are confined, the concentration of methane is high enough to burn. The gas leaks from these fractures suggest breach of seal and loss of sealing capacity of faults and the impervious shale overlying geological structures, in particular the flanks of Wunut anticline that contain gas accumulations.

The occurrence of gas bubbles also suggests that subsidence is not merely a shallow near-surface phenomenon as a result of surface loading by the weight of the mud or soft soil layer compaction, but instead also affecting deep horizons as the gas comes from a deep source. Gas chromatography of sampled gas bubbles near the main eruption vent in July 2006 indicates it primarily consists of methane but the presence of some heavier gases in small quantities including ethane, propane, butane, and pentane suggest a deep thermogenic origin and long extended fractures.

Fig. 28. (A)&(B). Photos before and after the collapse of levees west of the main vent. The levee has decreased in height by approximately 1 meter along a 150 m interval. (C)&(D). Photos before and after the collapse of the levee north of the main vent. The levee has decreased in height by 1.5 m along a 200 m interval.

Fig. 29. A mosque located in the Renokenongo village, to the east main of the vent is damaged due to fault reactivation. At this location a lot of fractures were oriented NE-SW, and these are patterns of the Watukosek fault syatem.

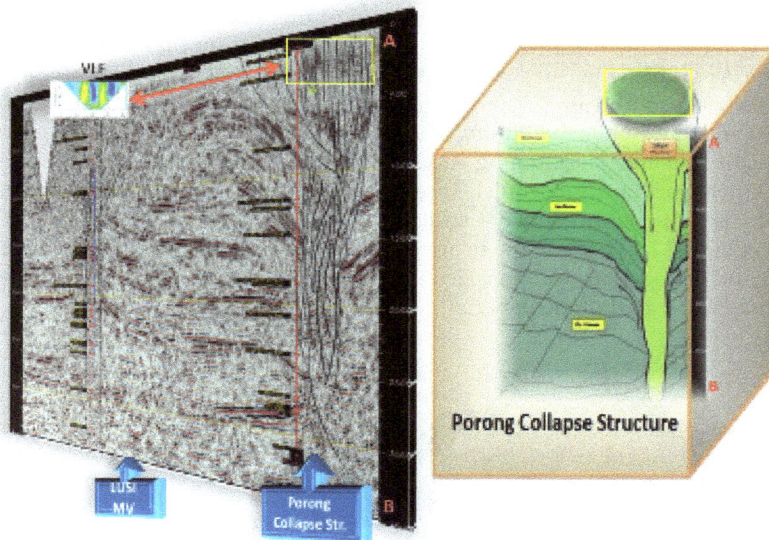

Fig. 30. The Porong paleo collapse structure was used as an analogy to predict the deformation pattern that will occur in LUSI. VLF measurement results indicate the existence of patterns of subsidence on the surface which is indicative of the onset of collapse

In terms of geohazard risks, the evidence and areas of concern include i) rupture of gas and water pipelines (shear and subsidence); ii) railroad bending (shear/faulting and subsidence); iii) road cracks (subsidence); iv) relief wells casing integrity (subsidence and shear); v) dyke collapse (subsidence); v) gas bubbles which appear along fractures and zones of weakness (shear/faulting and subsidence). (Istadi et al. 2009).

LUSI's continuing subsidence forms a depression bowl or funnel shaped structure. The subsidence forms an accommodation space, a natural basin to contain the mud. However, the high water content of the mud means it has a low viscosity and therefore cannot accumulate vertically to form a high and steep mountain-like structure. The mud, in particular the separated water tends to spread sideways which increases pressure on the mud retaining dykes that collapsed on a number of occasions and caused flooding. If the mud eruption continues with a high rate, then the potential flood prone areas will expand. The accumulated water at the peripheries away from the main eruption, which are held by the retaining dykes exert increasing hydrostatic pressure on the dyke walls. This increases the risk of retaining dyke failures.

Attempts to stop the mud flow should be implemented only after the likely causes of LUSI are studied and explained. The other alternative is to let nature find its own equilibrium and take care of itself, as LUSI mud volcano appears to be the result of a natural phenomena.

4. Other Mud volcanoes in East Java

Mud volcanoes are common in the northern part of Java and Madura Island (Satyana, 2008). Like elsewhere in the world, mud volcanoes in Java and Madura typically are located at the top of anticlines or along faults in the area. This phenomenon is demonstrated by the Sangiran mud volcano which is located at the top of the truncated dome on an up-thrown fault block, while the Bleduk Kuwu mud volcano is located on the top of the Purwodadi anticline, The Api Kayangan (means Fire of Heaven, or Eternal Fire) mud volcano is at the top of the Bojonegoro anticline. The Pengangson mud volcano located at the top of the Kedungwaru anticline, while the Pulungan and Kalang Anyar mud volcanoes are on the top of the Pulungan anticline. The Gunung Anyar (means New mountain) mud volcano is at the top of the Guyangan anticline, Bujel Tasek (Madura) and finally LUSI erupted on the extension of the Sekarputih anticlinal structure. Most of the mud volcanoes in East Java, with the exception of LUSI, are in the relative quiescence period and some can be considered in the dormant period with minimal activity. The existence of these mud volcanoes are described in the latter part of the paper

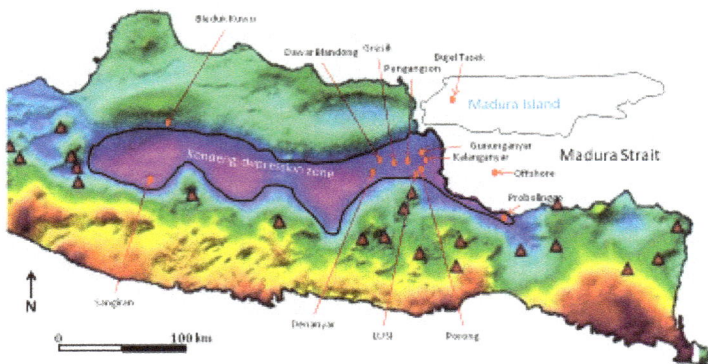

Fig. 31. Gravity map of East Java showing East Java Basin's depositional centers (blue) in the Kendeng depression zone. Red dots are the identified mud volcanoes, while the triangles are magmatic volcano locations.

The presence of mud volcanoes in the northern East Java Basin, especially along the Kendeng depression zone, is a common phenomenon. Their presence reflect the overpressure condition due to the very rapid deposition of the bluish-green mudstone and marlstone of the Sonde Formation during the Pliocene in a back arc basin setting that is folded and faulted (Dickinson, 1974).

In Central and East Java mud volcanoes are found within the Kendeng depression zone, except Bujel Tasek which appears in Rembang zone. The Kalang Anyar, Gunung Anyar, Pulungan, Bujel Tasek (Madura) and LUSI are in a straight line trending NE-SW contiguous with the regional fault trend, originating from the crater of Mt Penanggungan of the Arjuno–Welirang volcanic complex, following the Watukosek fault escarpment in the southern mountain ranges northward to Bujel Tasek in Madura island (see figure 32).

4.1 Kalanganyar

Kalanganyar mud volcano is located approximately 3 km south of Juanda Airport and approximately 15 km North-East of LUSI. The phenomenon of mud volcanoes in this area has been identified and mapped on the 1936 geological map of the Dutch era, suggesting it was formed long before the 1936 Duyfjes map. A temple known as Candi Tawangalun Majapahit (approximately 500 years ago) situated on the northern edge of the mud volcano suggest the significance of the mud volcano in the Majapahit era (see figure 33). Interestingly, some parts of the temple's material were made from material products of the Kalanganyar mud volcano.

The morphology of the Kalanganyar mud volcano forms a low relief hill overlying the alluvial plain deposit. The dimension of the main eruption cone is around 12x18 m. Some activities are still evident such as gas bubbles and some fresh mud around the main eruption vent (see figure 34) with a temperature of ± 38 °C. The unit is composed of silt-sized material and grains of fine sand and clay and saltwater that forms salt deposits.

The low relief of the mud volcano suggests that the mud has very low viscosity. The mud shrinks during dry season to form dessicated mud crack structures that are commonly found in the area. The mud material is derived from older rocks than the surrounding alluvial plains, correlatable with the mud at LUSI, the Upper Kalibeng Formation of Plio-Pleistocene age.

In the area around the active gas bubbles, rock fragments, such as siderite and salt deposits, are always found. The gas bursts are typically mild consisting of mixed gas, and formation fluids mainly connate water. Microbial activites are also found near the gas bubbles.

Mud breccias that are ejected from the Kalanganyar mud volcano are in the form of mud supported mudstone fragments which are light brown in color, composed of abundant carbonate mud, and quartz with opaque minerals in small quantities. Rudstone (Embry and Klovan, 1971) which is gray - brownish gray, grain supported, with fragment components consisting of shells of molluscs (Gastropoda and Pelecypoda, with a dominance of Ostrea shells) measuring 10-20 cm are abundant (> 10%) and bound by the matrix and carbonate cement (Figure 35 A&B). Balanus fossils were largely found freed from the carbonate (Figure 35 C) although some were still found to be bound by carbonate cement.

The mud eruption carried younger sediments and boulders which contain mud breccias and limestones. The stratigraphic position of the mud breccia based on Balanus fossil found on limestone within the mud breccia, indicate that this was sourced from the Sonde Formation of Pliocene age. The existence of mollusc and balanus contained in rudstone limestones suggest deposition in a shallow marine environment to the coastal littoral zone with strong

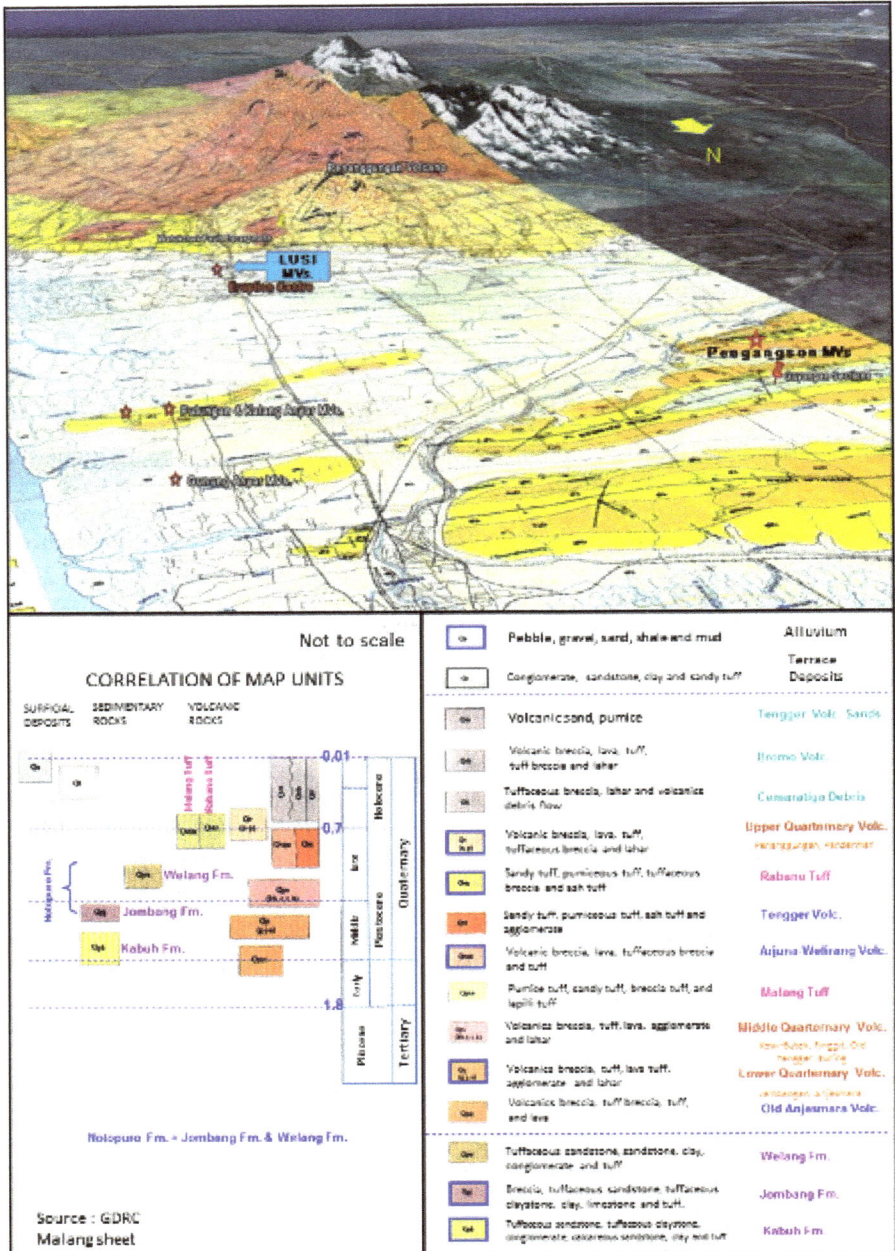

Fig. 32. Geological map overlaid with Google earth. Red stars are the identified mud volcano locations. The mud volcanoes are located across the top of anticlines and form a lineament.

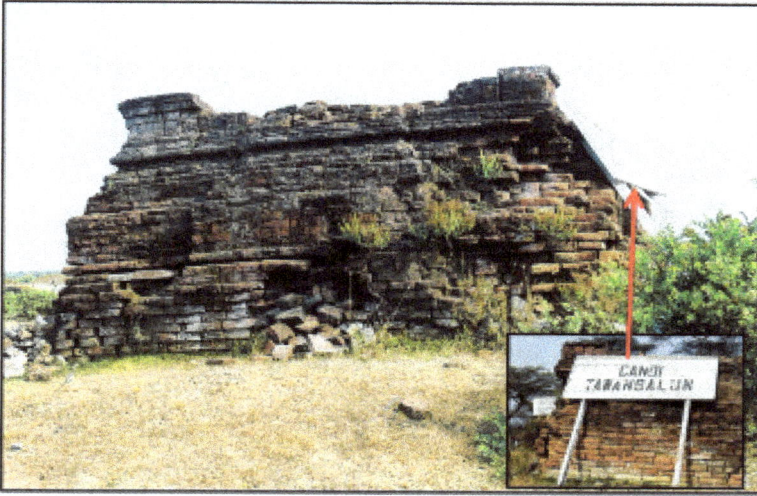

Fig. 33. Tawangalun temple was built during the Hindu kingdom (500 years ago) situated in the Northern edge of Kalanganyar mud volcano. Some materials of the temple were made from fragments from the Kalanganyar mud volcano.

Fig. 34. Kalang Anyar mud volcano in Kalanganyar village, Sidoarjo, East Java

energy currents. The dominant grain supported matrix of rudstone further suggests a high energy depositional environment. A shallow marine environment is located on the continental shelf with the fore reef sea conditions that are less affected by the supply of silisiclastics. The existence of mudstone with mud supported textures indicate deposition below wave base conditions on the back reef, rocks consequently do not experience the washing process (winnowing) by wave activity. The older mud volcano material is thought to have been sourced from the deeper Upper Miocene Kalibeng Formation, suggesting a regressive sequence in this part of the Kendeng zone of the East Java Basin.

Fig. 35. (A)&(B) Sandstone with mollusc shells dominated by Ostrea , (C). Balanus fossils among carbonates and siderite.

4.2 Gunung Anyar

The Gunung Anyar mud volcano (means "new mountain") is located 8 km to the west-northwest of Kalang Anyar mud volcano and is surrounded by densely populated residential area in the Gununganyar village, Surabaya. The morphology of the Gunung Anyar mud volcano is a northeast orientation and elongated hill-shaped geometry on the surrounding flat alluvial plains. The dimension of the still active main eruption vent is approximately 8x9m. The ejected material is composed of silt-sized grains of predominantly fine sand and saltwater. The temperature of the vent is ± 37.2 °C. More solid content than fluid has erupted from this mud volcano.

The lithology is similar to Kalang Anyar mud volcano where the composition is predominantly silt with the physical characteristics of brownish-grey, fine sand-sized to clay. In dry conditions, the structure shrinks, typically forming desiccated mud cracks. Seepage of crude oil of black color is also visible among the small bubbles that are still active.

Fig. 36. Gunung Anyar Mud Volcano morphology at Gunung Anyar Village in the Southern part of Surabaya, East Java.

Based on rocks and fragments carried by the mud eruption, the erupted materials were sourced from the following stratigraphic layers:

a. Marl

Brown marl with some weathered surfaces, white in color, with clay-sized fossil planktonic foraminifera and sand-sized bentonic fragments. Fresh outcrops of limestone fragments that appear to be newly ejected by the mud volcano. Stratigraphic position of the marl based on the physical properties and fossil content, indicate that this was derived from the Kalibeng and Sonde Formations.

b. Limestone

The limestone contains balanus fossils bounded by carbonate cement. The limestone based on fossil is thought to have been sourced from the Sonde Formation. The presence of balanus fossils typically suggest the coastal litoral zone depositional environment with the strong currents, while the texture of the limestone, suggest the low - moderate flow of energy environment with warm, calm and shallow waters possibly positioned on the back reef.

c. Calcareous sandstone

Weathered brown sandstone, poorly sorted, sub-rounded, fine sand-sized, composed of quartz, feldspar, biotite, and calcite. Black mudstone associated with these calcareous sandstones is also found in small quantities. The sandstone may have been sourced from the Pucangan Formation, Pleistocene in age.

d. Molluscs sandstone (grenzbank)

Sandstone containing molluscs of freshwater and seawater were found in this location. The freshwater molluscs which are characterized by the thin shell bi-valves are more dominant than the seawater molluscs. The presence of mixed seawater and freshwater molluscs suggest shallow to transitional marine deposition environment.

Fig. 37. Crude oil seepage coming out with salty connate water along with bubbles in Gunung Anyar Village.

e. Silt

Silt makes up most of the mud volcanic area. It is composed of silt-sized grains of predominantly silisiclastic brownish-grey colored clay materials and the salt water. The temperature is around 37 °C at the mud conduit. The mud contains more solid silt and clay materials than fluid. Silts were derived from underlying older rocks of possibly Upper Kalibeng Formation of late Miocene. The presence of rock fragments, siderite, and seepage of black crude oil are almost always found within the vicinity of active bubbles.

4.3 Pengangson

The location of Pengangson mud volcano is in the village of Kepuhklagen, Wringinanom, Gresik Regency. Compared to other mud volcanoes in East Java, the Pengangson mud volcano is the most ideal example of a mud volcano. The younger sediments outcropped and exposed at nearby excavated cliffs to the west of the mud volcano, and older rocks are found as fragments or clast carried by the mud volcano eruption. The geological structural components at this site are also clearly visible, such as folds, fractures, fault lines and sedimentary structures.

The morphology is formed of low hills between the alluvial plain surrounding. The unit is composed of silt sediment material with dominant clay-sized grains. The low temperature mud of ± 39.5 °C is typical of other mud volcanoes in other parts of East Java. The mud is thick and the liquid consists of a mix of formation fluid, crude oil seeps, salt deposits and gas bubbles.

Fig. 38. Pengangson Mud Volcano in Kepuhklagen village, Gresik, East Java

The materials ejected by the mud volcano are the following:

a. Sandy Mudstone - Sandy Siltstone

Rock outcrops are generally brownish-grey in fresh and partially weathered condition. Physical characteristics of the rock suggest that it is from the Sonde Formation. Based on the deposition enviroment, lithology, sedimentary structure, texture, mineralogical composition, and fossils suggest that this unit was deposited in a middle shelf environment - lower delta plain.

b. Tuffaceous Sandstone unit, Sonde Formation

The naming of the sandstone unit is based on the condition of the rock outcrop on the cliffs with tuffaceous sandstone lithology that shows the layered structure and planar crossbed. Rock samples in this unit do not contain foraminifera because of the dominance of volcanic material in the rocks.

The composition of the constituent material of plagioclase and the presence of volcanic glass which tends to be wacke, suggests that the source is not far from the sedimentation basin. The similarity of physical properties, and texture of rocks suggest this unit is part of the Sonde Formation.

c. Sandstone unit, Pucangan Formation

This unit is characterized by the presence of sedimentary structures such as grading, parallel lamination and slump structures. Outcrops of fresh rocks generally show somewhat weathered condition. The bottom of the unit is dominated by volcanic sandstone that contains calcareous mudstone layers. This unit is part of an eroded top of anticline.

Interpretation of the environment based on the lithology data, sedimentary structure, texture, mineralogical composition, and fossils indicates that this unit was deposited on the inner shelf environment - lower delta plain with traction and suspension flow mechanism. The sequence of lithologies indicates deposition in increasingly shallow conditions with the initial deposition on the inner shelf. The similarity of physical properties, texture, structure, age and environment of deposition of rocks suggests that this unit is part of the Pucangan Formation.

d. Tuffaceous Mudstone

The outcrop of rocks is generally fresh or slightly weathered. Overall this unit is dominated by massive mudstone and tuff. The bedding trends west - east and slopes to the north and south. Interpretation of the environment of deposition based on the lithology, texture, and mineralogical composition is that this unit was deposited in a braided stream environment or sub flood plain and stream sediment was transported through the mechanism of suspension. The similarity of physical properties and the texture of rocks suggests that this unit is part of the Kabuh Formation.

e. Volcanic sandstone

This unit is characterized by the presence of sedimentary structures such as grading, cross-lamination and parallel lamination. The outcrop of rocks generally show a somewhat weathered condition but rock structure is still visible. Interpretation of the environment of deposition based on the lithology, sedimentary structure, texture, and mineralogical composition suggests that this unit was deposited in a braided stream environment (minor channel) with traction flow mechanism. The similarity of physical properties, texture and structure of rocks suggests that this unit is part of the Kabuh Formation.

f. Silt unit

The naming of this rock unit is based upon the existence of a silt dominated mud volcano. The morphology of mud volcanoes forms a low hill between the alluvial plains. The unit is composed of silt sediment material dominant by clay-sized grains with a temperature 39.5 ° C. From the main vent fluid, gas bubbles and salty water are released.

Fig. 39. Crude oil seepage, brownish-black in color together with mud and gas that comes out of a gryphon.

In the vicinity, bubble-shaped Gryphons, siderite, and salt deposits are commonly found (see figure 39). Rock fragments are found in a limited number that consist of calcarenite (Sandy micrite), calcareous sandstone, calcareous mudstone (Micritic mudrock in the Mount, 1985) and sandstones with molluscs.

This mud deposit in Wringinanom contains foraminifera planktonic fossils suggesting a middle Pliocene age to late Pliocene (N20-N21) mud source, while the content of bentonic neritic foraminifera suggests the bathymetry position in the middle neritic.

4.4 Bujel Tasek

Bujel Tasek mud volcano is found in the Katol Barat village, Bangkalan district of Madura island. The morphology of the Bujel Tasek mud volcano is very different from other mud volcanoes in East Java. The shape is a cone edifice with a height of approximately 12 m with a diameter of approximately 5 m. The material that came out is a mixture of viscous mud, water and gas with a highly viscous mud. This cone shape is actually a giant gryphon formed by the high viscosity mud of a larger mud volcano that covers the area. Nearby this conic structure is a mud lake that represents an ancient mud caldera with mud conduits that are no longer active (see figure 40).

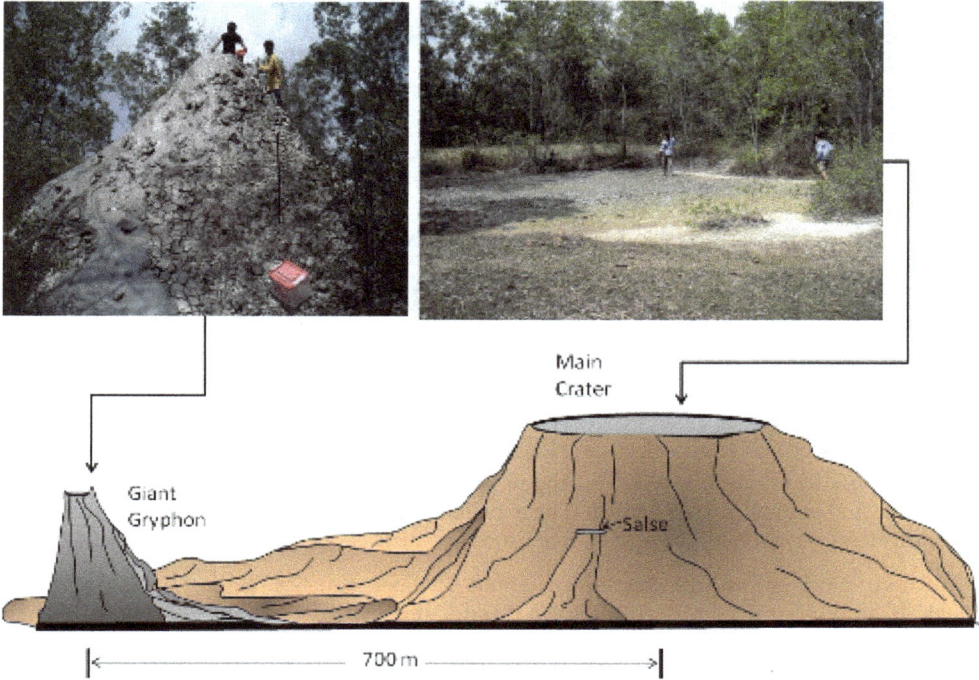

Fig. 40. Bujel Tasek mud volcano forms a cone morphology in the village of Katol Barat, Bangkalan, Madura island, East Java.

The mud sediment that came from the Lidah formation is characterized by its brownish-grey, fine clay. The mud breccia found is gravel to boulder sized, very abundant, with fragments ejected in the form of mudstone, calcareous sandstone, siderite and calcite.

The reddish-brown and gravel- pebble sized siderite mineral is found exposed. It is widely distributed, from the mud volcano to the valley. In addition, calcite is trapped in sandstones that fills the pores of the wood that look like silisified wood.

4.5 Bleduk Kuwu

This mud volcano is located in the Village Kuwu, Kradenan, Grobogan district, approximately 20 km south of Purwodadi in Central Java. The object of interest in Bleduk is the mud flow containing gas and salty water that takes place almost continuously in an area with a diameter of approximately 650 m (see figure 41). Etymologically, the name comes from Kuwu Bleduk. In the Javanese language 'Bleduk' means 'blast/burst' and 'kuwu' is derived from the word 'kuwur' which means 'run/scramble'.

Fig. 41. Bleduk Kuwu mud volcano during its almost continuous eruption. The gas is flammable and sometimes self-ignites. The expelled water is commercially used to extract salt. Eruptions generally occur four or five times a minute, as a burst of warm mud and gas.

The Kuwu mud volcano cluster covers about 45 hectares. The biggest vent can erupt materials as high as 5 meters with expelled mud in a diameter of about 9 meters. At the main Kuwu site the mud volcano usually erupts four or five times a minute consisting of mud accompanied by the release of gas and water (sometimes oil). Often the eruptions are accompanied by an explosion as the gas self-ignites. The temperature of the mud ranges from 28-30°C, while the smaller mud volcano is slightly cooler.

Bleduk Kuwu is surrounded by other mud volcanoes within a radius of approximately 1-2 km to the southwest, northeast and south with varying dimensional extents. To the southwest is Cangkring Bleduk mud volcano that occupies a larger area than the Bleduk Kuwu, while to the south is the Bleduk Banjarsari mud volcano, and to the East is Bleduk Crewek, and to the northeast is Medang Kamolan (figure 42). Other minor mud volcanoes in the area include Bledug Kesongo and Bledug Kropak. Geologically, these mud volcanoes are located at the boundary between North Serayu and Kendeng Depressions. Seismic sections across these mud volcanoes show disturbed zones from the top of the Kujung Formation, the top of Wonocolo Formation to the surface. The Bledug Kuwu disturbed zone is a chaotic mixture of upward convex and concave reflectors. Bledug Kesongo is characterized by a collapsed structure with upward concave horizons along the disturbed zone indicating a subsidence. The lower part of Late Miocene Wonocolo shales is believed to be the source of mud based on its fossil content. Seismic sections, however, show that the source of mud may also come from Early Miocene Tuban shales. Some diapirs also occur in this area and they are generally below the top of Wonocolo Formation. Folds in this area are considered to form diapirs as suggested by some seismic sections (Satyana and Asnidar, 2008).

Fig. 42. Medang Kamolan mud volcano, located approximately 3 km northeast of Bleduk Kuwu mud volcano.

4.6 Offshore mud volcanoes

The existence of submarine mud volcanoes and mud diapirs in the Madura Strait, offshore areas of East Java is visible in the seismic profiles. The cross sectional appearance looks like an upwards dipping strata around a venting system of seafloor-piercing shale diapir cutting the overlying sediment and forming a conic volcanic edifice (see figure 43).

The Madura Strait is an offshore extension of the Kendeng Depression. Thick Pliocene to Pleistocene sediments were deposited rapidly and compressed elisional system in the Madura Strait depression. Deepwater sedimentation is still taking place in this portion of the Kendeng zone, and it has not been uplifted. In the Madura Strait area, east-west trending left lateral wrench faulting triggered mobilization of Miocene basinal shales during the Plio-Pleistocene, resulting in a series of shale diapirs. Further south, the impact of on-going subduction along the Java Trench becomes increasingly significant and structures are dominated by north-directed thrusting, which may be independent of basement faulting (Satyana and Asnidar, 2008).

On the basis of structural style and the tectonic events, Widjonarko (1990) divided the Madura Strait block into five structural domains: wrench domain, slide domain, western basinal domain, eastern basinal domain, and southeastern fault block domain. Wrench and slide domains bound the Madura Strait to the Madura-Kangean High in the north. Southeastern fault block becomes the southern border of offshore Madura Strait. The main parts of the Madura Strait where mud diapirs and volcanoes exist are composed by western and eastern basinal domains. The Madura Strait Depression or Sub-Basin is one of the two deepest and thickest basins in Indonesia. In western basinal domain, very rapid sedimentation since the Late Miocene time resulted in the development of more than 3000 meters of Plio-Peistocene section. Eastern basinal domain is similar to western domain, the only difference is that the eastern basinal domain began to subside in the late Oligocene – early Miocene, much earlier than the western domain (Satyana and Asnidar, 2008).

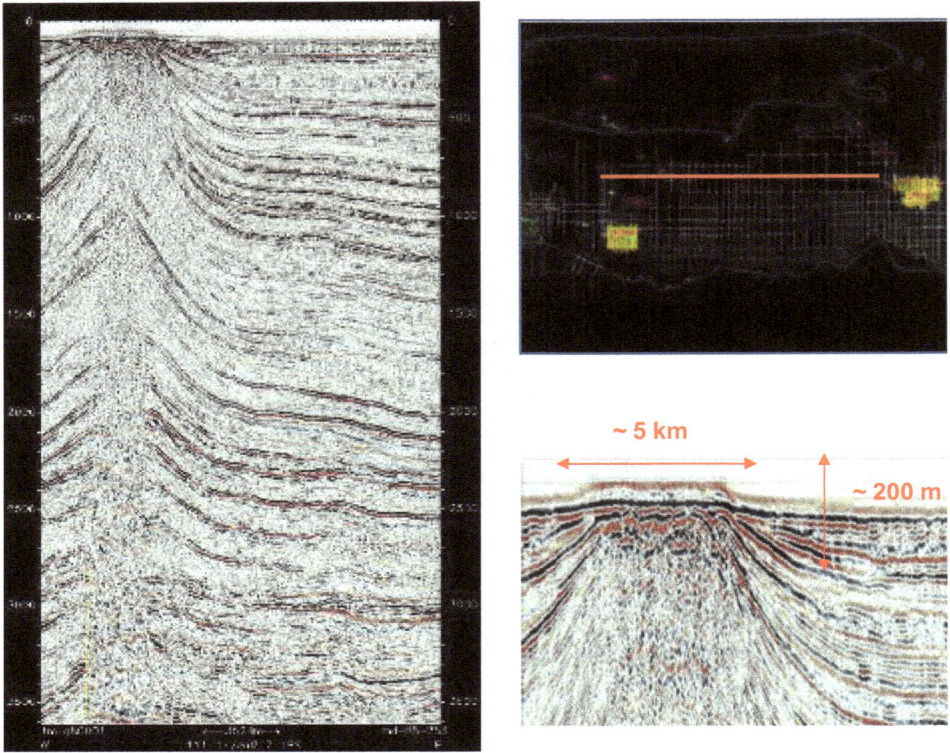

Fig. 43. Water Depth ~40 – 50 m in the offshore Madura Strait

Stratigraphy of the Madura Strait started in Middle Eocene time by deposition of transgressive clastics unconformably on top of pre-Tertiary basement. The deposition was terminated by a local uplift at the end of Eocene time. Subsidence during the Oligocene resulted in deposition of deep marine sediments. An uplift at the end of the Oligocene resulted in a regional unconformity throughout the basin. During the Early Miocene time the rapid subsidence resulted in deposition of deep marine sediments in the area. In the mid-Late Miocene time, the basin was filled and another uplift took place. After a short subsidence to the end of Late Miocene, sedimentation interrupted again by an uplift in Early Pliocene time. The rapid subsidence in the late Pliocene time is characterized by the deposition of overpressured thick clays. The area subsided again into a shallow marine environment after the Plio-Pleistocene regional uplift (Widjonarko, 1990; Satyana and Asnidar, 2008).

5. Conclusions

- Mud diapir and mud volcano are piercement structures showing the release of overpressured sediments piercing upward from subsurface to the Earth's surface due to buoyancy and differential pressure.

- The Kendeng-Madura Strait Zone is an ancient axial depression of Java to Madura Islands with elisional basin characteristics. The Mio-Pliocene and Pleistocene sediments were rapidly deposited into the depression and compressed as it was at the front of converging plate boundaries with high seismic activity. This resulted in numerous mud diapirs and mud volcanoes in the area. Sixteen mud volcanoes have so far been documented in East Java with the six mud volcanoes found along the Watukosek fault.
- LUSI, a new mud volcano, was born at the vicinity of the Watukosek fault. This geological phenomena as well as others occurring in the area such as the appearance of gas bubbles, cracks, subsidence, and vertical and horizontal displacement is believed to be due to a reactivation of existing faults in this area. The Watukosek fault system appeared to play a role in the existence of other six mud volcanoes at its vicinity.
- Early technical papers, such as Davies et al. (2007, 2008), Rubiandini et al. (2008) and Tingay et al. (2008) suggested a connection between the Banjarpanji-1 well and the mud volcano. These papers were based much on unverified and partial dataset. When the full dataset is integrated as in Sawolo et al. (2008, 2009 and 2010), it is evident that the well did not trigger LUSI mud volcano. Future analysis on the trigger of the mud volcano must consider this pitfall and integrate all available dataset. One must decipher and make use of the entire mud logger Real Time Data as it is the most reliable dataset, being automated, continuous and quantitative data that captures key operating parameters of the rig.
- The distance between the nearest volcanic complex, the Arjuno - Welirang volcanoes, to LUSI is about 10 km. This close proximity may have influenced the geothermal properties of LUSI and affected its temperature and geochemistry.
- Studies suggest that LUSI is likely to continue to flow for many years to come. Better understanding of its plumbing system and detailed subsurface studies must be conducted as part of the hazard mitigation effort. The more than 16 East Java mud volcanoes in the vicinity must be used as an analogy of mud volcano processes and its morphology.

6. Acknowledgement

The Authors wishes to express appreciation to the management of MIGAS, BPMIGAS, EMP, Lapindo Brantas Inc for the permission to publish the paper. Constructive discussions and inputs from our colleagues in particular Peter Adam and Awang H. Satyana are also appreciated.

7. References

Abidin, H.Z., Davies, R.J., Kusuma, M.A., Andreas, H., Deguchi, T., 2008., Subsidence and uplift of Sidoarjo (East Java) due to the eruption of the LUSI mud volcano (2006 present). Environmental Geology doi:10.1007/s00254-008-1363-4.

Akhmanov, G.G. and Mazzini, A., 2007, Mud volcanism in elisional basin, In: Proceedings of the International Geological Workshop on Sidoarjo Mud Volcano, Jakarta, IAGI-BPPT-LIPI, February 20–21, 2007. Indonesia Agency for the Assessment and Application of Technology, Jakarta.

Amelung, F., Galloway, D.L., Bell, J.W., Zebker, H.A., Laczniak, R.J., 1999. Sensing the ups and downs of Las Vegas: InSAR reveals structural control of land subsidence and aquifer-system deformation. Geology 27 (6), 483–486.

Davies, R.J., Brumm, M., Manga, M., Rubiandini, R., Swarbrick, R., Tingay, M., 2008. The East Java mud volcano (2006 to present): an earthquake or drilling trigger? Earth and Planetary Science Letters 272 (3–4), 627–638.

Davies, R.J., Swarbrick, R.E., Evans, R.J. and Huuse, M., 2007, Birth of a mud volcano: East Java, 29 May 2006: GSA Today, v. 17, p. 4-9.

De Genevraye, P., Samuel, L., 1972. Geology of the Kendeng zone (Central and East Java). In: Proceedings of the Indonesian Petroleum Association, 1st Annual Convention, pp. 17–30.

Deguchi, T., Maruyama, Y., Kato, M., Kobayashi, C., 2007a. Surface displacement around mud volcano, East Java captured by Insar using Palsar data. In: Proceedings of the 28th Asian Conference on Remote Sensing.

Deguchi, T., Maruyama, Y., Kato, M., 2007b. Measurement of long-term deformation by InSAR using ALOS/PALSAR data. In: Presented at FRINGE 2007 workshop, November 2007, Frascati, Italy. Workshop presentation available from: http://earth.esa.int/workshops/fringe07/presentations.html Workshop proceedings are in press.

Desaunettes, J.R., 1977, Catalogue of Landsform for Indonesia, Trust Fund of The Goverment of Indonesia – FAO, Bogor.

Dimitrov, L.I., 2002. Mud volcanoes: the most important pathway for degassing deeply buried sediments. Earth Science Reviews 59 (1–4), 49–76.

Duyfjes, J., 1936, The geology and stratigraphy of the Kendeng area between Trinil and Surabaya (Java) (in German), De Mijningenieur, vol. 3, no. 8, August 1936, p. 136-149.

Embry, AF, and Klovan, JE, 1971, A Late Devonian reef tract on Northeastern Banks Island, NWT: Canadian Petroleum Geology Bulletin, v. 19, p. 730-781.

Etiope, G., Feyzullayev, A., Milkov, A.V., Waseda, A., Mizobe, K., Sun, C.H., 2009, Evidence of subsurface anaerobic biodegradation of hydrocarbons and potential secondary methanogenesis in terrestrial mud volcanoes, Marine and Petroleum Geology 26 (2009) pp. 1692–1703

Fielding, E.J., Blom, R.G., Goldstein, R.M., 1998. Rapid subsidence over oil fields measured by SAR interferometry. Geophysical Research Letters 25 (17), 3215–3218.

Fukushima, Y., Cayol, V., Durand, P., 2005. Finding realistic dike models from interferometric synthetic aperture radar data: the February 2000 eruption at Piton de la Fournaise. Journal of Geophysical Research 110 B03206. doi:10.1029/2004JB003268.

Fukushima, Y., Mori, J., Hashimoto, M., Kano, Y., 2009. Subsidence associated with the LUSI mud eruption, East Java, investigated by SAR interferometry. Journal Marine and Petroleum Geology 26, 1740–1750.

Gourmelen, N., Amelung, F., Casu, F., Manzo, M., Lanari, R., 2007. Mining-related ground deformation in Crescent Valley, Nevada: implications for sparse GPS networks. Geophysical Research Letters 34 L09309. doi:10.1029/2007GL029427.

Guliyiev, I.S., Feizullayev, A.A., 1998. All About Mud Volcanoes. Nafta Press, Azerbaijan Publishing House, 52 pp.

Hanssen, R.F., 2001. Radar Interferometry – Data Interpretation and Error Analysis. Kluwer Academic Publishers, Dordrecht, The Netherlands.

Higgins, G.E., Saunders, J.B., 1974. Mud volcanoes. Their nature and origin. Verhandlungen Naturforschenden Gesselschaft in Basel 84, 101–152.

Hochstein, M. P. and Sudarman, S., 2010, Monitoring of LUSI Mud-Volcano - a Geo-Pressured System, Java, Indonesia., Proceedings World Geothermal Congress 2010, Bali, Indonesia, 25-29 April 2010

Istadi, B., Pramono, G.H., Sumintadireja, P., Alam, S. Simulation on growth and potential Geohazard of East Java Mud Volcano, Indonesia. Marine & Petroleum Geology, Mud volcano special issue, doi: 10.1016/j.marpetgeo.2009.03.006.

Istadi, B.P., Kadar, A., Sawolo, N., 2008. Analysis & recent study results on East Java mud volcano. In: Subsurface Sediment Remobilization and Fluid Flow in Sedimentary Basin Conference, October 2008. The Geological Society, Burlington House, Piccadilly, London.

Kadar, A.P., Kadar, D., Aziz, F., 2007. Pleistocene stratigraphy of Banjarpanji#1 well and the surrounding area. In: Proceedings of the International Geological Workshop on Sidoarjo Mud Volcano, Jakarta, IAGI-BPPT-LIPI, February 20-21, 2007. Indonesia Agency for the Assessment and Application of Technology, Jakarta.

Kholodov, V.N., 1983, Postsedimentary Transformations in Elisional Basins (example from Eastern Pre-Caucasus) (in Russian), 150 ps.

Koesoemadinata, R.P., 1980. Geologi Minyak dan Gas Bumi. 2 jilid, ed ke-2. Penerbit ITB, Bandung.

Kopf, A., 2002. Significance of mud volcanism. Reviews of Geophysics 40 (2), 1005, 2.1–2.52. doi:10.1029/2000RG000093

Kumai, H. and Yamamoto, H., 2007, Earthquake, the major trigger of Mud Volcanism at Sidoarjo, East Java., In: Proceedings of the International Geological Workshop on Sidoarjo Mud Volcano, Jakarta, IAGI-BPPT-LIPI, February 20-21, 2007. Indonesia Agency for the Assessment and Application of Technology, Jakarta.

Link, W.K., 1952. Significance of oil and gas seeps in world oil exploration. Am. Assoc. Pet. Geol. Bull. 36, 1505–1540.

Manga, M. and Brodsky, E. (2006): Seismic triggering of eruptions in the far field: Volcanoes and Geysers. Annual Review of Earth and Planetary Sciences, 34, 263-291.

Manga, M., 2007. Did an earthquake trigger the May 2006 eruption of the LUSI mud volcano? EOS 88 (201).

Manga, M., 2007. Did an earthquake trigger the may 2006 eruption of the LUSI mud volcano? EOS 88 (18), 201.

Manga, M., Rudolph, M.L., Brumm, M., 2009. Earthquake triggering of mud volcanoes: a review 26, 1785–1798.

Massonnet, D., Feigl, K.L., 1998. Radar interferometry and its application to changes in the Earth's surface. Review of Geophysics 36, 441–500.

Mazzini, A., Nermoen, A., Krotkiewski, M., Podladchikov, Y.Y., Planke, S., Svensen, H., 2009. Strike-slip faulting as a trigger mechanism for overpressure release through piercement structures. Implications for the LUSI mud volcano, Indonesia. Marine and Petroleum Geology 26, 1751–1765.

Mazzini, A., Svensen, H., Akhmanov, G.G., Aloisi, G. Planke, S., Malthe-Serenssen, A and Istadi. B., 2007, Trigering and dynamic evolution of LUSI mud volcano, Indonesia: Earth and Planetary Sciences Letters, v. 261, p. 375-388.

Mellors, R., Kilb, D., Aliyev, A., Glasanov, A. and Yetirmishli, G. (2007): Correlations between earthquakes and large mud volcano eruptions. Journal of Geophysical Research, 112, doi:10.1029/2006JB004489.

Milkov, A.V., 2000, Worldwide distribution of submarine mud volcanoes and associated gas hydrates, Marine Geology, 167, p. 29–42.

Milkov, A.V., 2005. Global distribution of mud volcanoes and their significance in petroleum exploration, as a source of methane in the atmosphere and hydrosphere, and as geohazard. In: Martinelli, G., Panahi, B. (Eds.), Mud Volcanoes, Geodynamics and Seismicity. NATO Science Series, IV Earth and Environmental Sciences, vol. 51, Springer, pp. 29–34.

Mori, J. and Kano, Y., 2009, Is the 2006 Yogyakarta Earthquake Related to the Triggering of the Sidoarjo, Indonesia Mud Volcano? Journal of Geography 118 (3)492-498 2009

Nawangsidi, D., 2007. In search of theory on flow mechanism of wild mud blow. In: Proceedings of the International Geological Workshop on Sidoarjo Mud Volcano, Jakarta, IAGI-BPPT-LIPI, February 20–21, 2007, Indonesia Agency for the Assessment and Application of Technology, Jakarta.

Pitt, A.M., Hutchinson, R.A., 1982. Hydrothermal changes related to earthquake activity at mud volcano, Yellowstone National Park, Wyoming. Journal of Geophysical Research 87, 2762–2766.

Satyana, A.H. and Armandita, C., 2004, Deep-Water play of Java, Indonesia : regional evaluation on opportunities and risks, Proccedings International Geoscience Conference of Deepwater and Frontier Exploration in Asia and Australasia, Indonesian Petroleum Association (IPA) and American Association of Petroleum Geologists (AAPG), Jakarta, p. 293-320.

Satyana, A.H. and Asnidar, 2008, Mud Diapirs and Mud Volcanoes of Java to Madura: Origins, Natures, and Implications to Petroleum System, Proceedings Indonesian Petroleum Association (IPA), 32nd annual convention, Jakarta, 27-29 May 2008.

Satyana, A.H., 2007, Bencana Geologi dalam "Sandhyâkâla" Jenggala dan Majapahit : Hipotesis Erupsi Gununglumpur Historis Berdasarkan Kitab Pararaton, Serat Kanda, Babad Tanah Jawi; Folklor Timun Mas; Analogi Erupsi LUSI; dan Analisis Geologi Depresi Kendeng-Delta Brantas, Proceedings Joint Convention Bali 2007-HAGI, IAGI, and IATMI, 14-16 November 2007.

Satyana, A.H., 2008, Roles of Mud Volcanoes Eruptions in the Decline of the Jenggala and Majapahit Empires, East Java, Indonesia: Constraints from the Historical Chronicles, Folklore, and Geological Analysis of the Brantas Delta-Kendeng Depression, Majalah Geologi Indonesia, vol. 23, no. 1-2, p. 1-10.

Sawolo, N., Sutriono, E., Istadi, B.P., Darmoyo, A.B., 2008. East Java mud volcano (LUSI): drilling facts and analysis. In: Proceedings of the African Energy Global Impact, AAPG International Conference and Exhibition, October 2008, Cape Town, South Africa.

Sawolo, N., Sutriono, E., Istadi, B.P., Darmoyo, A.B., 2009. Mud volcano triggering controversy: was it caused by drilling? Journal Marine and Petroleum Geology 26, (2009) 1766–1784. doi:10.1016/j.marpetgeo.2009.04.002

Sawolo, N., Sutriono, E., Istadi, B.P., Darmoyo, A.B., 2010. Was LUSI caused by drilling? – Authors reply to discussion. Journal Marine and Petroleum Geology 27 (2010) 1658–1675. doi:10.1016/j.marpetgeo.2010.01.018

Schiller, D.M., Seubert, B.W., Musliki, S., Abdullah, M., 1994, The reservoir potential of globigerinid sands in Indonesia, *Indon. Petroleum Assoc. 23nd Ann. Conv.*

Sharaf, E.F., BouDagher-Fadel, M.K., Simo, J.A. (Toni) and Caroll, A.R., 2005, Biostratigraphy and strontium isotope dating of Oligocene-Miocene strata, East java, Indonesia., Stratigraphy, vol. 2, no. 3, pp. 1-19, text figure 1-4, tables 1, plate 1-5.

Sudarman and Hendrasto, 2007, Hot Mud Flow At Sidoarjo., In: Proceedings of the International Geological Workshop on Sidoarjo Mud Volcano, Jakarta, IAGI-BPPT-LIPI, February 20–21, 2007. Indonesia Agency for the Assessment and Application of Technology, Jakarta.

Sukarna, D., 2007, Mud Volcano In Indonesia: Distribution and its geological Phenomenon., In: Proceedings of the International Geological Workshop on Sidoarjo Mud Volcano, Jakarta, IAGI-BPPT-LIPI, February 20–21, 2007. Indonesia Agency for the Assessment and Application of Technology, Jakarta.

Sumintadireja, P., Purwaman, I., Istadi, B., Darmoyo, A.B., 2007. Geology and geophysics study in revealing subsurface condition of Banjarpanji mud extrusion, Sidoarjo, East Java, Indonesia. In: Proceedings of the International Union of Geodesy and Geophysics (IUGG) XXIV, Perugia, Italy.

Sunardi, E., Guntoro, A., Alam, S., Koesoema, M.A., Hadi, S., Budiman, A., 2007. Studi Geologi Dan Geofisika Semburan Lumpur Di Daerah Porong Sidoarjo, Jawa Timur., In: Proceedings of the International Geological Workshop on Sidoarjo Mud Volcano, Jakarta, IAGI-BPPT-LIPI, February 20–21, 2007. Indonesia Agency for the Assessment and Application of Technology, Jakarta.

Sutaningsih, N.E., Humaida, H., Zaennudin, A., Suryono, Primulyana, S., 2010. Indikasi Sistim Geotermal Pada Semburan Lumpur Sidoarjo Ditinjau Dari Karakteristik Kimia, Proceedings PIT IAGI Lombok 2010, The 39th IAGI Annual Convention and Exhibition

Tingay, M.R.P., Heidbach, O., Davies, R., Swarbrick, R., 2008. Triggering of the LUSI Mud Eruption: Earthquake Versus Drilling Initiation. Geology, vol. 36(8), pp. 639–642.

Tingay, M.R.P., 2010. Anatomy of the Lusi Mud Eruption, East Java. Australian Society of Exploration Geophysicists 21st International Conference and Exhibition, 23-26 August 2010, Sydney, Australia, doi:10.1071/ASEG2010ab241.

Walter, T.R., Wang, R., Luehr, B.G., Wassermann, J., Behr, Y., Parolai, S., Anggraini, A., Günther, E., Sobiesiak, M., Grosser, H., Wetzel, H.U., Milkereit, C., Sri Brotopuspito, P.J.K., Harjadi, P., Zschau, J., The 26 May 2006 magnitude 6.4 Yogyakarta earthquake south of Mt. Merapi volcano: did lahar deposits amplify ground shaking and thus lead to the disaster?. G-cubed: Geochemistry Geophysics Geosystems, 9, Q05006, doi: 10.1029/2007GC001810.

Watanabe, N., dan Kadar, D., 1985, Quaternary Geology of the Hominids Fósil Bearing Formations in Java : Report of the Indonesia – Japan Research Project, CTA – 41, 1976 – 1979, Geological Research and Development Centre Special Publication, Bandung,P. 378

Widjonarko, R., 1990, BD field – a case history, Proceedings Indonesian Petroleum Association (IPA), 19th Annu. Conv., p. 161-182.

Willumsen, P., Schiller, D.M., 1994. High quality volcaniclastic sandstone reservoirs in East Java, Indonesia. In: 23rd Annual Convention, vol. I. IPA, pp. 101–111.

3

Xujiaweizi Rift Lower Cretaceous Yingcheng Group Volcanic Sequence Stratigraphic Features[1]

Zhang Yuangao, Chen Shumin, Feng Zhiqiang, Jiang
Chuanjin Zhang Erhua, Xin Zhaokun and Dai Shili
Daqing Oilfield Company Ltd.,Heilongjiang Daqing
China

1. Introduction

Sequence stratigraphy is a discipline that is developed on the seismic stratigraphy (Vail, 1987). The concept of sequence stratigraphy is born Since 1950'(Sloss, 1959), which is composed of LST, TST and HST. The theoretical system of sequence stratigraphy has been widely used by geologists after half a century of development. It is also developing from the classical three kinds system tracts to four systems tracts(Catuneanu,2006). Although the schemes of sequence boundary division are different(Catuneanu,2006; Catuneanu,2009).but they have stressed the fact is that the inherent mechanism of sequence genesis and phase distribution is controlled by sea level change(Arimoto,1997; Saydam,2000; Caquineay,2008). The theoretical framework of Chinese modern sequence stratigraphy is mainly composed of continental sequence stratigraphy(Shanley,1994; Liu ZJ,2002;Ji YL,2002) and high resolution sequence stratigraphy (Deng HW,2002). at the moment, the application fields of sequence stratigraphy have been developed from the research of the whole basin layers to a intrabasinal measure detailed study(Xu YX,2001; Chen F,2010; Wang QC,2010). With the development of exploration technology, the division of the sequence is more and more detailed, the control factor research of sequence development is more in-depth. However, with the breakthrough of exploration field, a large number of oil and gas resources are found in the volcanic rocks. Sequence stratigraphy is facing new challenges, how to create a volcanic sequence is our new research hotspot (Fig.1).

Yingcheng formation is made up of volcanic and sedimentary rocks in Xujiaweizi depression. volcanic rocks include volcanic lava and pyroclastic rocks that is a direct result of volcanism (Wang PJ,2008). sedimentary rocks include two sections: the one is the normal sedimentary rocks that occurred in the intermittent period of volcanism; the other is composed of weather worn volcanic which transport by water. Scholars(Stewart A L,2006; Busby CJ,2007; Wang P J,2006) have carried out intensive research about the volcanic lithology and lithofacies. But the targeted research of volcanic action filling model has not been carried out. Volcanic eruption and accumulation can take place in the high position. so the volcanic occurrence are not controlled by the accommodation space of water bodies, which is the biggest difference between sedimentary strata. Lead to the base level of

[1]Foundation Project ： National Basic Research Program of China (2009CB219307)

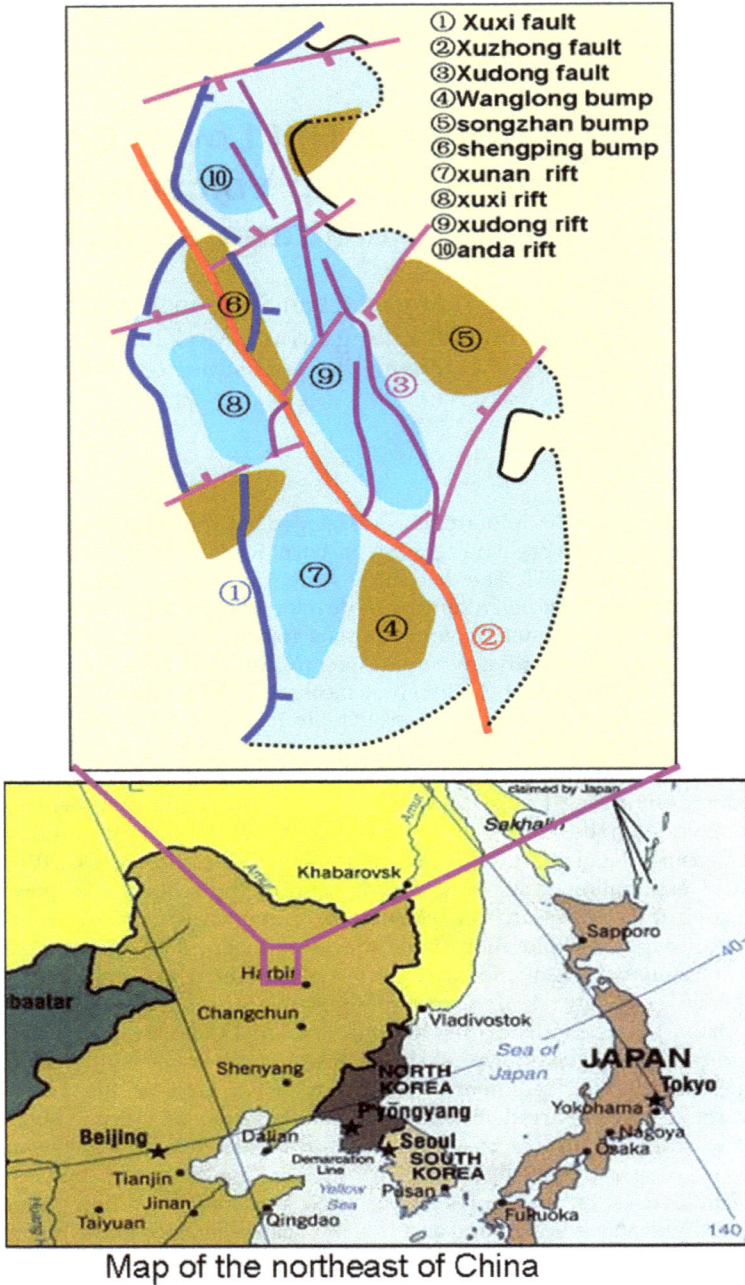

Map of the northeast of China

Fig. 1. Map of northeast of china ,in the upper shows locations of the fields mentioned in this article

traditional sequence stratigraphy is difficult to find and compare. In particular, when the volcanic eruption is strong and multi-stage, the phase sequence changes further complicated in the horizontal and vertical. It is more difficulty to find a unified datum in volcanic rock formation.

As the features of the volcanic eruption localized, the volcanism product is not the same in different parts of the sequence boundaries. Volcanic strata exploration must to face this problem. It is also the bottleneck restricting the development of volcanic exploration. The stratigraphic sequence of volcanic rocks is still in the exploratory stage for global scientists. The scholars of home and abroad (Gamberi F,2001; Schmincke H U,2004; Wang P J,2010) focus on the type, characteristics, causes of volcanic ejecta, etc. The volcanic rocks is usually considered to be filling sequence part of the sedimentary basin for more reseach, that the volcanic sequence develop in the ultra-sequence sets, super-sequence and the top of the bottom of the third sequence(Meng QA,2005; Dong GC,2005; Qiu C G,2006). With further research, the time and location of volcanic development is controlled by tectonic activity has been recognized (Cheng R H,2005;Tang H F,2007). However, these studies mentioned above have ignored the fact that volcanic sequence has the specific formation mechanism and controlling factors relative to sedimentary sequence.

Volcanic formation is an important part of fault basin. The rapid accumulation volcanic rock has important implications for the formation and evolution of stratigraphic sequence with volcanic rocks. In order to obtain new ideas for volcanic sequence research and establish a interpretation contrast mode of volcanic sequence strata, volcanism and its products should be analyzed together in this paper. Based on the new research results of construction and volcanic eruption mechanisms, We propose a new development model of volcanic sequence strata, in order to find out the distribution of volcanic rocks and the formation mechanism of volcanic reservoir.

2. Volcanism features of Yingcheng formation

Because the multi-center multiphase volcanic eruption, there is widely developed volcanic rocks in Yingcheng formation of Xujiaweizi fault depression (Fig.2). Volcanic strata have the feature of multiphase superposition in the vertical and migration in the horizontal. These kinds of feature indicate volcanic eruption with cycling and direction.

2.1 Multiphase volcanic rock superposition in the vertical

This appearance is common that volcanic rocks is vertical superposition in the wholly Xujiaweizi fault depression. The formation process that alternating layers of different volcanic rocks is more complex, which is also a record of volcanism processes. There should be two reasons for different volcanic rocks superimposed. The one is that the eruption of magma from different craters at different times and different distances superimposed in the same region. The other is that the volcanic rocks come from the same crater, but the volcanic rocks are different in different parts of the same volcanic edifice (Fig.3). This appearance is asynchronous volcanic action with the different features of eruptive material, effusive activity and affected area. The characteristic of volcanic rocks vertical development, resulting in a single well by volcanic sequences in different parts of the sequence have different characteristics. So it is not good to divide the volcanic sequence by the traditional way, we also can not expect to find a unified interface feature in the regional area.

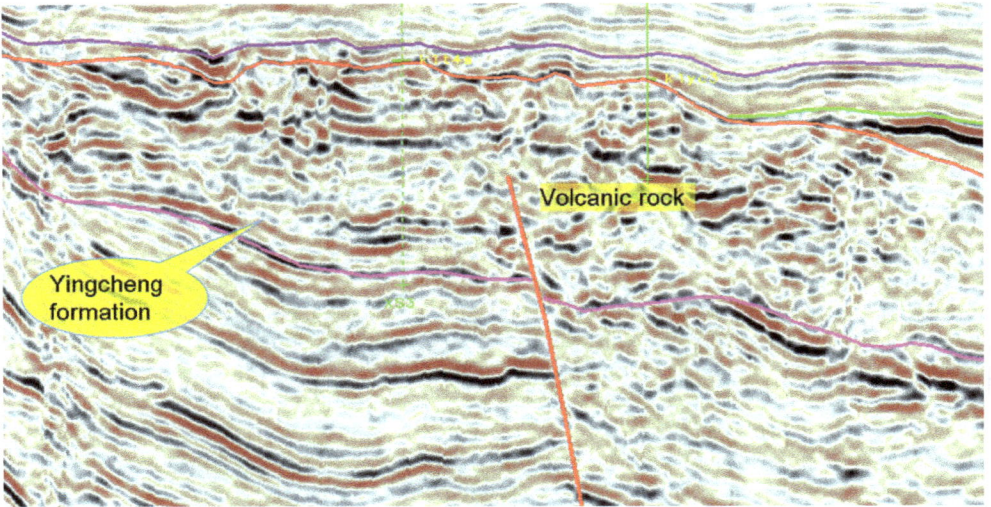

Fig. 2. Map of volcanic rocks development in Yingcheng formation

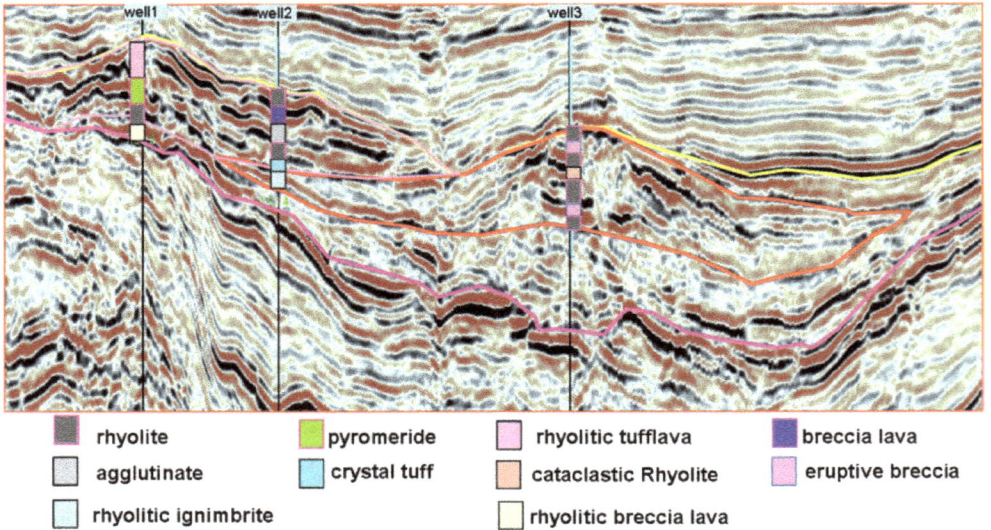

Fig. 3. Characteristic pattern of volcanic rock vertical superposition and horizontal change in Yingcheng formation

2.2 Volcanic eruption controlled by faults

Volcanic activity is related with the extensional movement of the mantle. When the unusual mantle bulge, causing supracrustal formation broken ground result in a volcanic eruption. The discordogenic fault that cut through the basement and earth's crust is the channel of magma upwelling. The Xuzhong and Xudong two major strike-slip fault systems (SSFS) and theirs connected subsidiary fractures are the channels of magma eruption in Xujiaweizi fault depression. The thick volcanic rocks distribute along the strike-slip fault systems of Xuzhong and Xudong (Fig.4). According to drilling and geophysical data, the craters are mainly located in the transition location and intersection of fractures. Based on the correlation theory of rock fracture mechanics and structural geology, stress is concentrated in the end. However, the two endpoints of fracture are the stress concentration points, but also it is the weakest area. The volcanism is firstly happened in the weakest zone (Fig.5).

Because of the characteristics of SSFS, the strata are subjected to compressional and extensional coincident movement. There will be the phenomenon that magma is not eruption in various parts of the same fault at the same time along the strike-slip fault zone. The characteristics of the same fault activity are often different in different locations. There is a gradual process of change. The volcanic eruption first happened in the extensional location form the feature of multi-center fissure eruption. Because the alternating tension and compression effects. The characteristics of volcanic activity are different in different fault segments. The existence of such differences is the main reason for the volcanic superposition and migration in the horizontal (Fig.6).

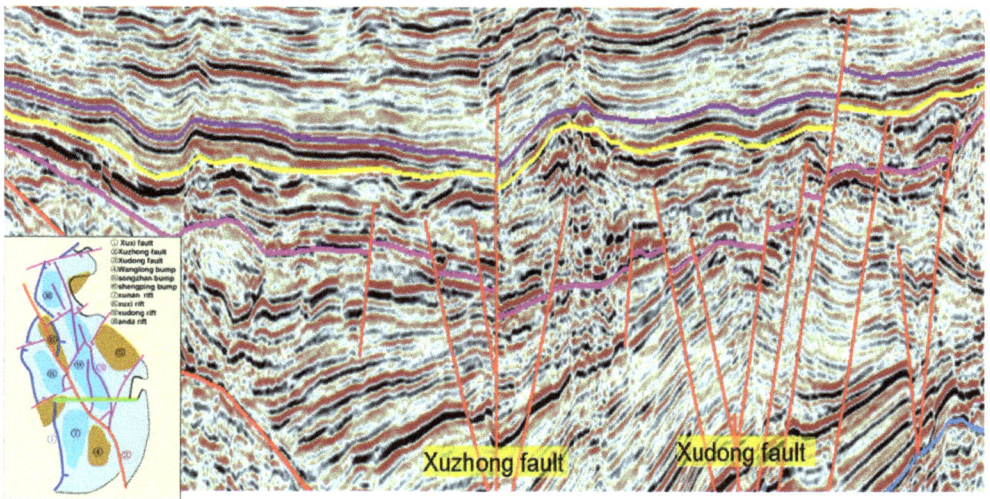

Fig. 4. The matched relationship of volcanic rocks thickness and fracture

2.3 The coexisting of volcanic and sedimentary rocks

The process of volcanic activity from start to finish is not continuous, but intermittent. The depositional interbedded stratum is the representative lithology for a quiet period of volcanic activity. Such as: there is about thickness of 45m deposition interlayer in the area of xs6 well volcanic rocks; there are also two sets of sedimentary sandwich thickness of 20m in the area of xs401 well. Regional deposition interlayer represents the interval of regional volcanic activity. There's a regional comparison deposition interlayer in the area of xs6-xs4-xs2-xs14-xs12 wells, which is the most obvious signs of volcanic cycle surface (Tang H F,2010).

The interbedded features of volcanic and sedimentary rocks, it is both on the evidence of intermittent volcanic activity, but also the evidence of underlying volcanic rocks below the

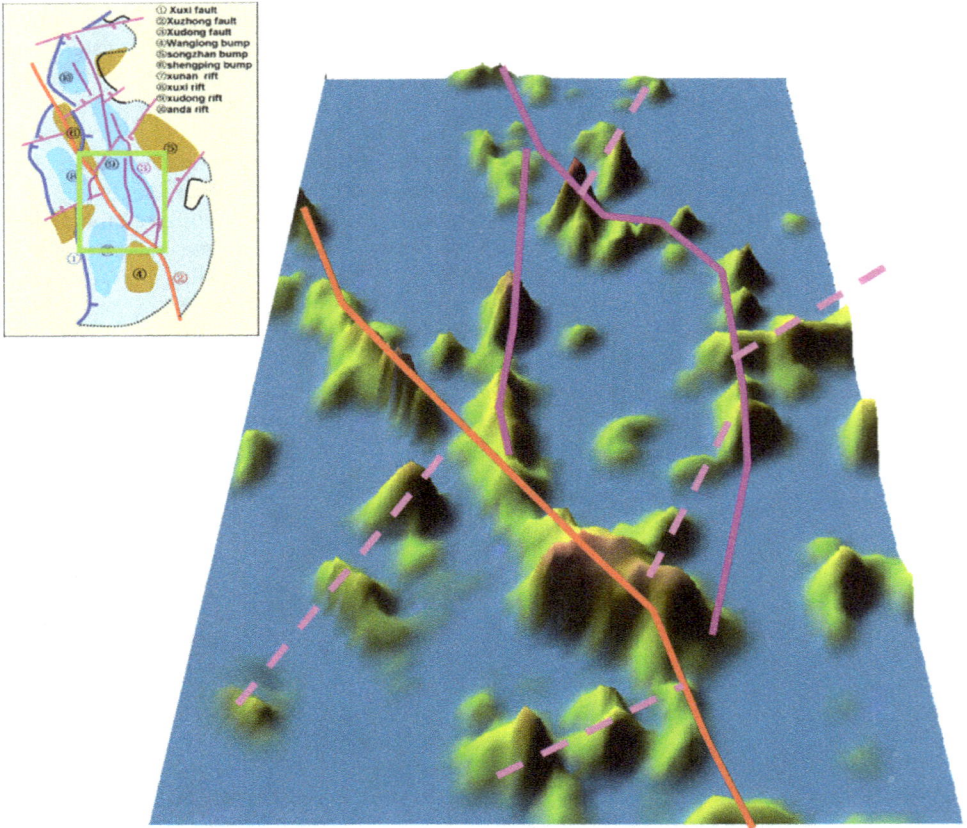

Fig. 5. The matched relationship of volcanic explosion vent and fracture

water surface a long period. This kind of sedimentary rocks indicates that the Yingcheng volcanic has the characteristics of subaquatic eruption. The pearlite of xs2, xs5 and deposit ash tuff of xs1-4 wells drilled are further indicates the presence of lacustrine environment. There is a complete set of clastic rock containing volcanic in the Yingcheng formation in the area of zs14 well, which confirms a certain amount of lake present at the same time of volcanic eruption. The contact relationship is diverse between sedimentary and volcanic rock, some sedimentary rock is located in the middle of volcanic rocks, some is also located at the bottom of volcanic rock. The difference of occurrence location is controlled by volcanic eruptive sequence and affected area for sedimentary rock.

3. The division of volcanic sequence for Yingcheng formation

The accurate division and attribution of layers is the basis to restore stratigraphic evolutionary sequence. It's also the basis for oil&gas exploration and development. Because the multiphase volcanic eruption in the same area, volcanic edifices occur reformed in the vertical and lateral migration. It's difficult to propose a simple evolution model of general

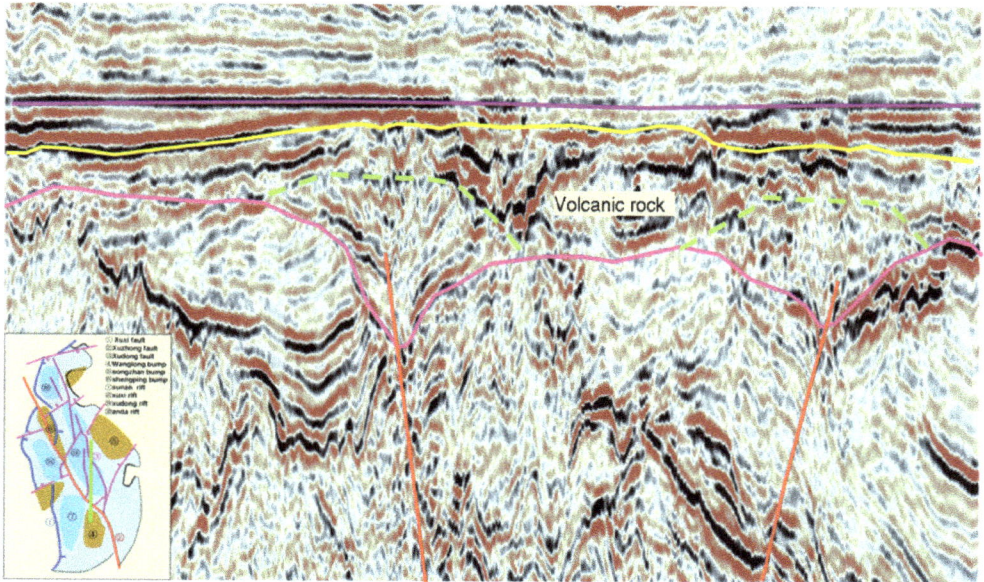

Fig. 6. The map of central vent eruption controlled by crack

application. It's also difficult to divide and compare volcanic sequence in accordance with this mode. Based on the summary of volcanic activity regular pattern, the authors analyze the volcanic sequence of the control factors and division and correlation in this article.

3.1 Distribution characteristics of Yingcheng strata

Based on the contrast from the regional stratigraphic features, there is main development of sedimentary rock strata containing volcanic material in the east and west sides of Xujiaweizi fault depression. In some areas of the rift center, such as xs801 well area, there is sedimentary rocks development in the lower part of Yingcheng formation. The set of sedimentary rocks is obviously different from the Shahezi sedimentary strata. The contact relationship of them is angle disconformity, this feature is very visible in the seismic profiles. It's further show that the volcanic rocks are not development in the whole Xujiaweizi rift in the period of volcanic eruption. It is jointly controlled by volcanic activities and sedimentary process for the temporal and spatial distribution of Yingcheng strata. It has the characteristics of multiphase volcanic eruption in the primary zone of volcanic eruption. Thus there is a large area of volcanic lava and athrogenic rocks development. There is the area of sedimentary rocks development in the outside of volcanic affected area in the Yingcheng formation. The construction feature of Yingcheng formation is associated with volcanic action and deposition (Fig.7).

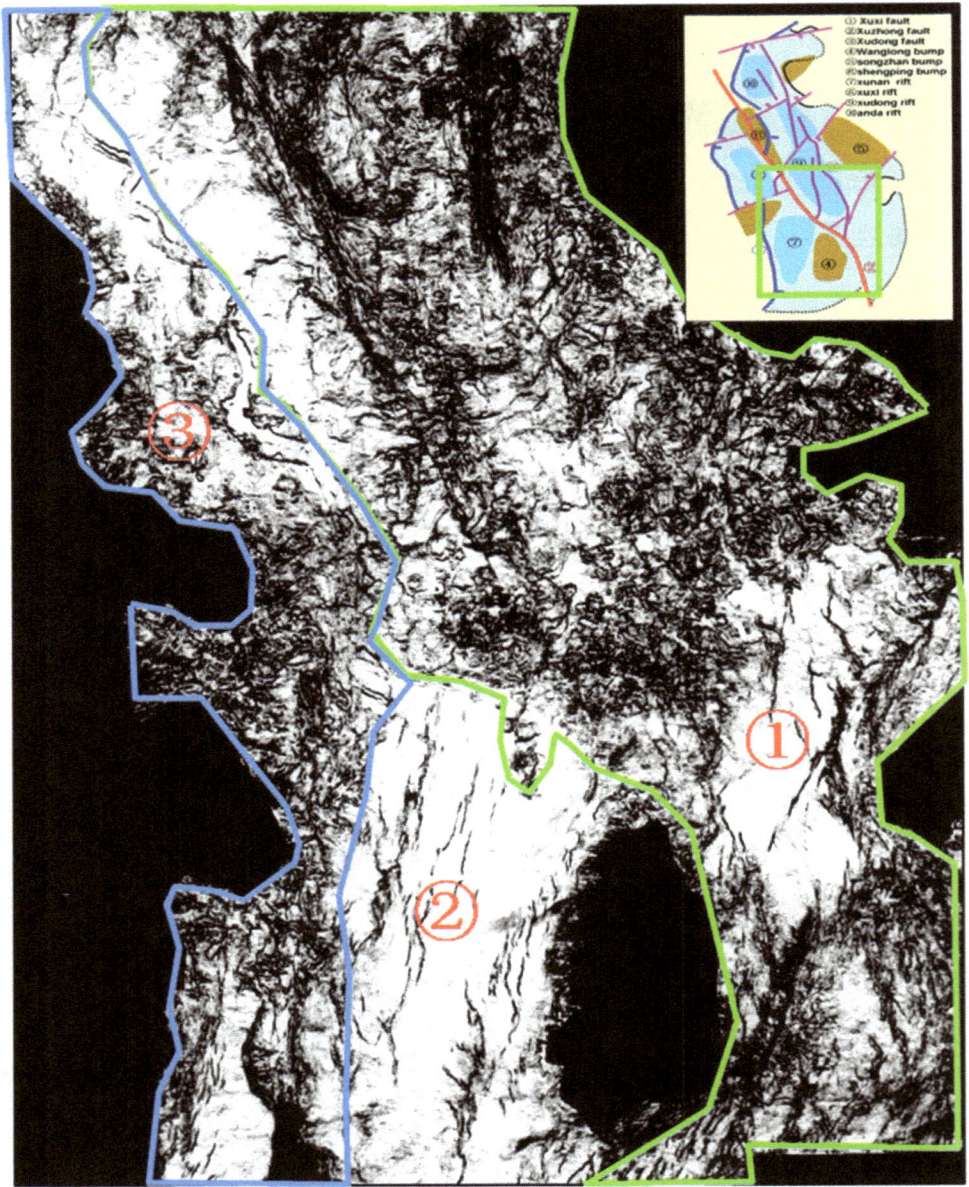

① Area of volcanic rock development
② Area of sedimentary rock development
③ Area of volcanic and sedimentary rock mixed development

Fig. 7. The forecasting map of lithology using coherence cube

3.2 Volcanic sequence classification foundation

It is to find the formation boundary that the upper and lower strata can be distinguished and correlatable in the horizontal for volcanic sequence division. For a concentrated continuous volcanic activity, the material composition, eruption methods and effusive activity will occur on a regular variation. which changes regularly must be inevitably form a set of genetic relationship of the volcanic sequence. Based on the single well detailed breakdown of volcanic rocks, Scholars (Wang PJ,etal, 2010) divided different numbers of volcanic rocks cycles in different regions in the Xujiaweizi rift. But these cycles explained is difficult to compare with others in the horizontal. Lithologic correlation is wrong correlation marker for volcanic sequence. It can reflects the variation of volcanic activity that superimposed array mode of volcanic apparatus(Chen SM,2011).Therefore, the directivity and superimposed mode of volcanic activity is different for diachronous volcano, volcanic sequence can be divided.

Based on comprehensive analysis of the volcanic rocks development characteristics of Yingcheng formation, the four correlation markers are proposed that is suited for volcanic rocks comparing in XJWZ rift. The first one is depositional interbedded stratum and ash tuff that can be regional compared. It is an important foundation for volcanic cycle dividing. The second one is the structural surface, in the upper and lower of the structural surface. There is distinguished difference for effusive activity, lithology, facies. So it is easy to form unconformity and the angular unconformity for volcanic formation, it is the good foundation for volcanic comparison (Fig.8). The third one is the superimposed mode of volcanic apparatus that diachronous volcano has different directivity of superimposition (Fig.9). The fourth one is depositional interface in the same period of volcano. Because the sedimentary formation records the evidence of volcanic event, classical infilling mode of fault depression has been changed when the volcanic goes into the rift. Each volcanic eruption correspond a tectonic movement, so the coupling of structure and sedimentary strata also recorded the volcanic activity cyclicity (Fig.10).

3.3 The division of volcanic sequence

Combination of the regional tectonic evolution, according to the macro-migration of volcanic activity, the volcanic rocks of Yingcheng formation can be divided into two three-stage sequences. The lower volcanic sequence is Y1 section, the upper volcanic sequence is Y3 section. The Y1 is mainly controlled by xuzhong strike-slip fault and its associated faults, the maximum thickness of volcanic rocks is along the fault zone. It is mainly distribution in the central and southern rift. The direction of volcanic eruption is mainly from north to south.As with the fault distance increases, the content of volcanic strata gradually reduced until it evolved into sedimentary strata. The Y3 is mainly controlled by xudong strike-slip fault and its associated faults. The direction of volcanic eruption is mainly from south to north. The xs22 well field is the starting point for the period of volcanic activity. Since the rapid release of volcanic energy, the xs22 Well field form a large-scale collapse crater, and form a thick pyroclastic filling. Because of the volcanic activity have the feature of direction and migration, there is a large area of debris deposition in the northern area that is the outside of volcanic activity sweep area. With the more large-scale violent volcanic eruptions and flooding, the larger and broader coverage of volcanic filling.

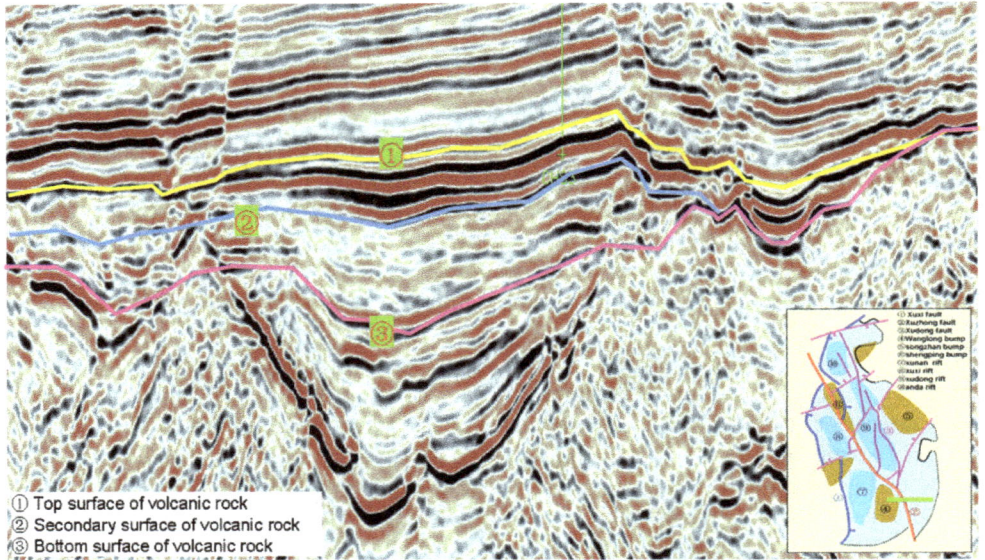

Fig. 8. The feature of sequence interface in volcanic rock

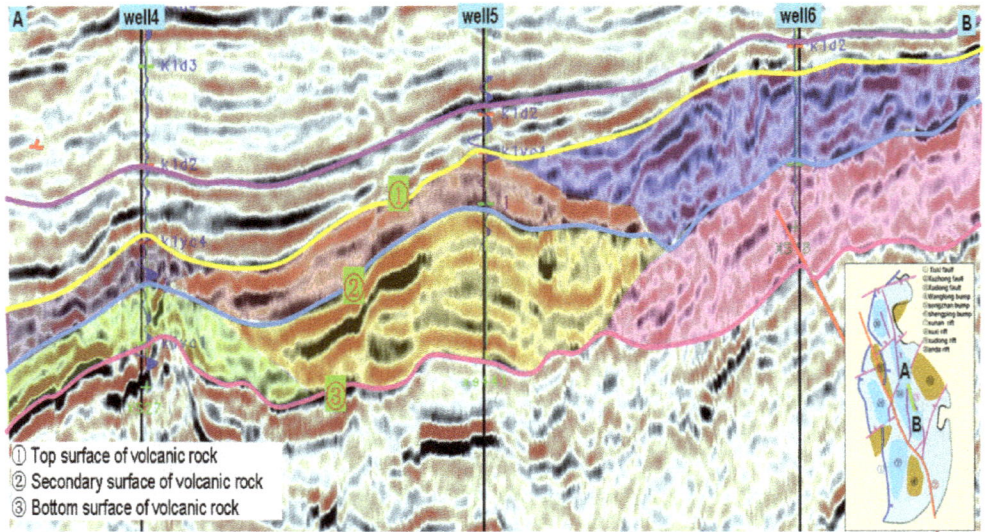

Fig. 9. The difference of volcanic edifice superimposed relationship nearby the volcanic sequence

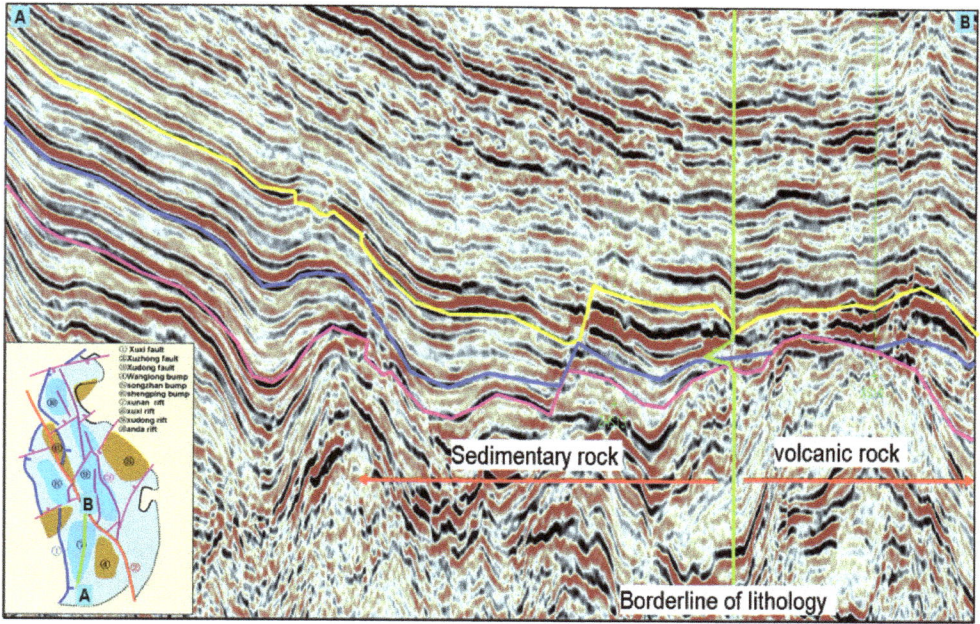

Fig. 10. The couple of sedimentary sequence and volcanic action

The volcanic sequence of Y1 and Y3 can be further identified three sub-sequences. they are the early volcanic eruption sequence(EVES), the strong volcanic eruption sequence(SVES) and the languid volcanic eruption sequence(LVES). Explosive eruption is primary in the EVES, there is also a certain amount of overflow phase developed. There is the area of sedimentary rocks development in the outside of volcanic affected area in the Yingcheng formation. The volcano shows the features of local central vent eruption along the cracks. Nowadays, the volcano of Wudalianchi has the same characteristics to EVES in the Heilongjiang province of China. The SVES has the typical feature that large-scale outbreak and the overflow happen simultaneously. There is the interbedded, and there is also some thick individual layer of lava flows and debris flows. The SVES is widely distributed in the large areas of sequence. It is not only the main volcanic sequence stratigraphy, but also the reservoir is the most development in the entire volcanic sequence. Local overflow and small-scale invasion is the main characterics of the LVES, because the size of LVES is small, the feature of sequence is no obvious. it is usually combined with the SVES as a sequence. But the LVES is the most favorable sequence for volcanic reservoir communication with deep fluid. it is also the favorable position for deep fluid easier to charging in the process of volcanic accumulation. It become the favorable accumulation area for carbon dioxide gas reservoir. The current exploration results have confirmed this characteristic in XJWZ rift (Fig.11).

4. Developmental patterns of Yingcheng sequence stratigraphy

The base level of volcanic sequence is unstable and not uniform in the region. Because of the existence of early volcanic eruption highland, the late sequence is not good overlay with the early. Therefore, the two-step volcanic sequence boundary is difficult to form a stable feature in the region. It is also hard to find the unified reflecting boundary for tracking.

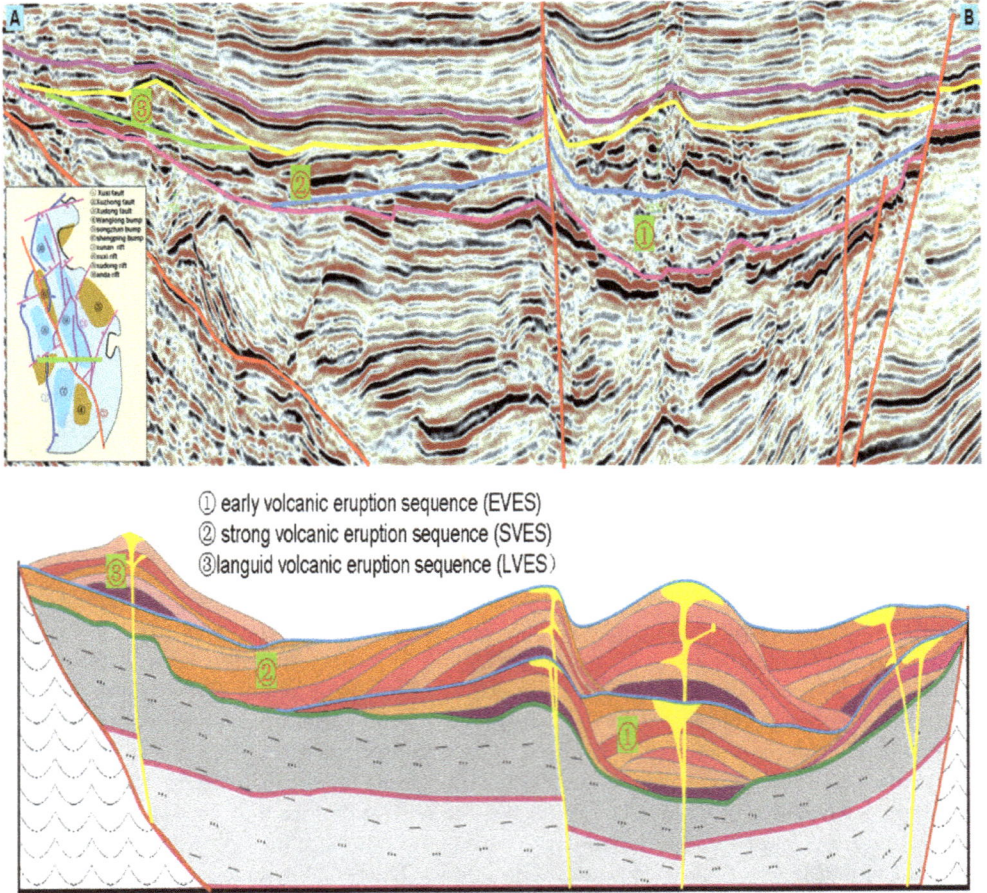

① early volcanic eruption sequence (EVES)
② strong volcanic eruption sequence (SVES)
③ languid volcanic eruption sequence (LVES)

Fig. 11. The map of volcanic sequence division

4.1 The Y1 developmental patterns

The size of Y1 volcanic apparatus is certain scale, the general current occurrence height is above 300 meters. The early eruption of the volcano can be located above water level. Therefore, compared to the late period volcanic apparatus, the weather-worn extent of early is strong. such as: the transformation feature of volcano is relatively obvious in the north well field, there is thick layer weathering crust development; but the transformation feature of volcano is significantly decreased in the south well field. This change indicates the order of volcanic action that is from north to south.

Volcanic activity is also accompanied by tectonic movement, resulting in frequent lake level fluctuation. When volcanic material quickly build-in the rift basin, the accommodation space is changed, resulting in level surface rising. The volcanic eruption early is located under the level surface, received clastic sediments. With the thickness of cumulo-volcano continuous increase, the late crater outcrop above the water level. Therefore, the interbedded phenomenon of sedimentary and volcanic rocks significantly reduced, there is mostly volcanic with high purity development in the upper volcanic rocks. There is the area of sedimentary rocks development in the outside of volcanic affected area. These sediments containing volcanic material is high, because the highland formed by volcanic eruption suffered weather worn activity, provide a filling material for the basin (Fig.12). Thus the unique geological feature is formed for Yingcheng formation. where the volcanic eruption, volcanic intermittent, erosion, deposition is interaction. The ancient landscape of Y1 is not flat where the north is lower, the south is higher.

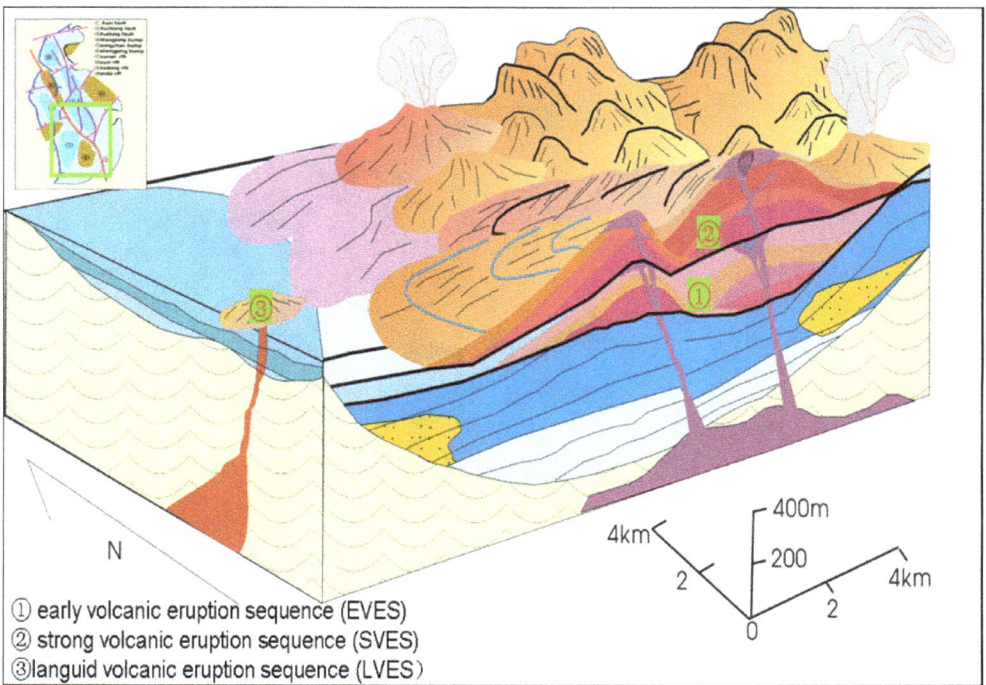

① early volcanic eruption sequence (EVES)
② strong volcanic eruption sequence (SVES)
③languid volcanic eruption sequence (LVES)

Fig. 12. The map of Y1 developmental patterns

4.2 The Y3 developmental patterns

Because of the different nature of the magma, the Y3 development pattern is completely different from Y1. The Y1 is mainly composed of acidic volcanic, but the basic volcanic rock is mainly development in Y3. Compared with the acid volcanic rocks, the basic volcanic rocks has different vents feature. Such as the volcanic vents is smaller, the plane position is relatively stable, the volcanic rocks is mainly volcanic lava. The local outbreaks characteristic of EVES is accompanied by fault activities. It is easy to form a barrier lake. Thus the formation of volcanic rocks and sedimentary rocks in symbiosis is the typical characteristic (Fig.13). When the volcano erupted violently, volcanic sub-sequence is in the stage of SVES. The rift basin is filled with volcanic material quickly, volcanic rock becomes the main rocks, and the sedimentary rock is little development. A large area of lava delta is development, in the vent, often accompanied by some small-scale outbreaks (Fig.14). The ancient landscape of Y3 is not flat where the north is higher, the south is lower.

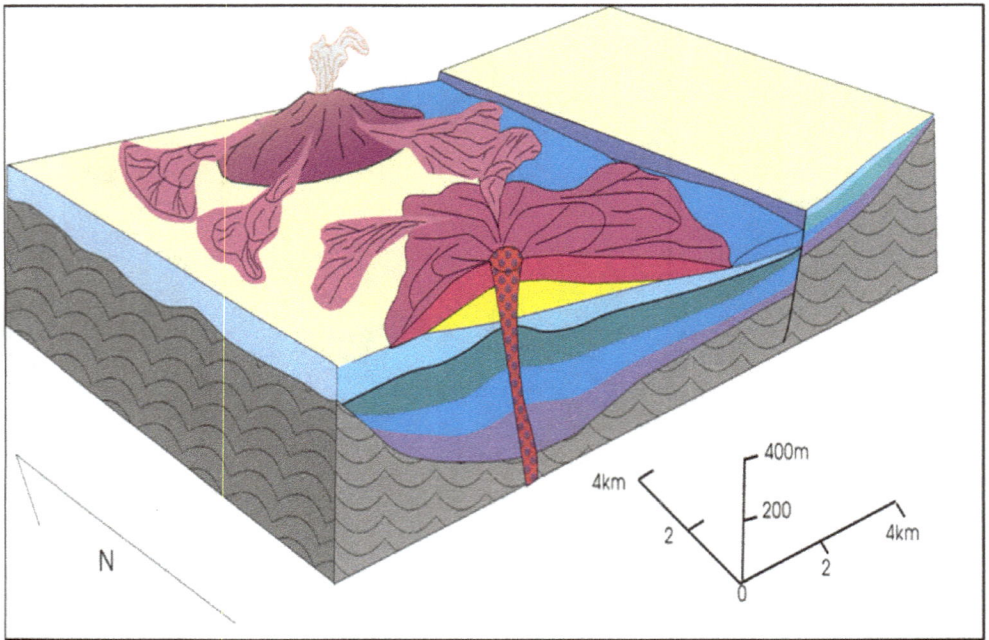

a. developmental patterns of EVES

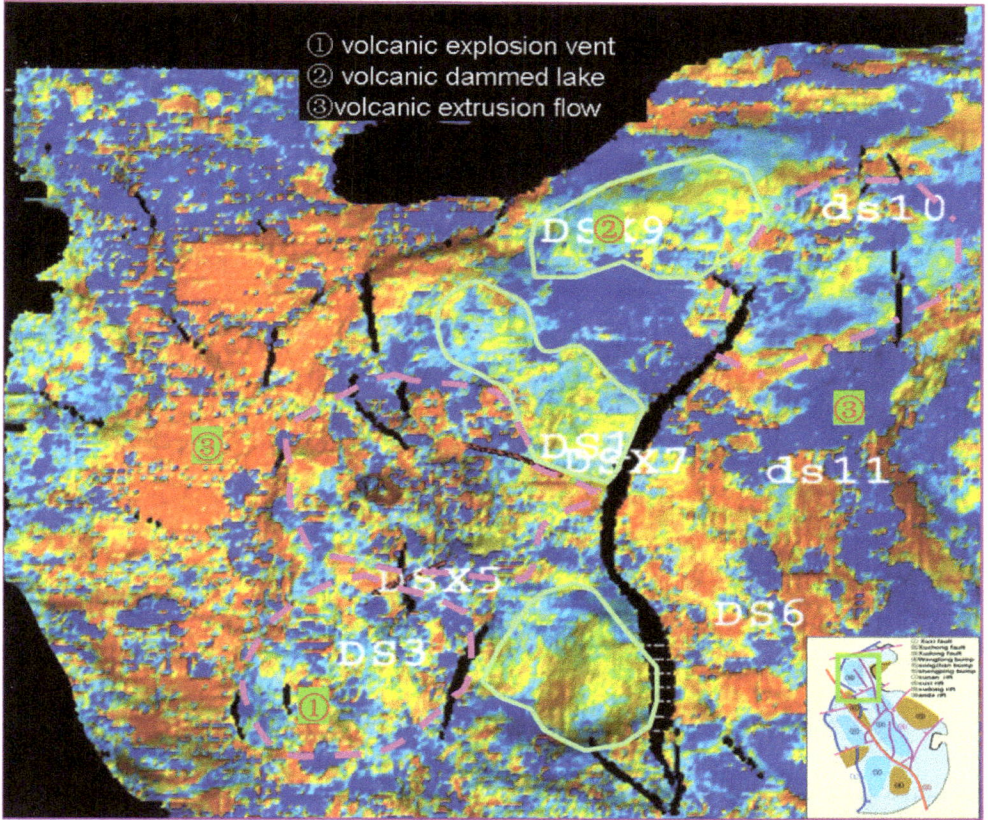

b. the example of EVES

Fig. 13. The map of Y3 developmental patterns(EVES)

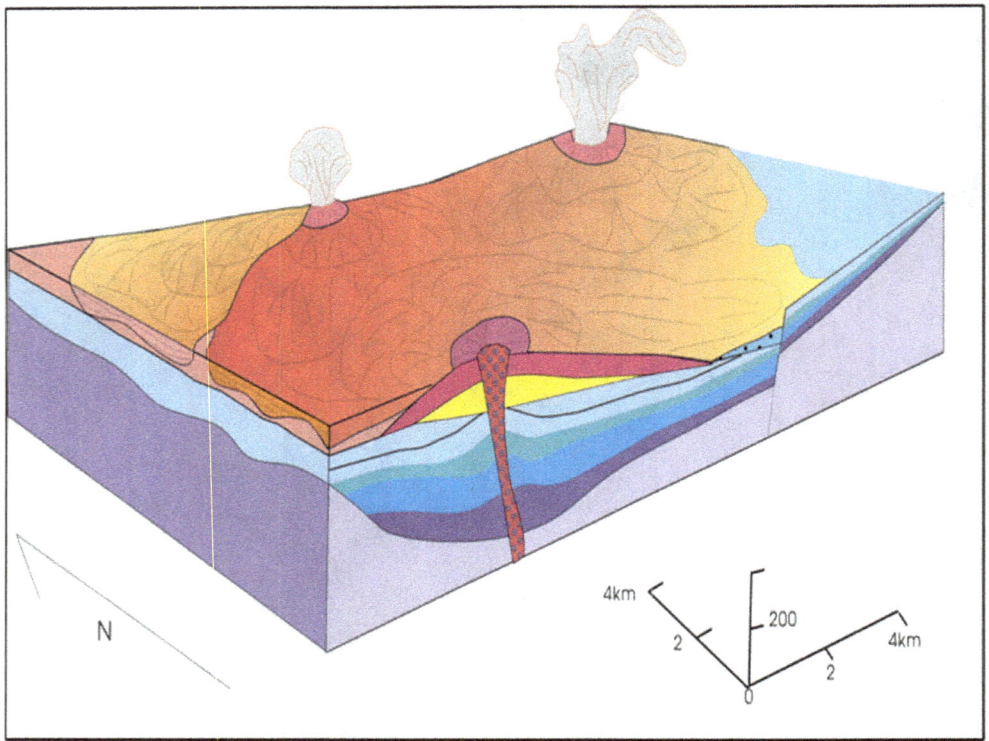

a. developmental patterns of SVES

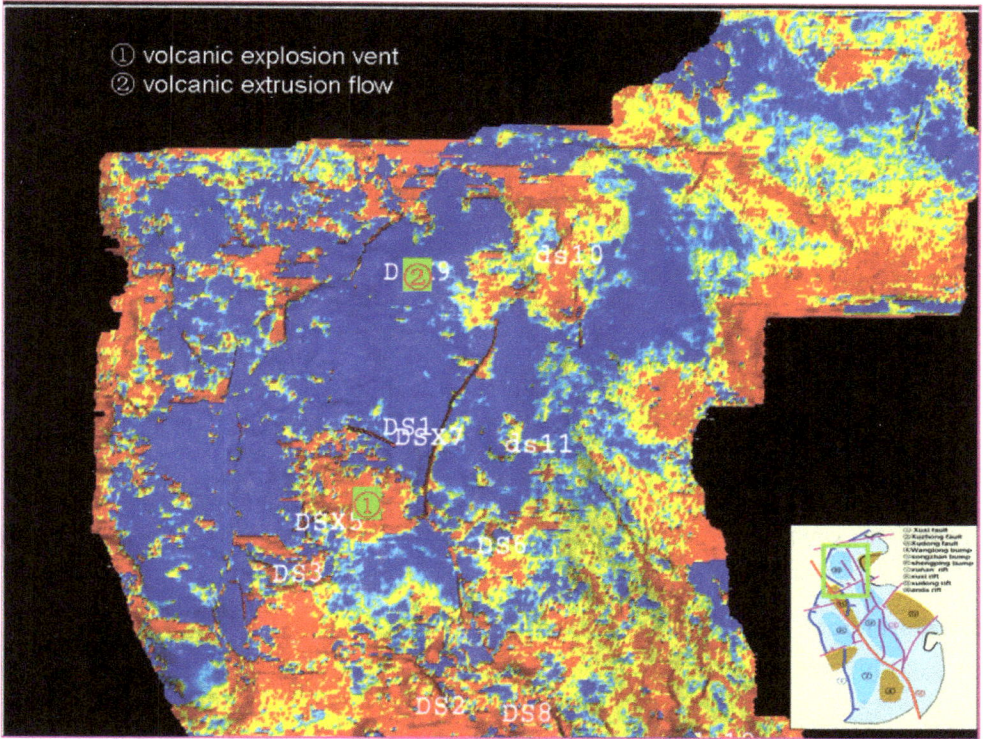

b. the example of SVES

Fig. 14. The map of Y3 developmental patterns(SVES)

5. Summary

1. The volcanic eruption is controlled by strike-slip fault systems(SSFS) and theirs connected subsidiary fractures. Because the order of fault activity is different, so that the volcanic rock of Yingcheng formation can be divided the upper and lower volcanic sequence. The the lower sequence(Y1) is mainly controlled by Xuzhong strike-slip fault and its associated faults, the upper sequence (Y3) is mainly controlled by strike-slip fault and its associated faults. So the research of volcanic sequence, it is very important to find the faults that control the volcanic eruption.

2. The regional sedimentary is present in volcanic rock strata. It is both on the evidence of intermittent volcanic activity, but also the evidence of underlying volcanic rocks below the water surface a long period. This kind of sedimentary rocks indicates that the Yingcheng volcanic has the characteristics of subaquatic eruption. The environment of volcanic eruption is shallow-water lacustrine facies. The interaction of fire and water, the volcanic eruption cycle can be found in sedimentary sequence.

3. Different parts of the same fault has different activities time, it also control the order of volcanic eruption. So it is the characteristic appearance that fissure flow multi-point

central vent eruption in XJWZ rift. The craters are mainly located in the transition location and intersection of fractures.

4. The difference of volcanic activity directional is the important basis of volcanic rock sequence division. It is also a reflection of plate movement. The direction change of volcanic activity is must be accompanied by the tectonic movement occurs.

5. Sedimentary and volcanic rocks are uniform in the strata, there is mainly volcanic sequence stratigraphy in the volcanic eruption area; There is the area of sedimentary rocks development in the outside of volcanic affected area . Sedimentary rock strata records the cyclicity of volcanism.

6. The volcanic sequence of Y1 and Y3 can be further identified three sub-sequences. They are the early volcanic eruption sequence(EVES), the strong volcanic eruption sequence(SVES) and the languid volcanic eruption sequence(LVES). It is mainly based on debris accumulation in EVES, it makes the reflection characteristics of chaotic and blank in seismic profile. The SVES usually have the reflection characteristics of strong energy in seismic profile. The LVES is local and sporadic development, it is difficult to find the regional characteristics compared in seismic profile. Therefore, it merged into the SVES. But it has the good indication for inorganic gas reservoir.

7. They are very important signs for volcanic sequence divided. Such as superimposed mode of volcanic apparatus,depositional interbedded stratum and ash tuff can be regional compared, structural surface, sedimentary sequence interface With the same period of volcanic activities.

6. References

[1] Vail P R.. 1987. Seismic stratigraphy interpretation using sequence stratigraphy. Part 1: Seismic stratigraphy interpretation procedure. American Association of Petroleum Geologists, Studies in Geology, 27:1-101

[2] Sloss L L. 1959. Sequences in the cratonic interior of North America, Part 2: Geological Society of America Bulletin, 70 (12) :1676-1677

[3] Catuneanu O, Khalifa M A, Wanas H A. Sequence stratigraphy of the Lower Cenomanian Bahariya Formation, Bahariya Oasis, Western Desert. Egypt Sedimentary Geology, 2006, 190: 121-137

[4] Catuneanu O.Principles of Sequence Stratigraphy. Elsevier, Amsterdam, 2006, 375

[5] Catuneanu O, Abreu V, Bhattacharya J P, et al. Towards the standardization of sequence stratigraphy. Earth-Science Reviews, 2009, 92: 1-33

[6] Arimoto R., B.J. Mass Particale Size Distribution of Atmospheric Dust and the Dry Deposition of Dust to the Remotr Ocean. Journal of Geopjyscical Research, 1997, 102: 15867-15874

[7] T., A.C., Saydam. Acidic and Alkaline Precipitation in the Cilician Basin, North-eastern Mediterranen Sea. Science of the Total Environment, 2000, 253: 93-109

[8] Caquineay, Gaudichet, Gomes. Saharan Dust: Clay Ratio as a Relevant Tracer to Assess the Origin of Soil-derived Aerosols. Geophysical Research Letters, 2008, 25: 983-986

[9] Shanley, K. W., McCabe, P. J., 1994. Perspectives on the sequence stratigraphy of continental strata. AAPG Bull., 78:544 - 568.

[10] Liu Z J, Dong Q S., Wang S. M., et al., 2002. Introduction and application to sequence stratigraphy of continental face. Petroleum Industry Publishing House, Beijing, 21 - 91 (in Chinese).

[11] Ji Y L. 2005. sequence stratigraphy. Tongji Univesity publishing house, Shanghai. (in Chinese)

[12] Deng H W, Wang H L, Zhu Y J, etal.2002. principle and application of high resolution sequence stratigraphy, Geological publishing house, Beijing (in Chinese).

[13] Hou M C, Chen H D, Tian J C. 2003. Sequence-filling dynamics——a new study direction on sequence on sequence stratigraphy. Journal of Stratigraphy, 27 (4) :358-364 (in Chinese)

[14] Xu Y X, Gao X L, Li Z Y, et al · Sequence stratigraphy and hydrocarbon distribution of the Eogene in the eastern slope of Chengdao area · Petroleum Exploration and Development, 2001, 28(4) :25-27 (in Chinese)

[15] Lin C S · Sequence and depositional architecture of sedimentary basin and process responses Acta Sedimentologica Sinica, 2009, 27(5) : 849-862 (in Chinese)

[16] Chen F, Luo P, Zhang X Y, et al. Stratigraphic architecture and sequence stratigraphy of upper Triassic Yanchang Formation in the eastern margin of Ordos Basin. Earth Science Frontiers, 2010, 17(1) :330-338 (in Chinese)

[17] Wang Q C, Bao Z D, He P. Sequence stratigraphic responses to the lacustrine basin deep-faulted period in the north area of the western sag, Liaohe Depression. Petroleum Exploration and Development, 2010, 37(1) :11-19 (in Chinese)

[18] Wang P J, Feng Z Q, Chen S M, et al. Basin Volcanic: Volcanic Rocks in petroliferous Basins: Lithology·Faces· Reservoir·Pool· Exploration. Science Press.2008, Beijing (in Chinese)

[19] Stewart A L, McPhie J. Facies architecture and LatePliocene-Pleistocene evolution of a felsic volcanic island, Milos, Greece. Bull Volcanol, 2006, 68:703-726.

[20] Busby C J, Bassett K N. Volcanic facies architecture of an intra-arc strike-slip basin, Santa Rita Mountains, Southern Arizona. Bull Volcanol, 2007, 69:85-103.

[21] Wang P J, Wu H Y, Pang Y M, et al. Volcanic facies of the Songliao Basin : sequence, model and the quantitative relationship with porosity& permeability of the volcanic reservoir. Journal of Jilin University (Earth Science Edition), 2006, 36(5) :805-812 (in Chinese)

[22] Gamberi F. Volcanic facies associations in a modern volcaniclastic apron (lipari and vulcano off shore, Aeolian island arc. Bull Volcanol, 2001, 63: 264-273.

[23] Schmincke H U. Volcanism. Berlin, Heidelberg, New York: Springer, 2004.

[24] Wang P J, Yin C H, Zhu R K, et al. Classification, description and interpretation of the volcanic products: ancient and modern examples from China. Journal of Jilin University(Earth Science Edition), 2010, 40(3) :469-481 (in Chinese)

[25] Meng Q A, Wang P J, Yang B J. Geological signatures of sequence boundary of the Songliao basin: new interpretation and their relation to gas accumulation. Geological Review, 2005, 51(1) :46-54 (in Chinese)

[26] Dong G C, Mo X X, Zhao Z D, et al. A new understanding of the stratigraphic successions of the Linzizong volcanic rocks in the Lhünzhub basin, northern Lhasa, Tibet, China. Regional Geology of China, 2005, 24(6) :549-557 (in Chinese)

[27] Qiu C G, Wang P J. Prelimimary study on sequence's location of volcanic stratigraphy in sedimentary basin-an example from Xujiaweizi depression of Songliao basion. Xinjiang Oil & Gas. 2006, 2(1):13-18 (in Chinese)

[28] Cheng R H, Wang P J, Liu W Z, et al. Sequence stratigraphy with fills of volcanic rocks in Xujiaweizi Faulted Depression of Songliao basin, Northeast China. Journal of Jilin University(Earth Science Edition), 2005, 35(4):469-474 (in Chinese)

[29] Tang H H, Wang P J, Jiang C J, et al. Seismic characters of volcanic facies and their distribution relation to deep faults in Songliao basin. Journal of Jilin University(Earth Science Edition), 2007, 37(1): 73-78 (in Chinese)

[30] Zhang Y G, Chen S M, Zhang E H. The new progress of Xujiaweizi fault depression characteristics of structural geology research. Acta Petrologica Sinica, 2010, 26(1):142-148 (in chinese)

[31] Zhang E H, Jiang C J, Zhang Y G, et al. Study on the formation and evolution of deep structure of Xujiaweizi fault depression. Acta Petrologica Sinica, 2010, 26(1):149-157 (in chinese)

[32] Jiang C J, Chen S M, Chu L L, et al. A new understanding about the volcanic distribution characteristics and eruption mechanism of Yingchen formation in Xujiaweizi fault depression. Acta Petrologica Sinica, 2010, 26(1):63-72 (in chinese)

[33] Tang H F, Bian W H, Wang P J, et al. Characteristics of volcanic eruption cycles of the Yingcheng Formation in the Songliao Basin. Natural Gas Industry, 2010, 30(3): 35-39 (in Chinese)

[34] Chen S M, Zhang Y G, Jiang C J. The analysis of volcanic edifice superimposition and its digital model parameters establishment. Chinese J. Geophys, 2011, 54(2): 499-507 (in Chinese)

Part 3

Remote Sensing

4

Remote Predictive Mapping: An Approach for the Geological Mapping of Canada's Arctic

J. R. Harris, E. Schetselaar and P. Behnia

Geological Survey of Canada, Ottawa
Canada

1. Introduction

Due to its vast territory and world-class mineral and energy potential, efficient methods are required for upgrading the geoscience knowledge base of Canada's North. An important part of this endeavour involves updating geological map coverage. In the past, the coverage and publication of traditional geological maps of a limited region demanded multiple years of fieldwork. Presently more efficient approaches for mapping larger regions within shorter time spans are required. As a result, an approach termed Remote Predictive Mapping (RPM) has been implemented since 2004 in pilot projects by the Geological Survey of Canada. This project falls under the larger Geo-mapping for Energy and Minerals (GEM) program initiated by Natural Resources Canada.

Remote predictive mapping comprises the compilation and interpretation (visual or computer-assisted) of a variety of geoscience data to produce predictive maps containing structural, lithological, geophysical, and surficial information to support field mapping. Predictive geological maps may be iteratively revised and upgraded to publishable geological maps on the basis of evolving insight by repeatedly integrating newly acquired field and laboratory data in the interpretation process. The predictive map(s) can also serve as a first-order geological map in areas where field mapping is not feasible or in areas that are poorly mapped. The fundamental difference between RPM and traditional ground-based mapping is that in the latter, the compilation of units away from field control (current and legacy field observations) is largely based on geological inference while in RPM this geological inference is repeatedly tested and calibrated against remote sensing imagery.

Remote predictive mapping is of course not an entirely new philosophy for geological mapping. Geologists have long assembled diverse layers (primarily aerial photographs and aeromagnetic contour maps) of geoscience data to study the relationships between the spatial patterns for resource exploration and mapping endeavours. In the past this has been accomplished using an 'analog' approach, forcing maps printed on mylar to be portrayed on a uniform map scale on a light table. However, with the increasing availability of digital data sets and the routine use of geographic information systems (GIS), the task of studying relationships between data and producing innovative maps to assist field mapping has become easier and more versatile. Contrary to the 'light table' approach, GIS allow maps and image data to be combined, overlaid, and manipulated at any scale with any combination of layers and subjected to any integrated enhancement.

A predictive map does not represent geological *truth* but rather a best estimate of what that area may represent on the ground based on the signatures derived from the interpreted data (geophysical, geochemical, remotely sensed). For that matter, even a traditionally produced geological map may not represent the geological truth, as all maps, no matter how they are produced, may contain spatial and classification errors. Thus, geological features of a predictive map do not necessarily correspond to how these features would be classified on the ground by a field geologist. At the categorization level, the geological term attached to a unit or structural feature may even prove to be wrong; yet at the detection level, the identified feature may correspond to a hitherto unrecognized mappable unit or structure that can be targeted for follow-up fieldwork.

The amount, variety, and quality of data used are obviously key factors in how closely the predictive map matches the geological patterns obtained by field mapping. Another factor is the nature of the geological terrain being mapped as the data sets and associated processing and enhancement techniques being employed will vary depending on the bedrock, surficial, and topographic environment. A remote predictive map can assist the geologist in a number of ways: (1) by predicting map units that would tentatively be assigned to rock types and/or geological formations (bedrock and surficial). This is based on establishing critical relationships between imaged physical properties (magnetic intensity, gravity, gamma-ray spectrometry, spectral reflectance, radar backscatter) and patterns obtained from available geological maps and field data, (2) by predicting areas that appear to be characterized on remotely sensed images by more complex and spatially heterogeneous geological patterns, thus focusing and prioritizing field work in these areas; likewise, areas with more homogenous signatures and simpler patterns can also be identified as possibly requiring less field work to geologically calibrate, (3) by predicting a variety of structures (foliation traces, faults, dykes, lineaments, glacial flow directions, etc.). The structural information can be used in advance of field work, to supplement field observations or as stand-alone geological information and, (4) by predicting the distribution of bedrock outcrop and other physiographic features such as wetlands, areas of forest fire burns, vegetation cover, and infrastructure to support fieldwork planning.

Predictive maps can also result in a different paradigm for planning field traverses. Instead of regularly spaced traverse lines, more detailed traverses can be set up that are focused on more complex areas and on areas where bedrock outcrop has been identified. This is especially advantageous in Northern mapping campaigns where the territory is vast and mapping expenses are high.

The mechanics of producing interpretations from various geoscience data sets can be greatly facilitated by GIS and image analysis technology. For example, image interpretation can be accomplished directly on a computer touch-sensitive screen as opposed to interpreting on mylar overlays. The advantage of this screen digitization process (i.e. heads-up digitization or interpretation) is that various enhanced images can be displayed quickly to facilitate interpretation by virtually real-time comparison between different data types at any scale. Multiple iterations can be undertaken and each digital interpretation can be stored as a different GIS layer. This by-passes the cumbersome procedure of scanning and digitizing hard-copy interpretations followed by georeferencing, which can introduce spatial errors. Similar to field mapping, the successful recognition and extraction of geological information is a learning process based on experience in interpreting image data in a variety of physiographic and geological settings.

2. RPM approach

Remote predictive mapping involves the acquisition, processing, and geological interpretation of available remotely sensed data sets as well as legacy geological data. The results are predictive maps (or GIS layers) of interpreted bedrock and surficial units as well as geologic structures. Remote Predictive Mapping can be either completed in isolation from field-based mapping or can be intimately integrated with it in order to ground truth the interpretation as field mapping proceeds. **Figure 1** shows a summary of the RPM

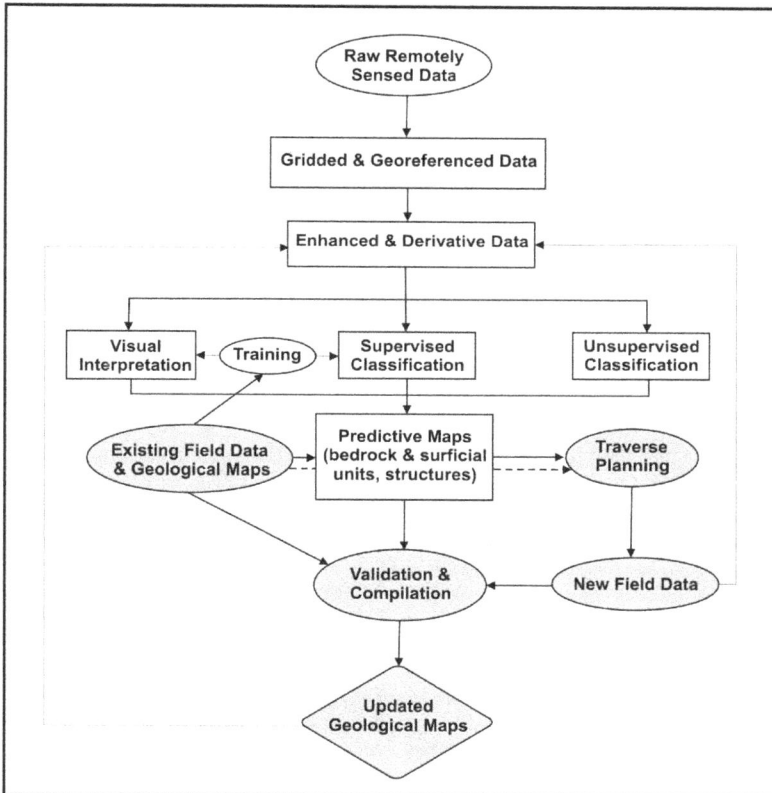

Fig. 1. Flow chart showing how RPM methods can be integrated in a geological mapping project. The grey area represents traditional field mapping methods whereas the white area represents remote predictive mapping methods. Predictive maps can be produced by enhancing and fusing various remotely sensed data and visually extracting geologic information from these products. Alternatively, a computer can be employed to automatically produce a predictive map (unsupervised approach) or by utilizing the geologist's expertise in concert with computer analysis (supervised approach). The geological interpretations are constrained or 'trained' by existing geological field data and existing geological maps. The arrow that loops back from *Updated Geological Maps* to *Enhanced and Derivative Data* emphasizes that the interpretation and map compilation process can be integrated over multiple iterations of field mapping.

process integrated into the work flow of a geological mapping project. The shaded portion represents the activities common to the *traditional* geological mapping process, whereas the portion that is not shaded represents the additional activities of the RPM approach. Regardless of whether the interpretation of remotely sensed data is fully integrated into a geological mapping project or not, the following provides a systematic outline of RPM work flow.

2.1 Mapping objectives

The first step in a RPM project is to define the mapping context, which includes the following:

- mapping focus (bedrock, surficial),
- nature of the geological terrain,
- surficial conditions and degree of exposure, physiography,
- data availability, quantity, and quality.

These factors will determine the data that will be most useful for bedrock mapping. Bedrock mapping projects that are planned in well exposed terrain and have thin residual till cover will benefit from the integration of magnetic, gamma-ray spectrometry, optical, and radar image data. In areas where sparse outcrops alternate with thick overburden, bedrock mapping will primarily profit from the interpretation of magnetic data.

In surficial mapping, optical and radar remote sensing techniques, together with gamma-ray spectrometry and digital terrain data, will contribute to distinguishing various types of surficial materials, identifying and mapping geomorphic features, and mapping streamlined glacial landforms that provide information on glacial movement. Geological setting and physiography of the terrain in combination with the spatial and spectral resolution, penetration depth, season of image acquisition, and aerial coverage of the remote sensing system (including airborne geophysics) are all important factors when choosing data sets for geological interpretation.

2.2 Data selection

Governments and private-sector contractors and/or vendors now provide much of the geoscience data in digital format that increasingly can be accessed through the internet. The core data types that are generally acquired and interpreted for RPM projects are listed in **Table 1** along with references to a sample list of websites to obtain them. In Canada, most of these data sets cover the complete landmass with the exception of gamma-ray spectrometry data. Nonproprietary, medium to low-resolution geophysical data, including magnetic and gamma-ray data were obtained from the Geological Survey of Canada's Geophysical Data Centre. LANDSAT 7 enhanced thematic mapper scenes of 180 x 180 km optical remotely sensed data with one 60 metre resolution thermal band, six 30 metre multispectral bands in the visible to mid-infrared range, and one 15 metre panchromatic band in the visible range can be obtained, free of charge, from the Geogratis website (http://www.geobase.ca). Radarsat data is obtained from the Canadian Space Agency (CSA). Digital elevation data (DEM) (CDED at 1:50,000 and/or 1:250,000 scale) can be downloaded from the Canadian Council on Geomatics (CCOG) website (http://www.geobase.ca). The internet providers of optical remotely sensed data often include a quick-look download service that allows for the inspection of cloud cover of the scenes before downloading.

Data	Data provider	Cost
LANDSAT TM	Geogratis web-site http://geogratis.cgdi.gc.ca/ MDA Geospatial Services	Free to download from Geogratis Cdn$720 per scene from MDA
RADARSAT	Geogratis web-site (100m pixel mosaic of Canada) MDA Geospatial Services (http://www.rsi.ca/)– for individual scenes from the archive or acquire new data – commercial users Canadian Space Agency (CSA) – as above – for government users	mosaic –free to download MDA - Cdn$3,000 – 4,000 /scene CSA - Cdn$300/ scene
Magnetic data	Geophysical Data center (GSC) http:// gdcinfo.agg.nrcan.gc.ca	Free to download
Gamma-ray spectrometry data	Geophysical data centre http:// gdcinfo.agg.nrcan.gc.ca	Free to download
DEM – CDED	Geobase http://www.geobase.ca/	Free to download
ASTER	USGS (http://edcdaac.usgs.gov/main.asp) Information can be found at: http://asterweb.jpl.nasa.gov/ http://asterweb.jpl.nasa.gov/gallery.asp	$40.0 US per scene
SPOT	IUNCTUS Geomatics Corp. http://www.terraengine.com/	$1200.0 per scene for SPOT 4 $1.00 - $6.00 per sq km – SPOT 5
IKONOS	MDA Geospatial Services (http://www.rsi.ca/) Information can be found at: http://www.infoterraglobal.com/ikonos.htm http://www.satimagingcorp.com/gallery-ikonos.html	$15.0 -$30.0 per sq km
QIUCKBIRD	MDA Geospatial Services (http://www.rsi.ca/) Information can be found at: http://www.satimagingcorp.com/gallery-quickbird.html http://www.ballaerospace.com/quickbird.html	See MDA web-site
ENVISAT	MDA Geospatial Services (http://www.rsi.ca/)	See MDA web-site
RESOURCESAT	MDA Geospatial Services (http://www.rsi.ca/)	$2750.0 per scene
IRS	MDA Geospatial Services (http://www.rsi.ca/)	$900.0 - $2,500.0
ERS-1 Radar	MDA Geospatial Services (http://www.rsi.ca/)	$660.0 per scene
Airborne hyperspectral data	- selected coverage of PROBE data – Baffin Island, Sudbury - Canada Centre for Remote Sensing (CCRS) and Geological Survey of Canada – Geophysical Data Centre	Selected scenes free to download

Table 1. Data sets used for the RPM projects discussed in this paper

There are also a number of other specialized remote sensing systems included in **Table 1** that do not yet provide complete coverage of the Canadian landmass. Optical sensors, including ASTER, SPOT, IKONOS, QUICKBIRD, WORLDVIEW I and II and airborne hyperspectral, can provide a wealth of geological information but these data are not available for all of Canada. However, these data can be acquired and when available their use should be considered, since they offer imagery with either higher spectral resolution (ASTER, 14 spectral bands) or higher spatial resolution (SPOT 5, IKONOS, QUICKBIRD, WORLDVIEW). The higher spatial resolution of the latter sensor systems with 4.0 to 2.4 metre multispectral and 1.0 to 0.4 metre panchromatic data acquisition is not only useful for mapping and logistical planning but also as a navigational guide in hand-held field computers.

Existing field and laboratory data and published geological maps can be integrated into the RPM process to guide, calibrate, and test interpretations (see Case Studies 5 and 9 in Harris, 2008 and Schetselaar et al., 2000). This can be accomplished by overlaying the field observations (lithological unit, strike and dip measurements) on the predictive map(s) in a GIS environment to calibrate the interpretation of geological units and structures. Field data can also be used in training computer classification algorithms. The statistical relationships between the numerical values of image data (representing spectral reflectance, magnetic field intensity, radar backscatter, etc) and lithological units can be computed at field stations and then used to predict other areas with similar signatures. Geological mapping is increasingly being supported by digital field-data capture technology using hand-held computers and global positioning systems (GPS). This is a revolutionary development in RPM as it allows the validation of remote predictive maps on the outcrop. Simultaneous display of remote predictive maps and GPS position in real time may lead the mapping geologist to make small deviations from planned traverses to inspect subtle anomalous patterns that appear to be geologically significant when analyzed in the context of the immediate surroundings of an outcrop. This may apply, for example, to confirming the presence of a dyke, when short-wavelength linear magnetic anomalies from near surface magnetic bodies appear to be in close proximity to the field site.

2.3 Data processing and enhancement

A wide range of processing and enhancement methods can be used to facilitate extraction of geological information from RPM data sets (**Table 2**). Harris (2008) provides many examples of enhanced image data (mainly from Canada's North). Generally the methods employed depend on the data type to be enhanced. Derivatives of potential field data include vertical derivatives, upward continuation, analytic signal, magnetic susceptibility, and pseudogravity, among others (Pilkington et al., 2008). Grids of measured magnetic and gravity data, as well as their derivatives, are improved by applying contrast-enhancement and relief-shading algorithms or both in combination (Milligan and Gunn, 1997). Spatial convolution-filters and colour-enhancement techniques, such as decorrelation stretch (Gillespie et al., 1986) and saturation enhancement (Kruse and Raines, 1994) may be applied to enhance optical remotely sensed (Chapter 5 in Harris, 2008), multibeam radar (Chapter 6 – Harris, 2008) and gamma-ray spectrometry data (Chapter 4 – Harris, 2008) , while band ratios or pairwise principal component analysis (Jensen, 1995; Jolliffe, 2004; Richards and Jia, 2006) are useful to enhance geological information on multispectral or multibeam radar imagery. Most of these enhancements can be generated semi-automatically using computer algorithms available with GIS and/or image analysis systems. User input, however, is always important to fine-tune the enhancement, since this is guided by insight on how the dynamic range and spatial frequency distribution of the imaged physical properties are associated to geology. In addition to the enhancement of individual data types, image fusion

(Harris et al., 1999) combines image data into single images to highlight features of interest and assist in the analysis of complementary geological information.

Data Source (including various enhancements)	RPM Product
Magnetics	Map of magnetic units (domains)
	Map of structures (faults (ductile, brittle), dykes, lineaments, foliation/ bedding traces, folds, potential lithologic contacts)
Gamma ray	Map of radioelement units (domains) that can provide insight into lithologies, different granitic phases and regional metamorphic conditions
Digital elevation data (DEM)	Map of terrain units (based on relief)
	Glacial landforms
	Map of structures (based on topographic expression) – bedrock or glacial (ice-flow features)
	Map of drainage basins (watersheds)
LANDSAT	Map of structures (faults (ductile, brittle), dykes, lineaments, foliation/ bedding traces, folds, potential lithologic contacts)
	Map of spectral units (spectral absorption features due to white mica, clay minerals (potentially associated to hydrothermal alteration) and carbonates) especially carbonates – may represent a combination of bedrock lithology and surficial units
	Fe –oxide map (3/1 – ratio)
	Clay-alteration map, Carbonate, white mica and other OH-group minerals (5/7 – ratio)
	Map of vegetation (4/3 ratio)
	Outcrop map (1+7/4 or 7/4 ratio)
	Map of wetlands (band 4)
	Map showing forest fire burns
	Map of snow and ice
	Drainage map (can provide more detail than topographic maps depending on scale)
Radarsat data	Map of terrain units that may represent surficial or lithologic units
	Map of structures (faults (ductile, brittle), dykes, lineaments, foliation/ bedding traces, folds, potential lithologic contacts)
Hyperspectral	Map of spectral units (can be calibrated to actual lithologic units or specific minerals in certain environments)
	Map of structures (as above)
	Alteration map (if good exposure)

Table 2. RPM data types and products (maps)

2.4 Data analysis

Interpretation can be undertaken visually, on various enhanced and fused images using the well-known principals of photo-geologic interpretation or by employing computer-assisted techniques that can lead to automatically generated maps or products that require some geologic interpretation and calibration by the geologist (**Fig.1**).

2.4.1 Visual interpretation

Visual interpretation of the enhanced and or fused remotely sensed data can be based either on making hard-copy images or by digitizing on a touch-sensitive computer screen. The latter method is more flexible as it allows for instantaneous display of different data sets, thus facilitating the extraction of complementary information while weighing the geological significance of image patterns in each of the data layers. It can provide interpretations of units, unit contacts, or faults that are automatically georeferenced to the database, can be virtually overlain on other data for comparison, and serve as a basis for geological map compilation once new field data are acquired.

Regardless of the data type being rendered, visual interpretation is based on recognizing geological features using seven diagnostic elements. These include tone and/or colour, texture, patterns, shape, size, shadow, and association (Lillesand and Kieffer, 2000; Drury, 2001). Depending on the type of geoscience data used for predictive mapping (including remotely sensed and geophysical) data one or more of these photo-geologic elements can be captured. *Tone* and/or *colour* refers to the relative brightness or colour of objects in an image. It is the most fundamental element of image interpretation, as its variation also allows appreciating other elements, such as texture, pattern, and shape. Tonal and/or colour response can be captured from optical sensors (i.e. LANDSAT and may others) sensitive to reflectance properties of the Earth's surface and entail the use of spectral signatures to characterize various earth materials. Magnetic data captures tonal response due to variations in magnetic susceptibility and these tonal variations often reflect underlying lithology and geologic structure. Gamma ray spectrometer tonal variations reflect radioelement emissions (eU, eTh and %K) from the surface and are useful for mapping geochemical variations at the surface. *Size, shape* and surface *texture* can be captured by both optical and microwave remote sensors as well as digital elevation models. Radar is particularly useful for capturing textural responses from the Earth's surface due to variations in surface roughness and moisture. *Pattern* refers to the repetitive arrangement of discernable features in an image and different patterns can be captured based on what each sensor responds to, as discussed above. *Shadow* refers to the part of an object that is obstructed from incoming radiation from a natural, active, or artificial energy source. Shadow provides a perception of the profile or relative height of a target. It, however, may also hamper the identification of an object since it lowers or completely obstructs the reflectance from that object. *Association* refers to the relationship of an object with other recognizable objects in the vicinity. The identification of features that one would expect to associate with other features may provide information to facilitate identification. Typical geological examples include radial drainage patterns around circular objects, such as those associated with impact structures, and intrusive and tectonic domes and volcanoes.

2.4.2 Computer-assisted (numerical methods)

In addition, or as a compliment to visual interpretation, numerical interpretation methods can be used to produce remote predictive maps (**Fig. 1**). Automated numerical methods can include supervised and unsupervised classification and image segmentation algorithms (Lillesand and Kieffer, 2000; Richards and Jia, 2006). These methods provide alternatives for extracting geological information in a systematic and unbiased manner, although visual interpretation is commonly judged to outperform methods of automated pattern recognition. However, numerical methods are superior to visual methods at simultaneously manipulating and interpreting multiple data sets having a large number of image variables. Supervised classification methods allow geologists to have input into the map-making process by using geological field data during the training stage of the classification

(Schetselaar and de Kemp, 2000; Schetselaar et al., 2000; also see Case Studies 2, 5, 6, and 7 in Harris, 2008). In supervised classification, decision rules for class allocation are derived from multivariate statistics computed from the relationships between classes and image variables at the sample sites (i.e. field sites considered representative for bedrock or surficial units). The decision rules are used in the classification stage to allocate all pixels or grid cells to particular classes. The available classification algorithms differ in the way probability density functions for each class are modelled and estimated from the training data. The classification algorithms can be broadly categorized into (1) parametric classifiers that model the class probability density functions with the estimated parameters of a multivariate normal distribution or (2) nonparametric classifiers that directly estimate the class probability density functions from the data.

2.5 Data Integration (making a predictive map)

Various aspects of the surface can be emphasized and enhanced on various geoscience datasets. The difficulty comes in how all this information can be integrated into a final geologic map. Firstly the concept of what constitutes a map has changed with the explosion of digital data and tools (i.e. GIS and image analysis systems) to manipulate, enhance, combine and analyse data. A map now can be defined on demand by extracting themes of interest from a geodatabase housed within a GIS comprising a series of geo-referenced layers. These layers can then be combined to create a customized, or in fact a *virtual* geologic map representing different aspects of a geologic terrain. Two examples, one dealing with bedrock geology and the other with surficial materials (surficial geology) are presented below to illustrate this concept.

2.6 Validation

All maps whether predictive or based on field measurements and observations are a generalized model of the Earth's surface. Both approaches (remote and ground-based) are complimentary. There are obviously geological features that can only be observed and mapped in the field, complimented by various laboratory analysis. However, *the view from above* using a variety of geoscience datasets offers a different geologic perspective of the terrain to be mapped, highlighting features and patterns not easily seen or evident when on the ground. Both methods of producing a geologic map, are characterized by different types of uncertainties and these should be (but not always are!) indicated on the map. These include uncertainties in what feature is being mapped, and the spatial location of these features. Capturing these uncertainties is an integral part of the map-making process and example 2, discussed below, illustrates how statistical and spatial uncertainty was quantified when producing a predictive surficial materials map.

3. Examples of predictive maps

Two examples are discussed demonstrating how the concepts discussed above can be applied to make a predictive geological map. The first example deals with the creation of a bedrock geology map which includes spectral/lithologic units as well as structural features over a small portion of the Hall peninsula, Baffin Island, in Canada's Arctic. Both visual and computer-assisted techniques will be presented, compared and contrasted. The second example deals with the creation of a predictive surficial materials map using computer-assisted techniques over a much broader region of the Hall peninsula, Baffin Island. Data used to create these predictive maps include freely available Canadian geoscience datasets including LANDSAT 7 TM, CDED, 1:50,000 DEMS, airborne magnetic geophysical data, hydrographic and geographic GIS layers and legacy field data (digital maps and GIS databases). Image processing software (ENVI™) in concert with GIS software (ArcGIS™) were used to produce the maps using touch-screen display technology.

The study areas for these two examples (Fig. 2) are from the Hall peninsula of south-central Baffin Island, Canada. This area has not been systematically mapped since the 1960's and thus requires updating for both bedrock and surficial information. The geology of the Hall Peninsula corridor can be divided into three principal lithological domains. An eastern domain of Archean tonalitic gneisses, monzogranite and minor metasedimentary rocks, a central domain of Paleoproterozoic siliciclastic metasedimentary rocks and subordinate Paleoproterozoic metaplutonic rocks, and a western domain dominated by orthopyroxene- and garnet bearing monzogranites of the Paleoproterozoic Cumberland batholith (Scott, 1997). The terrain is rough and rocky, with hills near the coast. The Hall peninsula has permanent ice; the Grinnel glacier calves icebergs into Frobisher Bay. The Hall Peninsula is part of the Arctic Tundra biome — the world's coldest and driest biome.

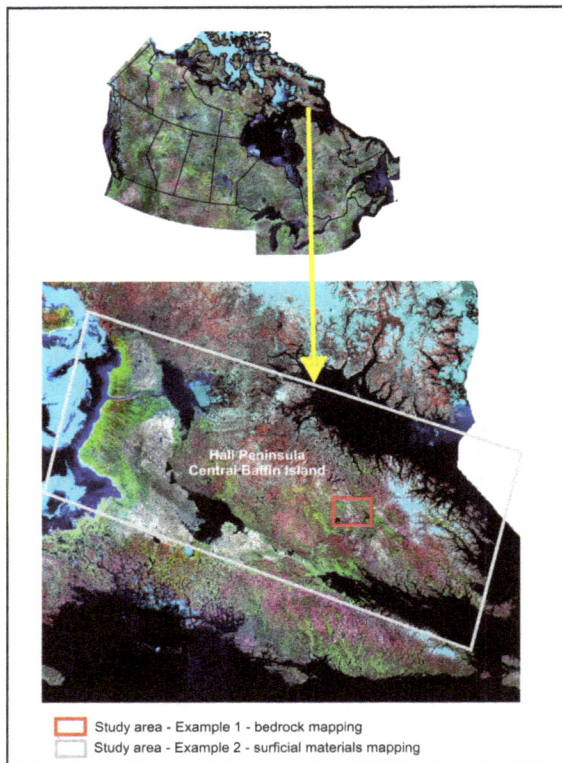

Fig. 2. Study areas for the two examples (bedrock and surficial) of predictive mapping – Hall Peninsula, south-central Baffin Island, Nunavut, Canada.

3.1 Example 1 – Bedrock mapping
3.1.1 Visual assessment
Figure 3 presents a generalized flow-chart summarizing the RPM protocol for producing a bedrock geology map by visual interpreting the enhanced LANDSAT (**Fig. 4**) and magnetic data (**Fig. 5**). Both structural form lines comprising potential lithological contacts, bedding and foliation trends, faults and lineaments (no dykes were evident in the area) and spectral units and magnetic domains were identified. A *heads-up* digitization (interpretation) process was utilized in which interpretations were undertaken directly on a touch-sensitive display

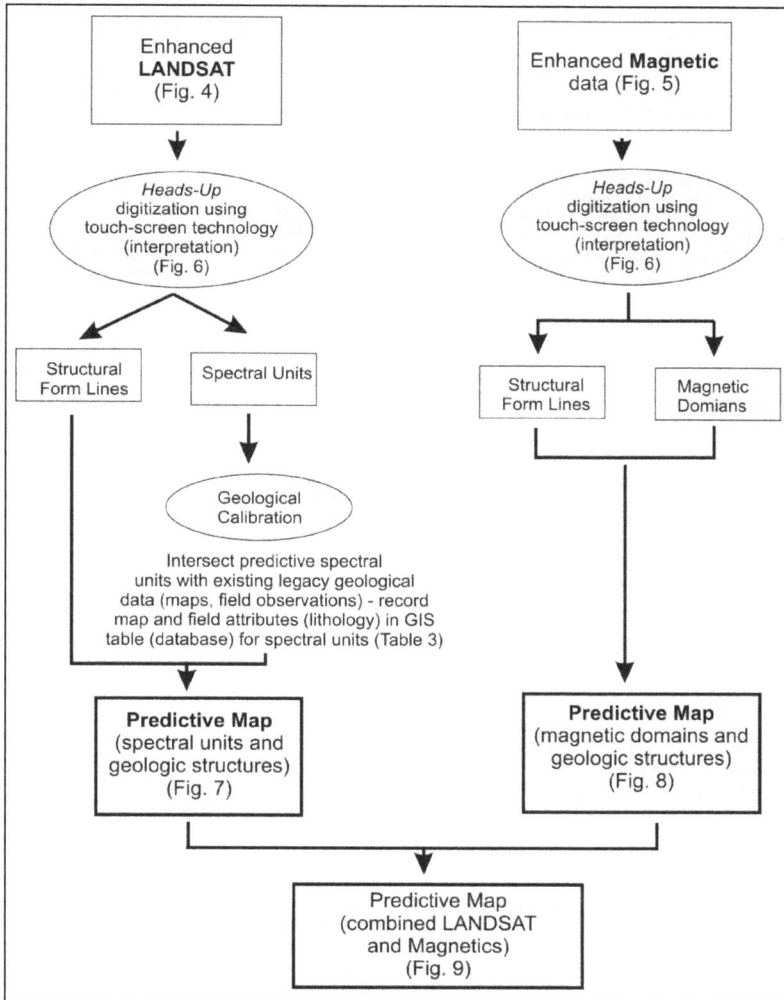

Fig. 3. Flow chart outlining the steps for producing a bedrock predictive map from LANDSAT and airborne magnetic data using visual interpretation techniques and the final integration of the two predictive maps.

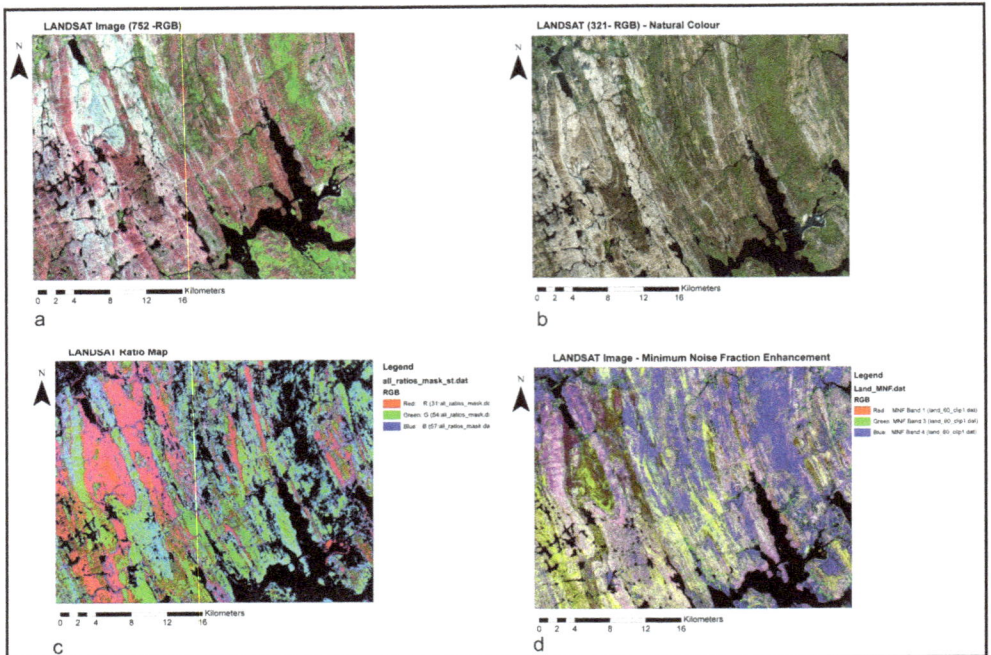

Fig. 4. Enhanced LANDSAT data used for predictive mapping (visual and computer-assisted) (a) band 7,5,2 (RGB) ternary composite image (contrast enhanced), (b) band 3,2,1 (RGB) ternary natural colour composite image (contrast enhanced), (c) LANDSAT ratio ternary composite image (R = ferric iron ratio - red/ blue wavelengths (bands 3/1); G = ferrous iron ratio – SWIR / NIR wavelengths (bands 5/4); B = clay ratio - SWIR / SWIR (bands 7/5). Red areas are higher in ferric iron content, green higher ferrous iron and blue, higher clay (possible sericite), (d) Minimum Noise Enhancement (transform), R = MNF component image 1, G =   MNF component image 2, R = MNF component image 3. In these images, note the good spectral separation leading to the identification of distinct spectral units.

screen (Cintiq screen) using a stylus pen (**Fig 6**). An ArcGIS geodatabase was first defined with pre-selected structural and lithological attributes and using the touch screen, all interpretations were immediately incorporated and attributed within feature classes of the geodatabase. The *heads-up* interpretation process is akin to overlaying transparent paper over a hard-copy image and conducting photo-geologic interpretation. However it offers the advantage of flexibility and efficiency as the enhanced image data displayed on the background can be interactively changed while interpretations are fully geo-referenced and are immediately incorporated within the geodatabase. **Table 3** shows the results (GIS attribute table) of geologically calibrating the spectral units by intersecting the polygon map of spectral units with legacy geological data (maps, field stations) thus assisting in assigning a lithological name to each spectral unit. This was accomplished within the GIS by comparing the interpretation of the spectral interpretations with lithological units displayed as polygons on the digital geology maps and field stations in which rock type was recorded

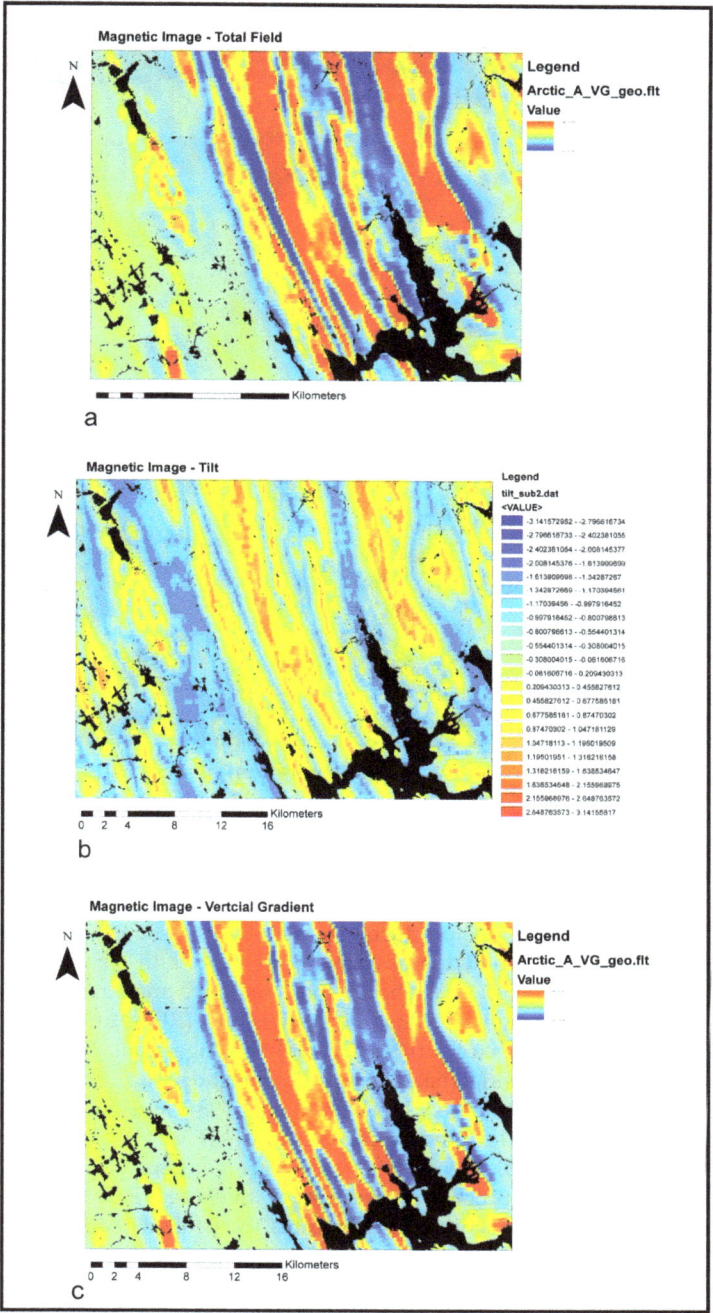

Fig. 5. Enhanced airborne magnetic data (a) total field, (b) tilt, (c) vertical gradient

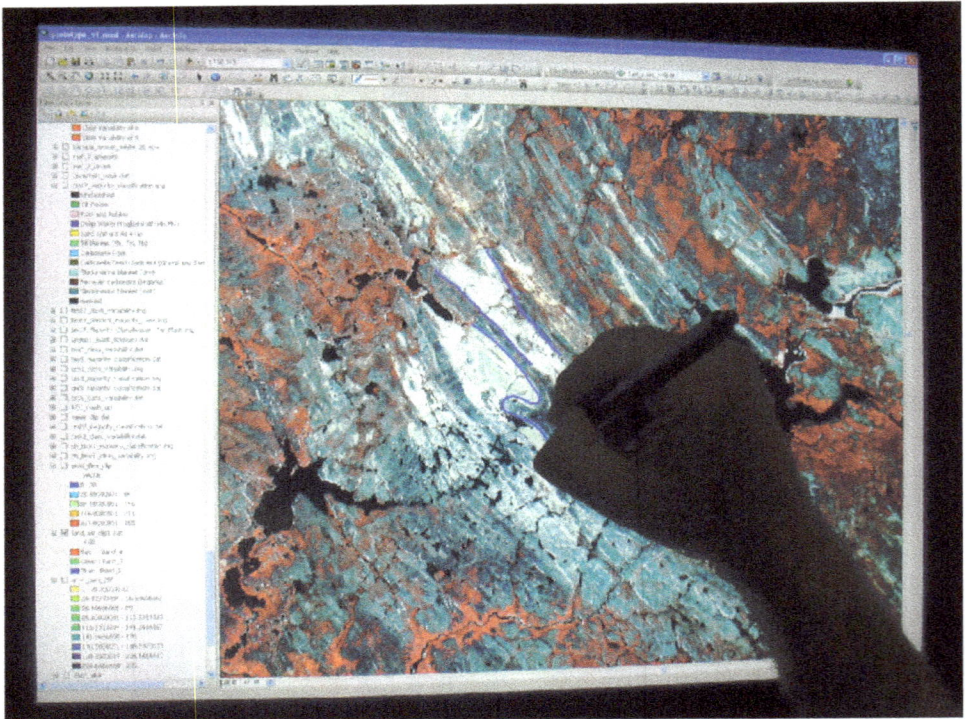

Fig. 6. Example of the *heads–up* digitization (interpretation) process using a touch-sensitive screen – geologist is drawing boundaries on an enhanced LANDSAT image.

in a point database. Note that initially a spectral unit was assigned based on interpretation of the LANDSAT data and after comparing these to the geological data (maps and field stations) a tentative rock unit was assigned. The tentative rock name of course requires field validation. The final predictive map produced by visually interpreting the LANDSAT data, which combines spectral units and the associated database with structural form lines, is shown in **Figure 7** whereas **Figure 8** shows the predictive map produced by visually interpreting the enhanced magnetic data. Five divisions (RPM units 1 - 1d) of the sedimentary rock assemblage (Lake Harbour Group –St- Onge et al., 1998)) , four intrusive units (RPM units 2a,b, comprising the Ramsey River orthogneiss assemblage and 4, 6 comprising the Cumberland Batholith (St- Onge et al., 1998)) and one gneissic unit (RPM unit 5), have been identified by differing spectral responses (**Fig.7**)

Components of these two predictive bedrock maps are combined in the final predictive map, shown in **Figure 9**. The process of overlay the interpretations is a crucial decision process in RPM that is often difficult as this requires the conflicts between interpretations from different image types to be resolved. One approach is to combine the interpretations after all are complete. An alternative approach is to combine the interpretations *on the fly* by dynamically changing the imagery on the computer screen during the interpretation process.

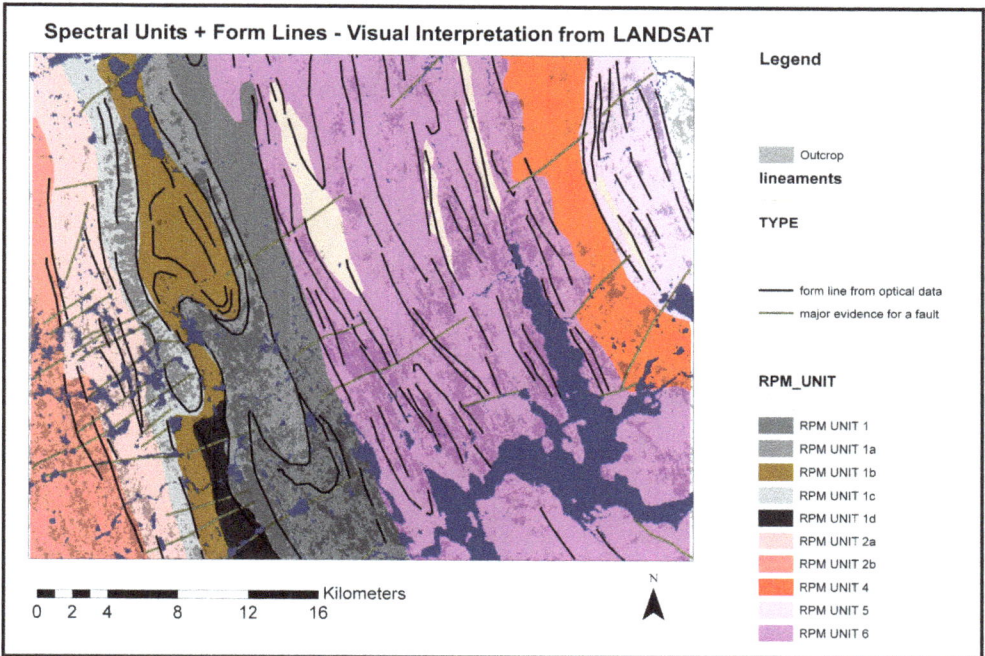

Fig. 7. Predictive bedrock geology map produced by visually interpreting enhanced LANDSAT data (Fig. 4) using a *head-up* digitization process (Fig. 6). The steps for producing such a map are outlined in Fig. 3. Note hat the grey shaded areas within each spectral unit are areas of bedrock outcrop identified on the LANDSAT data. This was accomplished by producing a Blue / NIR wavelength (1/4) ratio as exposed outcrop reflects blue energy and absorbs NIR energy. An upper threshold on the histogram of this ratio image was identified creating a binary raster map of outcrop and non outcrop areas that were included as part of the predictive map.

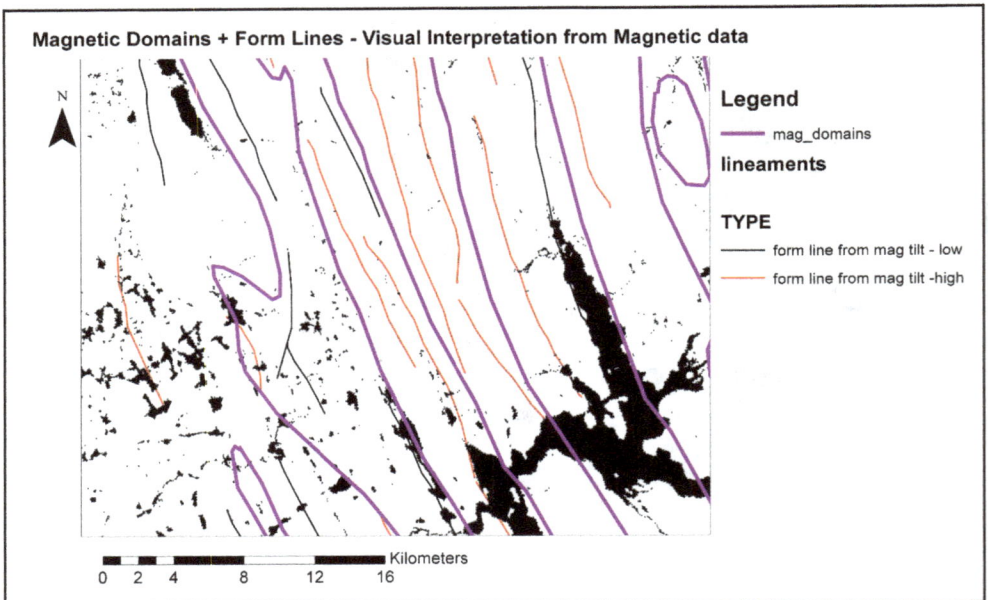

Fig. 8. Predictive bedrock geology map produced by visually interpreting enhanced airborne magnetic data (Fig. 5) using a *head-up* digitization process (Fig. 6). The steps for producing such a map are outlined in Fig. 3. The boundaries of each magnetic domains (which have not been polygonized and thus are not coloured as are the spectral units in Fig. 7) are shown in purple the structural form lines, interpreted largely form the tilt image (Fig.5) in black and red.

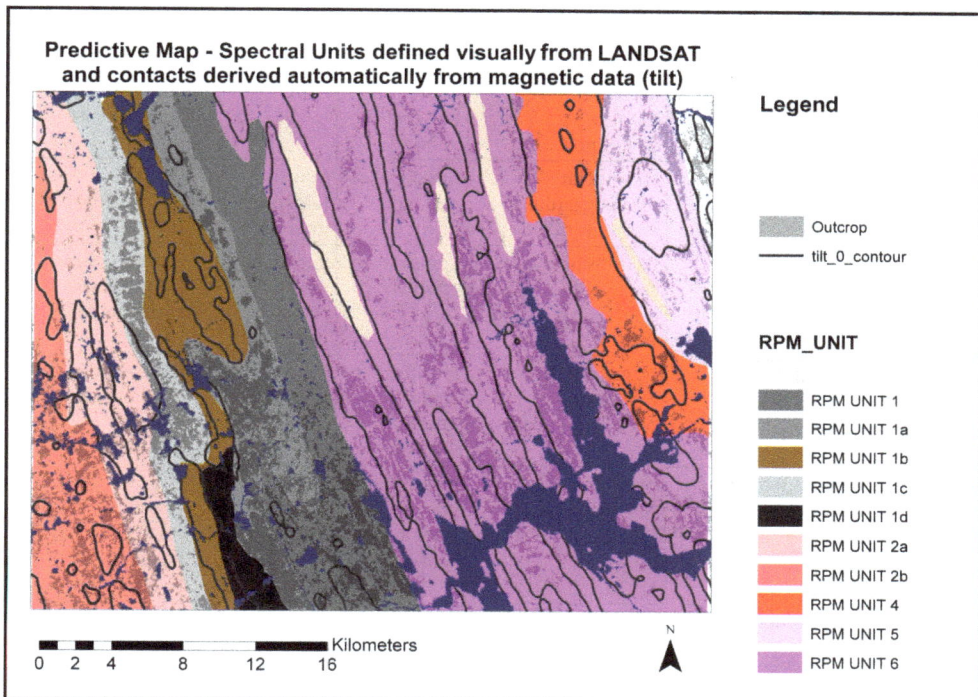

Fig. 9. Predictive map which combines spectral units (geologically calibrated – see Table 3) visually interpreted from the enhanced LANDSAT imagery and magnetic contacts extracted automatically from the magnetic tilt data (0 contour – see description in the text). Areas of bedrock, as described on Fig. 7 have been overlaid in grey. Note that there is good correspondence between the magnetic contacts and the boundaries of the spectral units. However, certain spectral units (RPM 6 for example) are characterized with more frequent and apparent magnetic contacts, perhaps representing significant differences in magnetic susceptibility contrast within each spectral unit, which may be due to metamorphic and / or tectonic processes (e.g. new growth and retrograde destruction of magnetite). This would, of course, benefit from field follow-up work.

SPECTRAL UNIT	MAP UNIT 1 (International Polar Map –IPY– not shown) (Harrison et al., 2011)	DESCRIPTION	FIELD UNIT (Fig 14)	MAP UNIT 2 (Fig. 14)	RPM UNIT
RPM5	orthogneiss	monzogranite-tonalite	gneiss (mafic enclaves)	drift	Orthogneiss – monzogranite-tonalite
RPM4	Igneous intrusive	monzogranite-tonalite orthogneiss	quartz feldspar gneiss	drift	Intrusive - orthogneiss
RPM6	Intrusive	charnockite – monzogranite to syenogranite	quartz feldspar gneiss	quartz-feldspar gneiss	Intrusive - charnokite
RPM1	Sedimentary	psammite - semipelite	gneiss -buff , grey	garnet biotite quartz feldspar gneiss	Meta-sediment 1- psammite - semipelite
RPM1a	Sedimentary	psammite -garnet-biotite-quartz-feldspar	granite, rusty gneiss, gneiss	rusty paragneiss	Meta-sediment 2 - psammite
RPM1b	Sedimentary	psammite, semipelite	rusty gneiss, gneiss, granite	rusty paragneiss - gneiss	Meta-sediment 3 - psammite – semipelite - (rusty – high Fe content))
RPM1c	Sedimentary	psammite garnet-biotite-quartz-feldspar	gneiss (buff)- granite	garnet-biotite -quartz-feldspar Gneiss + rusty paragneiss	Mea-sediment 4 - psammite (less rusty)
RPM1d	Sedimentary	psammite - semipelite	quartz feldspar gneiss	quartz-feldspar gneiss	Meta-sediment 5 -psammite - semipelite
RPM2a	Intrusive	monzogranite-tonalite orthogneiss	quartz feldspar gneiss – buff gneiss	quartz feldspar gneiss	Gneiss 1 – quartz feldspar
RPM2b	Intrusive	monzogranite-tonalite orthogneiss	quartz feldspar gneiss	quartz feldspar gneiss	Gneiss 2 – quartz feldspar

Table 3. Attribute table produced by intersecting the spectral (RPM) units visually interpreted from the LANDSAT data (see Fig. 7) with 2 legacy geological maps (note the column labeled *Map Unit 2* was derived from the geological map shown in Fig. 14 –) *Map Unit* 1 was derived from the International Polar Year Map (Harrison et al., 2011), the field data was derived from field stations shown on Fig. 14

3.1.2 Computer-assisted

The numerical power of an image analysis system in concert with a GIS can be leveraged to extract geological features automatically from remotely sensed imagery producing a stand-alone interpretive, GIS layer and/or a product that will facilitate visual photo-geologic interpretation. **Figure 10** presents a generalized flow-chart summarizing the RPM protocol for producing a bedrock geology map utilizing computer-assisted techniques. Spectral units that may or may not relate to underlying lithologic patterns can be extracted from optical data such as LANDSAT using unsupervised and/or supervised classification techniques in which the geologist provides *a priori* information on the spectral /lithologic features to be classified. Training areas, representing distinct spectral units, were identified on the

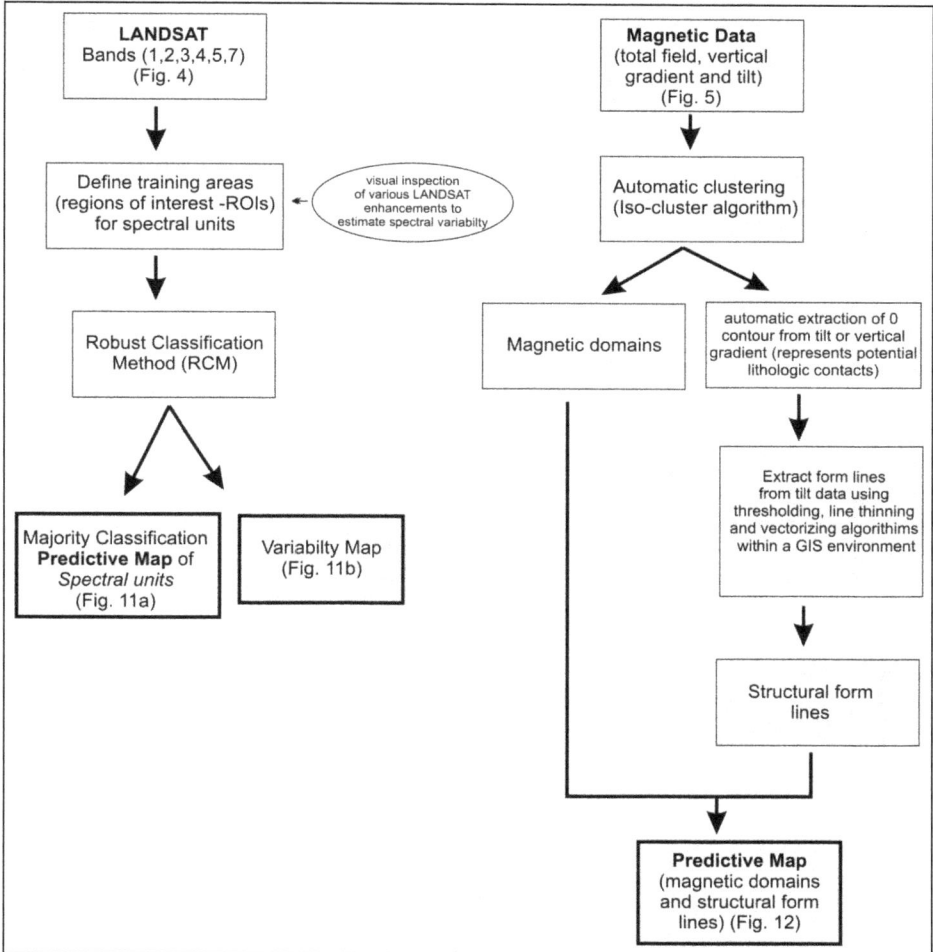

Fig. 10. Flow chart outlining the steps for producing a bedrock predictive map from LANDSAT and airborne magnetic data user computer-assisted (semi-automatic to automatic) techniques.

enhanced LANDSAT data (**Fig. 4**) and used to classify the entire image. The Robust Classification Method (RCM) was employed using the maximum likelihood algorithm to classify the data into spectral units. The RCM method involves a repetitive sampling of a training dataset in concert with cross validation to produce a user-specified number of predictions (classified maps) of spectral units. The RCM process provides a better classification result as the final map comprises a majority classification whereby each pixel is assigned the class that occurred most frequently over the user-specified number of repetitions and the spatial uncertainty of the process is captured by a variability map (cross-validation process). A majority classification map (**Fig. 11a**) for the 10 repetitions of RCM as well as a map that shows the spatial variability (uncertainty) (**Fig. 11b**) over the 10 repetitions are produced as part of the outputs from RCM. Interested readers can find more details on RCM in Harris et al., (2011).

A fair degree of correspondence between the automatically derived and visual derived spectral boundaries exist (**Fig. 7a vs. 11a**). The main difference is that the spectral map derived through supervised classification techniques provides more potential detail within the main visually derived spectral units, perhaps reflecting slightly different lithologic compositions and/ or weathering conditions. With respect to the classification variability map (**Fig. 11b**) no large areally extensive zones of classification uncertainty (variability) exist. However, a few NNW-SSE trending linear zones in the central portion of the study area (green and yellow) have been identified as uncertain using RCM.

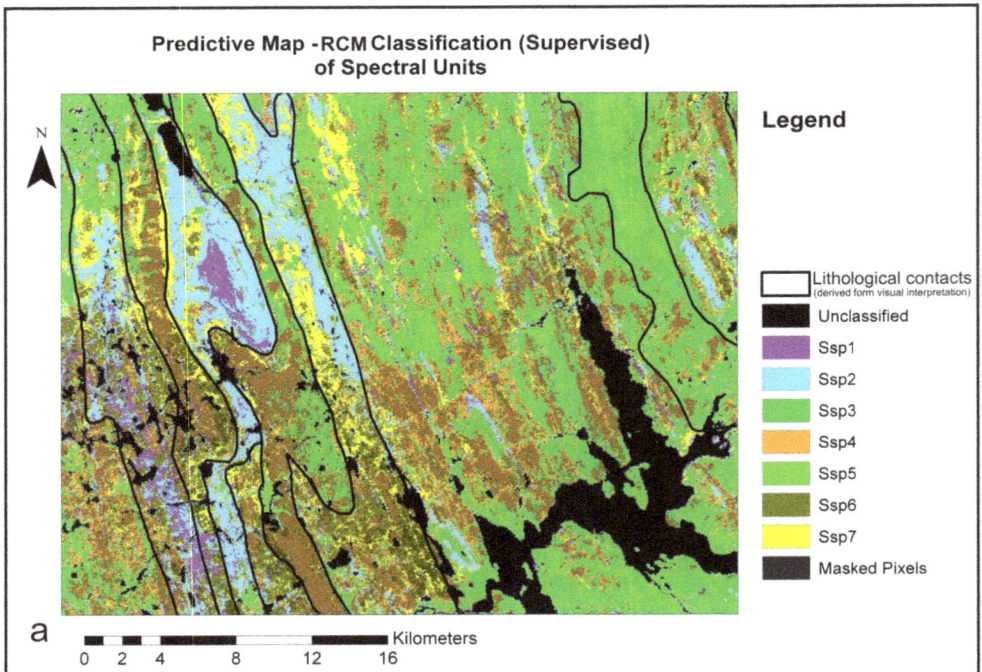

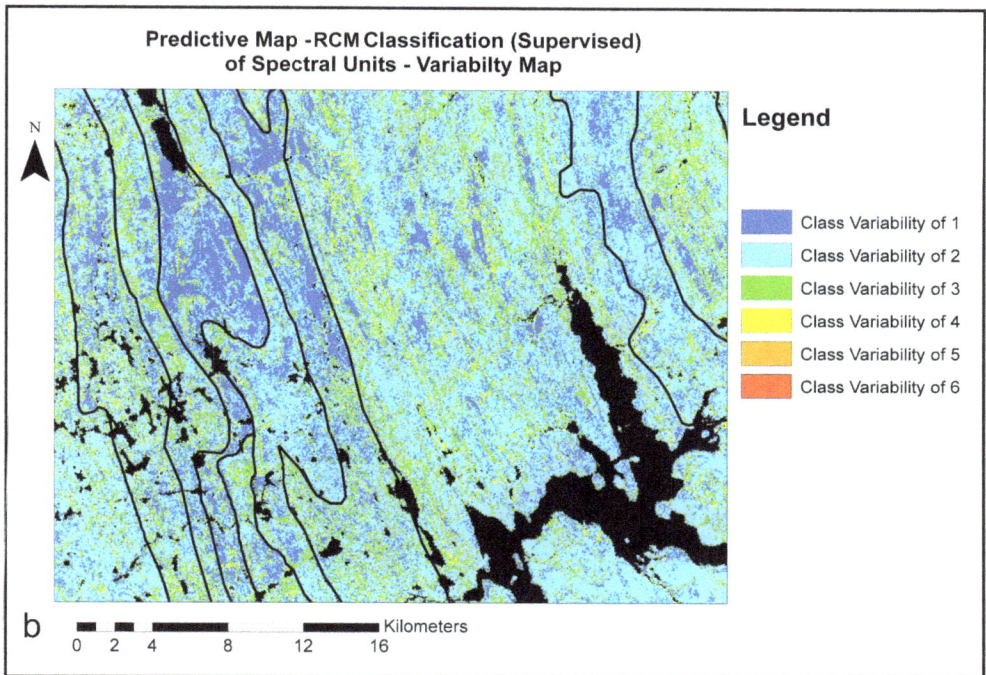

Fig. 11. Predictive bedrock maps of spectral units identified using a supervised classification technique referred to as the Robust Classification Method (RCM) (see description in text and Harris et. al., 2011 for more details on this algorithm). (a) Majority classification predictive map of spectral units. The main spectral boundaries identified through visual interpretation (Fig. 7) have been overlaid for comparison purposes. (b) associated map produced from RCM showing the spatial uncertainty in the spectral classification (i.e. spectral variability map).

Magnetic domains can be automatically produced from the multi-band magnetic dataset (total field, tilt and vertical gradient) by employing unsupervised clustering techniques. This processing involves identifying similar statistical clusters in N-dimensional space (in this example – 3 dimensions (i.e. 3 magnetic images)) based on magnetic susceptibility and then plotting these spatially creating a magnetic domain map (**Fig. 12**).
Potentially meaningful geologic structural features can be automatically extracted from magnetic data forming the basis of a structural map comprising form lines (**Fig 12**). Mapping the locations of lateral magnetization contrasts (i.e. the edges of magnetic bodies or sources) is one of the most useful applications of magnetic data for geological mapping (Pilkington et al., 2009). Contacts can be automatically extracted from magnet tilt data by selecting zero values (which exist over potential edges) and then contoured in the GIS environment creating a vector map of potential lithologic contacts. Furthermore, the linear high and low areas from a vertical gradient or tilt image can be extracted by simple density (thresholding) slicing, followed by thinning the binary map produced from thresholding to a single pixel and then vectorizing producing a vector map of structural form lines (**Fig. 12**).

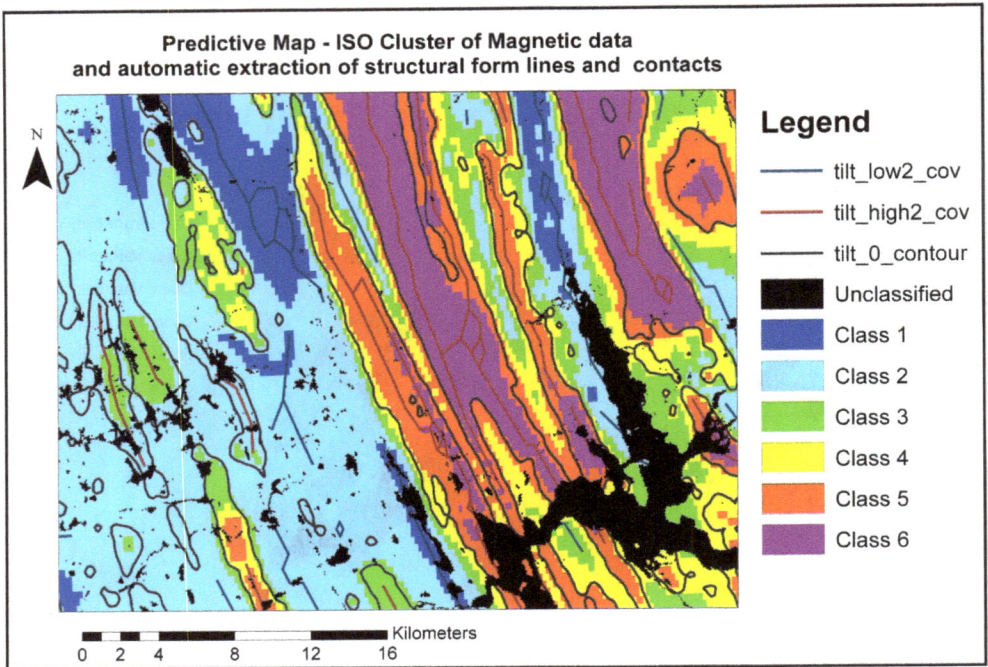

Fig. 12. Predictive bedrock map produced from automatic and semi-automatic processing of the airborne magnetic data. Magnetic domains have been identified and mapped by automatically clustering the total field, tilt and vertical gradient data and contacts and structural form lines have been extracted from the tilt data using semi-automatic methods (see Fig. 10 and descriptions in the text).

3.1.3 Evaluation of predictive bedrock maps

Selected components from the predictive bedrock maps produced from visual and computer-assisted techniques can be combined creating a predictive map which is a hybrid of both interpretation techniques (**Fig. 13**). Although this is a somewhat busy bedrock map it illustrates the power of using the GIS to compile and integrate various layers from the LANDSAT and magnetic data contained within a geodatabase. The various layers can then be combined producing a custom geologic map determined by the geologist and to meet the requirements of what the map is designed to highlight and display (i.e. be it for mapping, exploration etc). Thus the concept of a geologic map now is the geodatabse containing the various geological and geoscience information as points, lines, polygons and rasters as opposed to the traditional static paper map. This new paradigm of a geologic map now allows customization depending on the geological application and fully supports a print-on-demand concept.

There are some similarities in the patterns between the predictive and legacy geological map and in fact the legacy map (**Fig. 14**) was used to geologically calibrate the spectral RPM units as discussed above (see **Table 3**). However, on the legacy map the entire central-north area has been mapped as Quaternary cover. This is clearly not the case as evidenced (and mapped) on the LANDSAT in concert with the magnetic data, both of which offer more detailed geological information in this area. Of course the predictive map would benefit from field follow-up especially with respect to verifying and assigning rock names to each RPM unit.

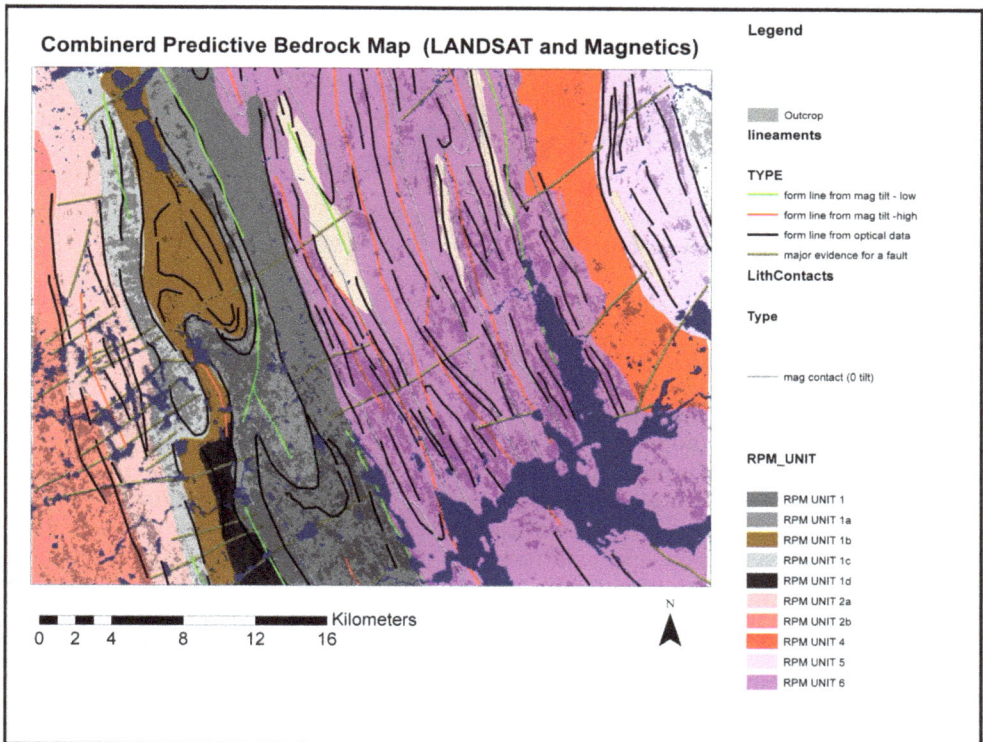

Fig. 13. Predictive bedrock map combining spectral units, bedrock outcrop and form lines derived from visual interpretation of the enhanced LANDSAT imagery with form lines and contacts extracted from semi-automatic interpretation of the magnetic (tilt) data.

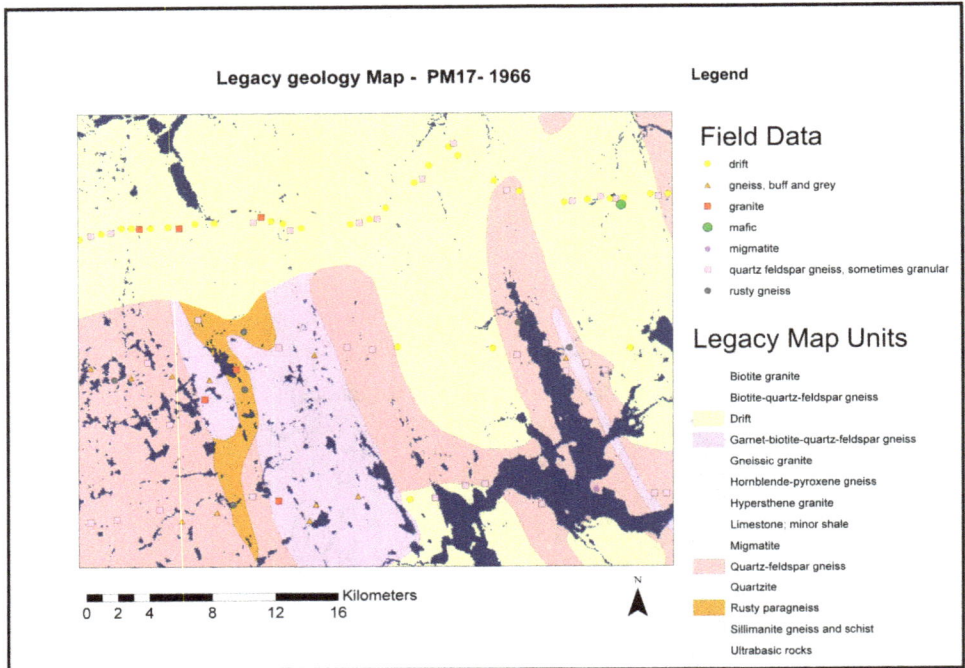

Fig. 14. Legacy geological map (Blackadar, 1966)

3.2 Example 2 – Surficial materials map
3.2.1 Computer-assisted (supervised classification)

The RPM protocol for producing a predictive map of surficial materials is presented as a processing flow-chart in **Figure 15**. This process involves selecting representative training areas (regions of interest) by an expert surficial geologist, knowledgeable about the area to be mapped, selection of geoscience and remotely sensed data to use and selection of an algorithm to perform the classification. In this example, the Robust Classification Method (RCM), discussed and used for bedrock mapping in example 1, was again employed. The data used to produce the predictive surficial materials map included LANDSAT, to capture spectral reflectance characteristics of surficial materials, derived textural derivatives of the LANDSAT bands (entropy and homogeneity) to capture spatial variations in surface texture and finally derivatives from a digital elevation model (DEM) designed to capture topographic characteristics of the terrain. The derivatives of the DEM were based on a 16 by 16 pixel neighbourhood filter which was passed over the DEM and at each pixel the difference from the mean, standard deviation and percent difference were calculated based on the total number of pixels in the neighbourhood. The difference from the mean was used as a measure of topographic position, the standard deviation as a measure of local relief and percent as the range in elevation (Wilson, 2000). Thus in this case both surface reflectance, textural and topographic properties were used to classify surficial materials.

The majority classification map (**Fig. 16**), as described above in example 1, shows the class that was most frequently assigned on a pixel-to-pixel basis over 10 repetitions of RCM whereas

Figure 17 shows a variability map in which the warmer colours represents pixels (areas) that showed much variability in the class each was assigned to through the repetitive classification process. In fact, these variable pixels could be excluded from the majority classification map, as they represent a high degree of uncertainty in the classification process.

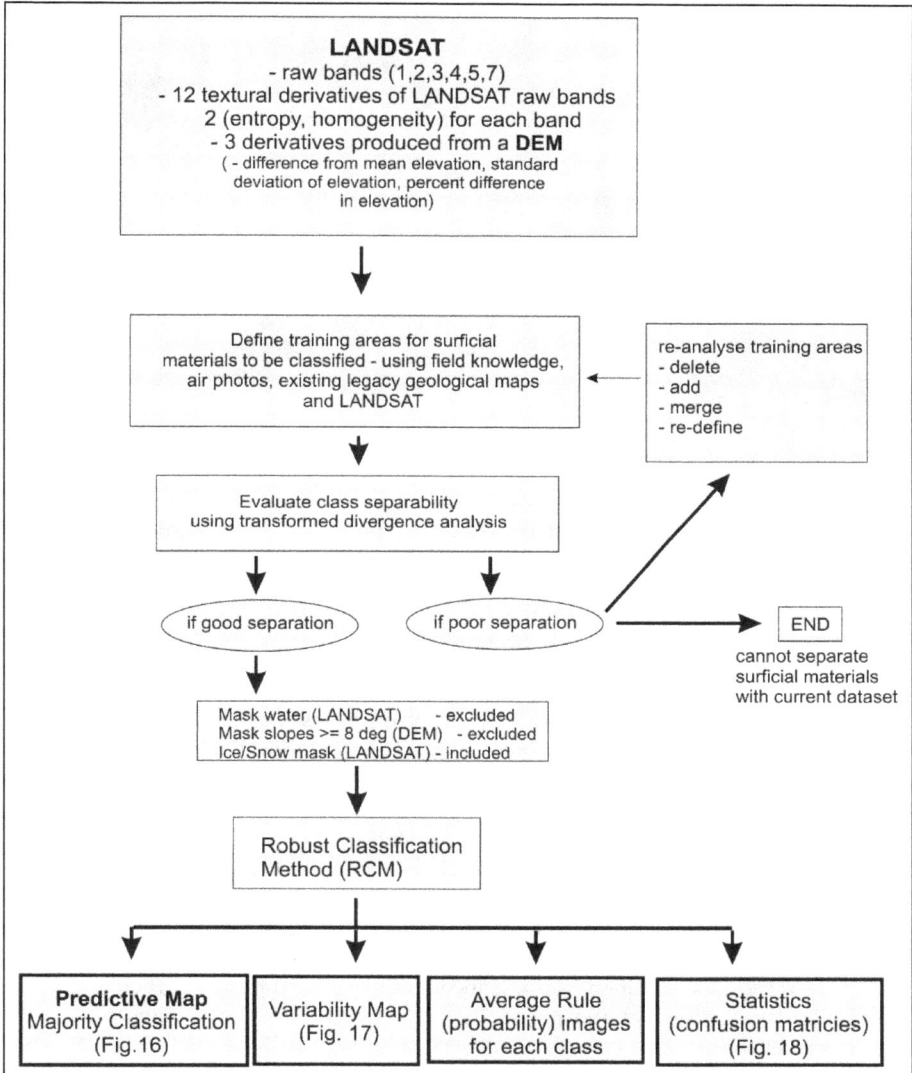

Fig. 15. Flow chart outlining the steps involved in producing a predictive map of surficial materials using a supervised classification technique referred to as the Robust Classification Method (RCM) (see description in text and Harris et. al, 2011 for more details on this algorithm).

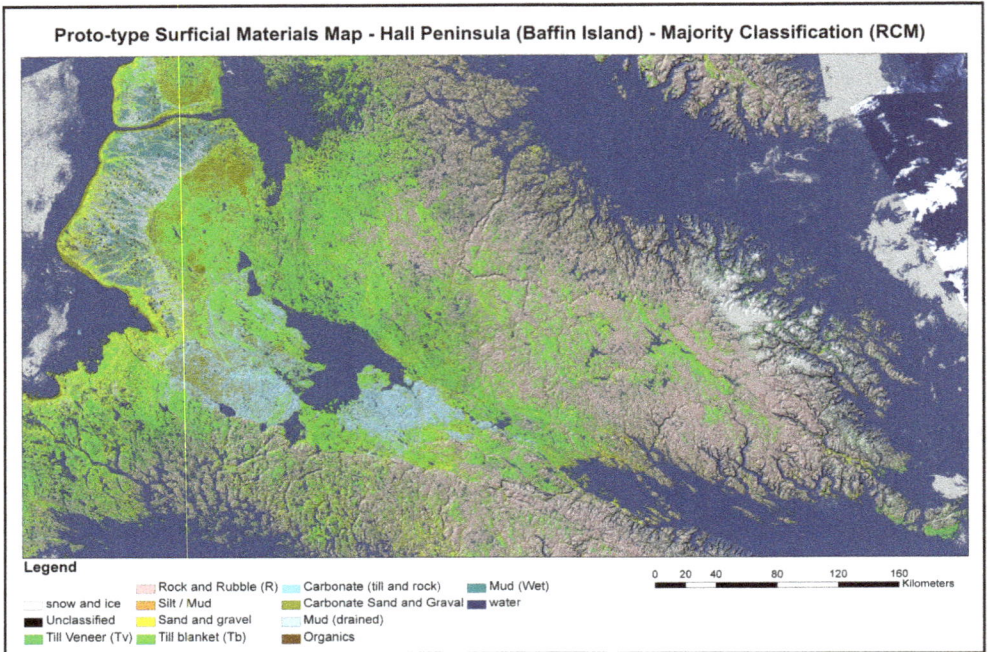

Fig. 16. Predictive surficial materials map – This map produced by RCM shows the majority classification of surficial material on a pixel-to-pixel basis for 10 iterations of the classification algorithm. This map was produced in the same manner as the predictive map of spectral units (Fig. 11). The classification has been combined with a shaded DEM (CDED data) to enhance topographic and geomorphologic variations in the landscape as they relate to the distribution of surficial materials.

The overall average classification accuracy of the majority classification map (**Fig. 16**) is 75.9% whereas the mean accuracy (based on the average of the producer's accuracy) is somewhat lower at 64%. These accuracies do not reveal whether the classification errors are evenly distributed over all classes. Thus, **Figure 18** shows plots of both user's and producer's accuracy for each surficial class which gives a better representation of error as a function of each class. Although the overall accuracy is good some classes are characterized by very poor user's accuracy yet good producer's accuracy and vice versa. Specifically, surface materials with poor producer's accuracy (errors of exclusion – pixels on the classified map that do not match the reference data (training pixels)) yet good user's accuracy (errors of inclusion – pixels on the map that are not the class specified or pixels incorrectly excluded from a particular class.) are : silt/ mud, till veneer and sand and gravel. Thus, pixels in these classes have a much lower probability of being classified correctly on the image, yet on the map they have a higher probability of being correct. Materials that have an opposite relationship (i.e. high producer's but low user's accuracy - pixels incorrectly assigned to a particular class that actually belong in other classes.) are carbonate (till and rock) and organics. Thus the materials that have the least uncertainty of being misclassified are rock and rubble, carbonate sand and gravel, both dry and wet mud and to a lesser extent, till blanket.

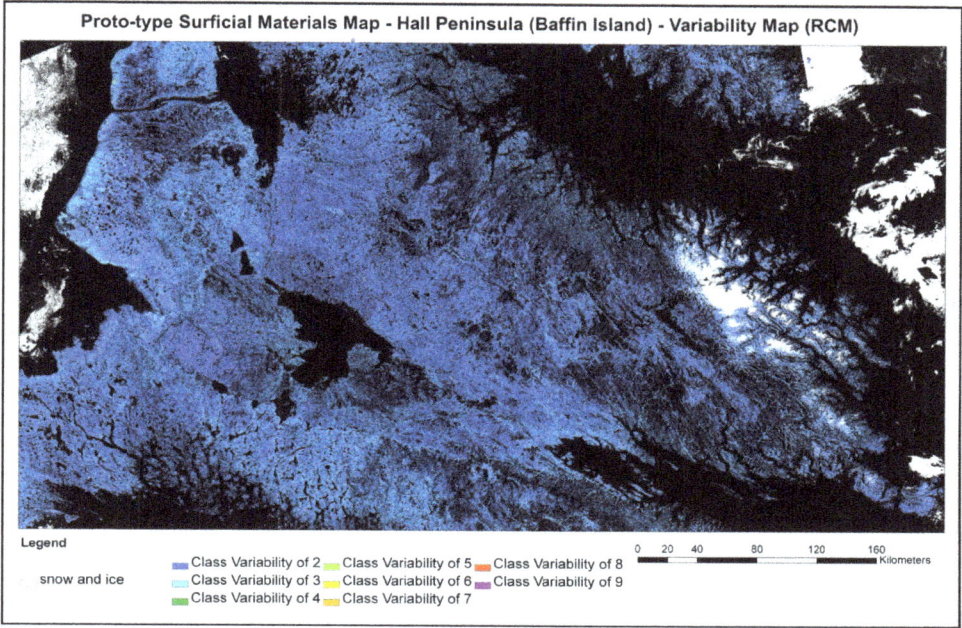

Fig. 17. RCM Variability showing the spatial variability in the surficial material majority classification map (Fig. 16). There is only a small to very moderate variability in the classification as indicted by the predominance of blue hues indicating a class variability of 3 or less through the 10 iterations of RCM.

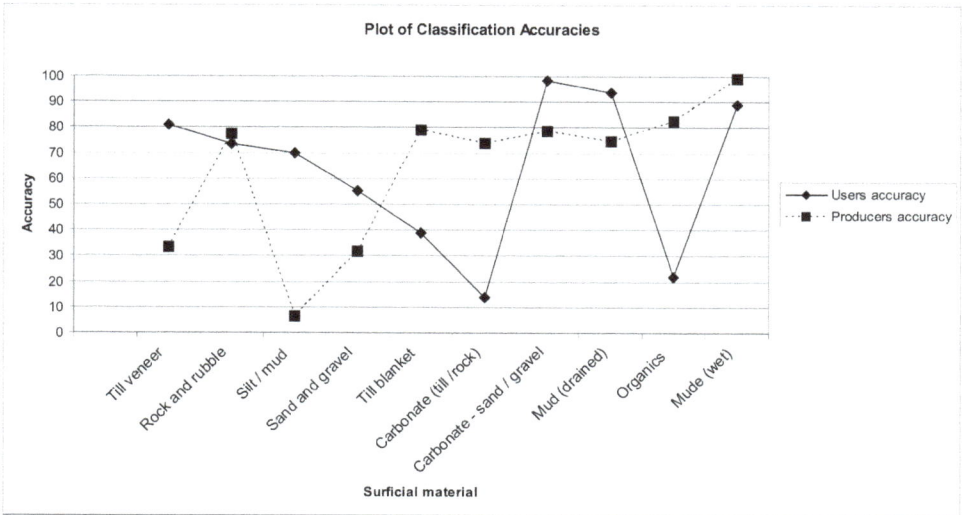

Fig. 18. Plot of user's and producer's accuracies for each surficial materials class shown on Fig. 16 – see text for discussion

Thus, with respect to a user of this map, a high percentage of silt/mud, till veneer and sand and gravel are classified as these materials on the ground. However, the producer's accuracy of these categories are quite low indicating much misclassification of the original training (reference) data. The opposite situation exists for carbonate (till and rock) and organics.

4. Discussion and conclusion

With respect to the best method for producing a predictive geological map, a number of factors, discussed in the introduction section, are important. Mapping bedrock geology is generally more difficult than mapping surficial materials as most remotely acquired data, with the exception of magnetic data, respond to surface parameters (spectral reflectance, backscatter, radioelement emission, topography) only. Capturing all the factors that comprise a bedrock map arguably is more easily done visually as a decision to draw a geological boundary often requires the geologist to integrate all the photo-geologic parameters in the interpretation process. This is difficult to do using computer assisted algorithms unless these photo-geologic parameters can be readily transformed into numerical variables that yield complementary discrimination potential in using multivariate image classification. Furthermore, even in Arctic terrains, the target (bedrock) is often covered by glacial deposits and lichen which can obscure important spectral, radar, backscatter and radioelement characteristics of the underlying bedrock. It is critical to note that the nature of the glacial overburden and whether it is residual or transported is an important factor in determining the effectiveness of remotely sensed data for mapping bedrock patterns. For example, if the glacial material is largely residual, the overburden often reflects the underlying bedrock composition and thus the bedrock can be mapped in part remotely using spectral reflectance, backscatter and radioelement characteristics of the surface. Glacial and vegetative cover, of course, is not a severe limitation with magnetic data. The Canadian Arctic islands and coastal areas are better environments for predictive bedrock mapping using optical remote sensors due to less lichen and vegetation cover whereas inland areas, even though bedrock outcrop is plentiful, are largely covered by lichen which suppresses spectral reflectance variations. This, however, does not apply to structural mapping as several types of geologic and glacial structures, regardless of whether the mapping area is inland, island or coastal, are often clearly expressed on optical, radar and topographic data. The only issue is separating glacial from bedrock structures. It is suggested the best method for producing a predictive bedrock map is to combine both visual and computer-assisted approaches. Automatic or semi-automatic methods can be employed and the results incorporated in the GIS database. The geologist is then free to screen, geologically calibrate and use these automatically derived results in whole or in part on a predictive bedrock map as shown on **Figure 13** which combines distinct spectral boundaries and units, derived through classification of optical data and automatically derived form lines from the magnetic data. Furthermore, the structural data can be screened based on attributes such as orientation, length and correlation with structural features interpreted from optical, topographic and microwave data.

Mapping of surficial materials is a somewhat easier endeavour than bedrock mapping using remotely sensed data as it is the surface material (which may be noise for bedrock mapping!) that forms the target for surficial mapping. Furthermore surficial materials mapping, as demonstrated in example 2 above, is more amenable to computer-assisted techniques for producing a predictive map. The key to producing meaningful predictive surficial material maps lies in the identification of representative training areas. The protocol being followed by RPM efforts in Canada is to establish a database of representative training areas by eco-region which are regions defined based on similar terrain, geologic and biophysical characteristics.

Validation of predictive maps is certainly a key issue. Statistical and spatial uncertainties can be quantified when using computer-assisted algorithms (i.e. classification) as demonstrated by both examples presented in this paper (variability maps, confusion analysis). However, the process of characterizing uncertainty is more subjective when creating a predictive map using visual interpretation techniques. This has traditionally been done by the geologist making the map by adding symbologies such as inferred contacts, extrapolated boundaries etc. as demonstrated in example 1. However, these types of uncertainties are not always included in the final map product and are dependent on the geologist making the map. Part of the Canadian RPM project is to develop these standard mapping protocols.

Canada's Arctic region (north of 60°) comprises a vast territory that is difficult to access and is extremely expensive to map by a traditional *"boots on the ground"* approach characterized by evenly spaced traverses (3- 5 km) that transect all rock and surficial material types, regardless of complexity and variability. This traditional approach often leads to under sampling areas of complex geology and oversampling areas that are characterized by less complex geology. It is often the more complex areas that are of interest from a mineral exploration point of view. Field work is an integral and absolute essential part of geological mapping and of course this will always be the case. No geologist would disagree with this! Remote Predictive Mapping protocols are not meant as a replacement for traditional mapping methods but as a compliment. In many case the view from above captures different geological information than that observed on the ground. The integration of the two approaches is essential in order to provide systematic geological data over large tracts of Canada's North. This combined style of mapping utilizing RPM protocols (and variations of) presented in this paper will provide consistent, efficient and broad coverage of Canada's North. Associated with predictive mapping is a different form of field work which relies on focused traverses in areas of complex geology, as indicated by the predictive map, and less dense field checks in areas characterized by more homogeneous signatures and patterns. Ultimately this will lead to a more complete geoscience database of Canada's northern territory.

5. References

Blackadar, R.G., 1966. Geology, Cumberland Sound, District of Franklin, Geological Survey of Canada, Preliminary Map 17-1966.

Drury, S.A., 2001. Image Interpretation in Geology, 3rd edition Cheltenham, UK: Nelson Thornes; Malden, MA : Blackwell Science, 304 p.

Gillespie, A.R., Kahle, A.B., and Walker, R.E. 1986. Colour enhancement of highly correlated images. I. Decorrelation and HSI contrast stretches; Remote Sensing of the Environment, v. 20, p. 209-235.

Harris, J.R. (ed), 2008. Remote Predictive Mapping: An Aid for Northern Mapping, geological Survey of Canada Open File 5643, DVD.

Harris, J.R., Viljoen, D., and Rencz, A. 1999. Integration and visualization of geoscience data, Chapter 6 *in* Manual of Remote Sensing, Volume 3: Remote Sensing for the Earth

Sciences, 3rd edition, (ed.) A. Rencz; John Wiley and Sons Inc., New York, v. 3, p. 307-354.

Harris, J.R., He, J., Grunsky, E. Gorodetsky and Brown, N., 2011. A Robust, Cross Validation Classification Method (RCM) for Improved Mapping Accuracy and Confidence Metrics– Canadian Journal of Remote Sensing –in press)

Harrison, J C; St-Onge, M R; Petrov, O V; Strelnikov, S I; Lopatin, B G; Wilson, F H; Tella, S; Paul, D; Lynds, T; Shokalsky, S P; Hults, C K; Bergman, S; Jepsen, H F; Solli, A., 2011. Geological map of the Arctic / Carte géologique de l'Arctique, Geological Survey of Canada, "A" Series Map 2159A, 9 sheets 1 DVD; Natural Resources Canada / Ressources naturelles Canada; 1:5,000,000.

Jensen J.R. 1995. Introductory Digital Image Processing: A Remote Sensing Perspective, 2rd edition; Prentice Hall, 316 p.

Jolliffe, I.T. 2004. Principal Component Analysis, 2nd edition; Springer-Verlag, New York, Springer Series in Statistics, 486 p.

Kruse, F. and Raines, G. 1994. A technique for enhancing digital color images by contrast stretching in Munsell color space, *in* Proceedings of the ERIM Third Thematic Conference, Environmental Research Institute of Michigan, Ann Arbor, MI, p. 755-760.

Lillesand, T.M. and Kieffer, R.W. 2000. Remote Sensing and Image Interpretation, 4th edition; John Wiley and Sons Inc., New York, 724 p.

Milligan, P. R. and Gunn, P. J., 1997. Enhancement and presentation of airborne geophysical data; AGSO Journal of Australian Geology and Geophysics, v. 17, p. 63-75.

Pilkington M., Keating, P.B., and Thomas, M.D., 2008. Chapter 3 –Geophysics, in Harris, J.R. (ed), Remote predictive Mapping: An Aid for Northern Mapping, Geological Survey of Canada Open File 5643, DVD.

Richards, J.A. and Jia, X. 2006. Remote Sensing Digital Image Analysis: An Introduction; Springer-Verlag, New York, 4th edition, 439 p.

Schetselaar, E.M., Chung, C.F, and Kim, K.E., 2000. Integration of Landsat TM, gamma-ray, magnetic, and field data to discriminate lithological units in vegetated granite-gneiss terrain, Remote Sensing of the Environment, v. 71, pp. 89-105.

Schetselaar, E.M., and deKemp, E.A., 2000. Image classification from Landsat TM, airborne magnetics and DEM data for mapping Paleoproterozoic bedrock units, Baffin Island, Nunavut, Canada. ISPRSS Amsterdam, July 2000.

Schetselaar, E. M., Harris, J.R., Lynds, T. and de Kemp, E. A. 2007. Remote Predictive Mapping (RPM): A strategy for geological mapping of Canada's North, Geoscience Canada, v. 34, no. 3/4, pp. 93 -111.

Scott, D. J., 1997. U–Pb geochronology of the eastern Hall Peninsula, southern Baffin Island, Canada: a northern link between the Archean of West Greenland and the Paleoproterozoic Torngat Orogen of northern Labrador. Precambrian Research, 93: 5-26.

St-Onge, M.R., Scott D J., and Corrigran, D., 1998. Geology, Central Baffin Island area, Nunavut, Geological Survey of Canada Open File Reports 3536 and 3537.

Wilson, J. P., amd Gallant, J.G., 2000. Terrain Analysis: Principles and Applications, John C Wiley and Sons Inc. New York, 479 p.

5

Laser Altimetry: What Can Be Learned About Geology and Surface Processes from Detailed Topography

Tim Webster
Applied Geomatics Research Group
Nova Scotia Community College, Middleton,
Canada

1. Introduction

Earth Science has utilized new remote sensing techniques for many years, weather it be airborne geophysics to sense the magnetic field or aerial photography and satellite imagery to obtain that ever important synoptic view that aids in our interpretation of the landscape and geology. The field of geomatics, which is the acquisition, analysis and mapping of the earth's surface, has emerged and drives the commonplace web applications like Google maps and Google earth. Geomatics is important in the earth science sector for many areas including: utilizing global positioning systems (GPS) for locating their property, infrastucture and geological samples, a geophysical-image analysis system for analyzing and display of their remote sensing data from geophysical (seismic, radiometric isotopes, electromagnetic, etc.) to imagery (airphotos, satellite) data, and a geographic information system (GIS) to house all of these data in addition to other geospatial data (points: wells, sample assays, etc.; lines: roads, streams, contours, etc.; and polygon: claim block, watershed, anomalies, etc.) and raster or grid cell based maps. Landscapes are influenced by several factors including geology, soils, climate, glaciations, topography, and vegetation cover, among others. In order to study geology and the influence on landscapes and their evolution, we attempt to map these different factors using geomatics.

2. Digital elevation models – lidar

One of the most critical layers to describe a landscape is the topography of the terrain, which is expressed as a digital elevation model (DEM) within a GIS environment. Most elevation models have been derived from stereo aerial photography, in which measurements of the ground are hampered by the tree canopy. The challenge to make accurate topographic measurements of the earth under the forest canopy has been a problem until recently. Airborne laser scanning has the ability to solve this problem and see through the vegetation, depending on the canopy density and closure. Light Detection and Ranging (lidar) is a technique that combines a laser ranging system with an inertial navigation system comprised of a survey grade GPS and an inertial measurement unit (IMU) in an aircraft (Fig. 1). Detailed technical overviews of lidar systems have been described by various authors (Flood et al., 1997; Gomes Pereira and Wicherson, 1999; When and Lohn, 1999; and Maune,

2001). Lidar has been used in a number of geoscience applications, including the analysis of river networks (Stock et al., 2005), the generation of cross-sections across rivers (Charlton et al., 2003), in general glaciology (Krabill et al., 1995, 2000), groundwater monitoring (Harding and Berghoff, 2000), investigation of landslides (McKean and Roering, 2003), and in the mapping of tectonic fault scarps and geomorphic features (Haugerud et al., 2003) and examining coastal processes (Brock et al., 2002). Lidar has been used to demonstrate improvements in mapping bedrock and surficial geology as well as landscape metrics such as stream incision, and to resolve and map the individual volcanic flow units of the North Mountain Basalt and the identification of crater structures within the lower flow unit (Webster et al., 2006, 2006 A). Lidar has been merged with geophysical data to revise the geological boundaries along the Avalon-Meguma terrain boundary in Nova Scotia, Canada (Webster, Murphy and Quinn, 2009). Webster et al. (2009) used lidar and drill logs to estimate the thickness of aggregate deposits in the Annapolis Valley, NS.

The detail and resolution of DEMs derived from lidar are ten times better than previous available data for these areas. Generally, DEMs derived from aerial photography or other remote sensing systems such as the Shuttle Radar Topography Mission (SRTM) have degraded accuracies in forested areas and have horizontal resolutions of ca. 20 – 30 m. The benefit of lidar is that a narrow laser beam is directed from the aircraft towards the earth's surface and reflected back in order to measure the range or distance from the aircraft to the ground. The beam divergence is typically very small (0.3 rmad), resulting in a laser footprint diameter of 30 cm on the ground at 1000 m flying height. The system is mounted in an aircraft and the laser fires hundreds of thousands of shots per second that are directed across a swath toward the earth's surface by an oscillating mirror (Fig. 1).

The laser pulse is reflected back to the sensor, which records the two-way travel time that is then converted into a range or distance based on the speed of light (Fig. 1). Since the laser pulse can partially hit several targets (top of canopy, branches, tree trunk, buildings, shrubs, and ground) the lidar sensor can record several returns. Earlier Lidar sensors, ca 2003, could record a single return, first or last. Today's sensors are capable of capturing multiple returns, for example the Optech ALTM-3100 model is capable of recording up to four returns per emitted pulse. For many surveys there is no requirement for these intermediate laser returns so only the first and last returns are recorded during the survey. The laser range distances are combined with the angular and trajectory data from the scan mirror, GPS and IMU to determine the three-dimensional location of the targets in the GPS World Geodetic System of 1984 (WGS84) mapping system. A local GPS base station is setup over a known monument to establish geodetic control for the survey (Fig. 1). Ground check points should also be acquired along road surfaces within the survey area in order to validate the lidar elevations as part of the vertical accuracy assessment process. The GPS from the aircraft (rover) is combined with that of the base station (reference) in order to obtain the position of the aircraft every second. The GPS information is combined with the angular measurements from the IMU that are 200 times per second (Fig 1). The lidar points are generally output to a map projection coordinate system, typically Universal Transverse Mercator (UTM) in meters east and north. The lidar elevations are referenced to the GRS80 ellipsoid and not above mean sea level or a local national vertical datum. A geoid-ellipsoid model can be used to convert the elevations from ellipsoidal to orthometric heights above the geoid. In Canada we currently use the HT_2 model supplied by the Canadian Geodetic Survey of Natural Resources Canada to relate ellipsoidal heights to the Canadian Geodetic Vertical Datum of 1928 (CGVD28). The lidar surveys are typically acquired in swaths along overlapping flight lines or strips (Fig. 2).

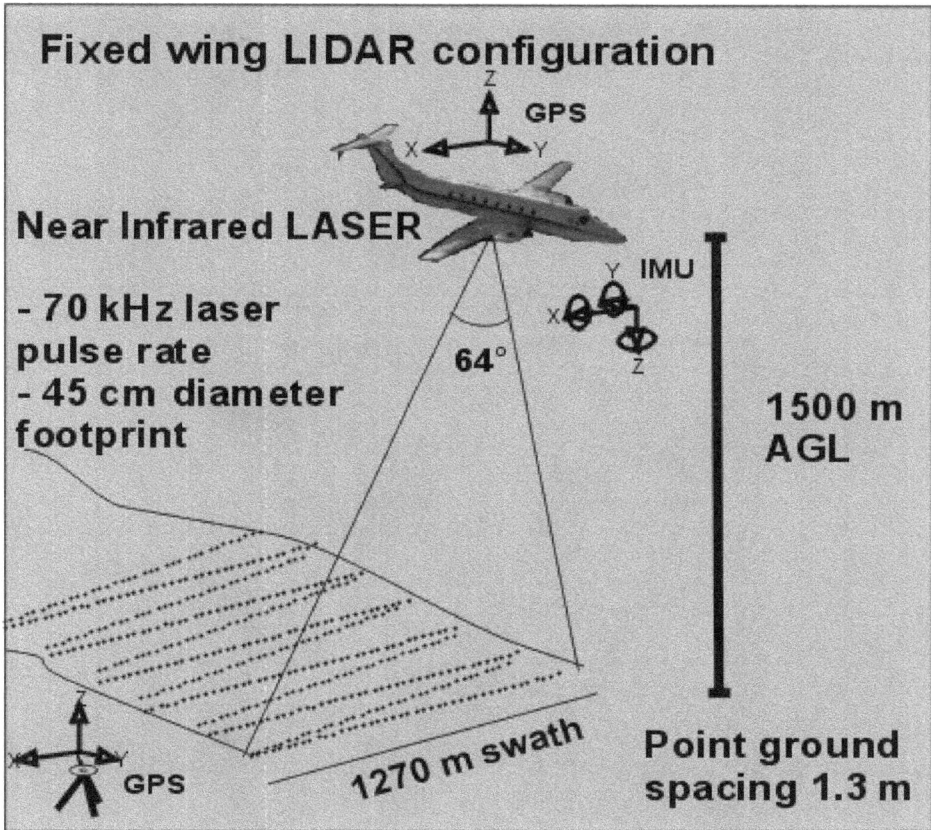

Fig. 1. Typical wide area lidar survey configuration. The laser firing at 70 kHz, with the pulses directed across the swath at 25 Hz at a height of 1500 m.

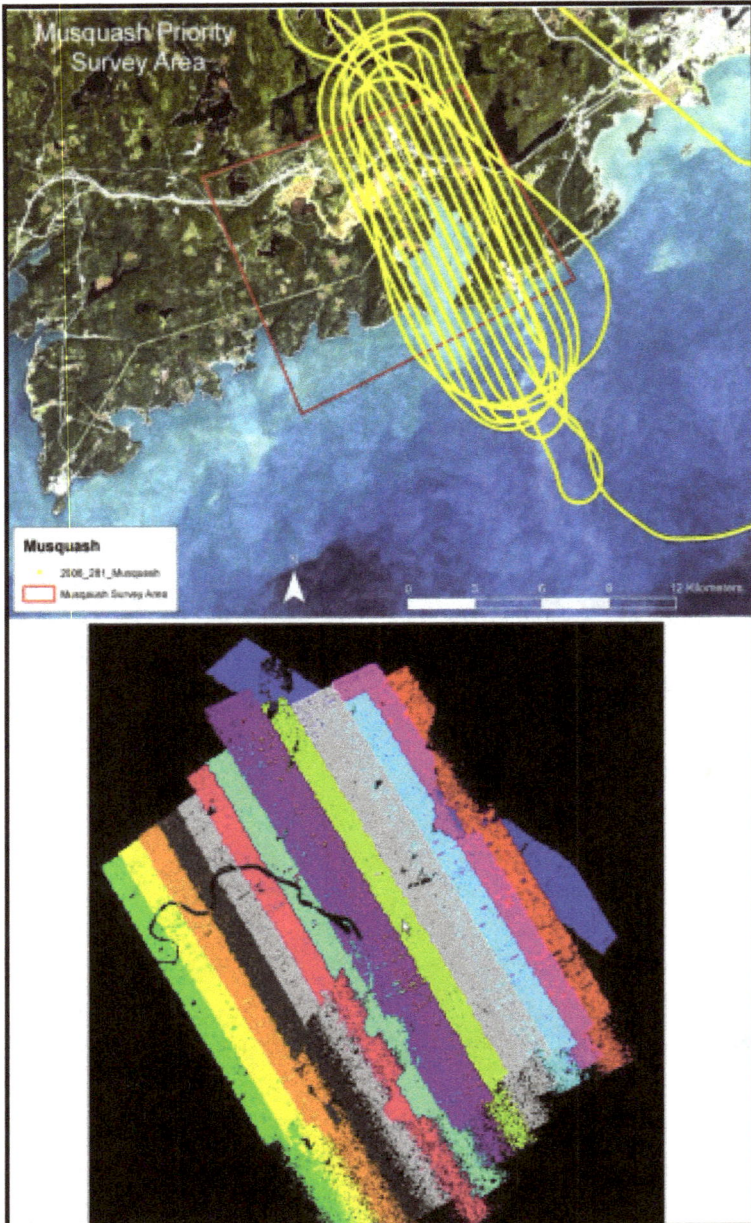

Fig. 2. Top map shows the flight trajectory of the aircraft during a lidar survey, yellow dots. The bottom map shows the lidar swaths associated with the above flight lines. Note the black outline of the river where no returns were detected because of the smooth mirror like surface of the water. Musquash, New Brunswick, Canada.

The trajectory is solved from the blend of GPS and IMU data to position the aircraft, then the laser ranges and scan mirror angles are used to compute the target position in space. The results are a set of high-density points known as a 'point cloud' that represent the ground and other targets, such as vegetation or anthropogenic features e.g. roads, buildings, bridges (Fig. 3).

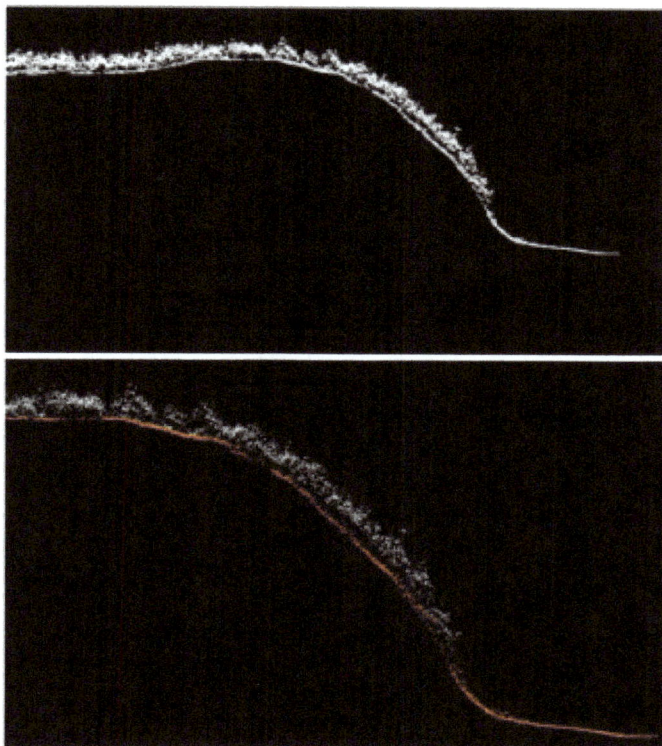

Fig. 3. Cross section of a lidar point cloud along a coastal area. Top cross-section of unclassified points. Bottom cross-section of classified ground points in orange. The remaining points are vegetation.

In order to derive an accurate DEM, the lidar points are classified or filtered into 'ground' and 'non-ground' target classes (Fig. 3). The point cloud is classified using specialized software where the points for each strip are merged together and broken down into a series of tiles based on a map projection grid system and processed individually. The classification algorithms can have problems producing accurate results in rough terrain or discontinuous slopes, dense forest areas where the beam cannot penetrate to the ground, and low vegetation being confused with the ground. Generally the lowest points are used to construct an initial surface from a Triangular Irregular Network (TIN). Then each additional point is added to the TIN if the parameters are below the threshold settings. The problem is that different thresholds are required for different terrain conditions. The two sets of lidar points, 'ground' and 'non-ground' are integrated into a GIS that can be used to interpolate different types of surfaces from the combination of points. Surfaces, such as a Digital Surface

Model (DSM) by using all of the lidar points (including those representing vegetation and ground) and the DEM using only the 'ground' points are constructed using the interpolation routines (Fig. 4). In addition to recording the time of the near-infrared 1064 nm laser pulse, the lidar system also records the amplitude of the returning pulse, known as the intensity. The intensity will vary depending on the material of the target; low reflective materials like asphalt will have very little energy returned and a low intensity compared to grass which will have a high intensity. The intensity of the points are interpolated to form a black and white type of photograph, since all of the points are used we refer to it as DSM intensity or DSMI (Fig. 4). Because the DSMI is grey-scale and is related to land cover rather than elevation, it can be combined with the DEM and DSM to form a hybrid image (Fig.4). The manipulation of lidar surface models in a GIS allows for the construction of maps that can preferentially highlight subtle geomorphic features (e.g. artificial sun illumination and vertical exaggeration). Such features are often not readily observed in traditional DEMs, or from stereoscopic inspection of aerial photographs, because of their low relief and obstructions from vegetation. Because of the scale of the many geological features being studied, regional-scale lidar surveys are required in order to assess its applicability to geomorphic research, such that features with a topographic expression can be detected and traced over long distances.

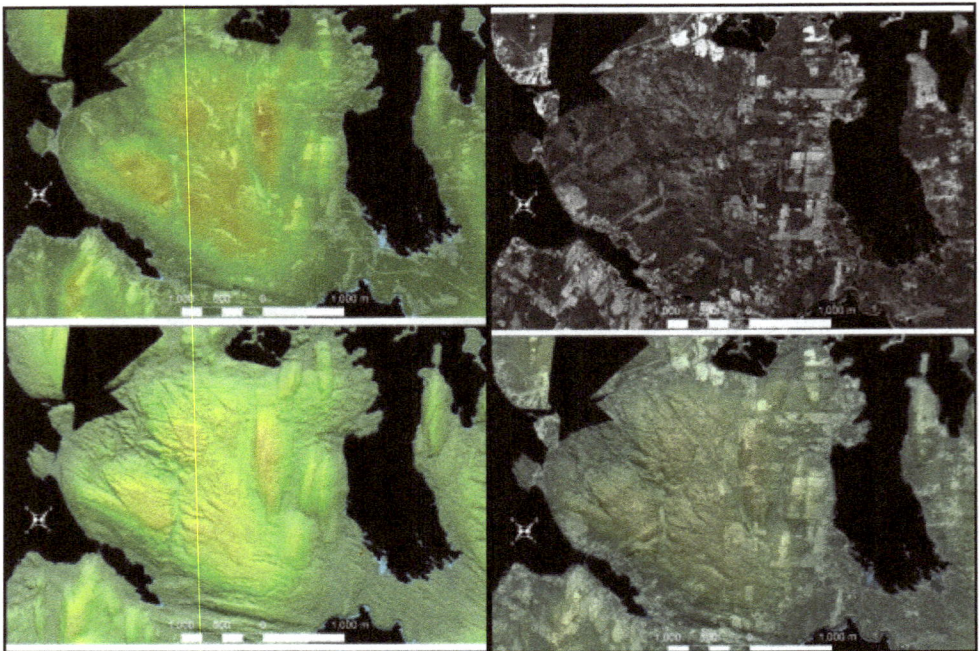

Fig. 4. Lidar surface models: Upper left map - Digital Surface model DSM using all the lidar points; Lower right map is a Digital Elevation Model DEM of ground only points; Upper Right map is the grey scale lidar intensity (land cover); Lower right map is the hybrid of the intensity plus the DEM. The folds and faults in the bedrock and drumlins are visible in the 2nd map of the DEM. Riverport, Nova Scotia, Canada.

3. Visualization

Grey-scale shaded relief maps can be constructed by illuminating the DEM from azimuth angles to highlight topographic features which trend in the perpendicular direction, for example if a topographic high trends east-west, then an illumination azimuth angle of 0 or 180 degrees would highlight the features. In addition to an azimuth illumination angle, a zenith angle can be expressed to denote where the light source is in the sky relative to the horizon. The cartographic convention for shaded relief maps is to illuminate the terrain (DEM) from the northwest or 315 degrees and with a zenith angle of 45 degrees. In order to enhance the terrain and have enough contrast between bright and darker surfaces, depending on the relief of the study area, a vertical exaggeration can be applied. A lidar DEM has been processed for a section of the Annapolis Valley, Nova Scotia and used to demonstrate various processing techniques. In the following example, the DEM has been illuminated from the eight cardinal directions 0, 45, 90, 135, 180, 225, 270 and 315 degrees at a constant zenith angle of 45 degrees and a 5 time vertical exaggeration applied (Fig. 5). The resultant grey-scale images reveal the texture or relief of the terrain, however they do not indicate the absolute elevation of features, for example a sloping surface will look the same regardless of the absolute elevation. By illuminating the terrain from different azimuth angles, different topographic features are revealed, depending on their orientation (Fig. 5). The ability to vary the sun angle and apply a vertical exaggeration is very useful in geology to enhance subtle topographic features such as lineaments, contact ridges or surficial deposits. These maps reveal two distinct morphological characteristics of the terrain with respect to the roughness of the topography. Some areas of the terrain are rough and ridges represent different volcanic flow units in contrast to smoother sections that represent areas that have a glacial till blanket covering the bedrock (Fig 5). This reflects differences in glacial history; areas to the west consist of glacially scoured bedrock with a thin till veneer, and areas to the east have a thick blanket of glacial till, known locally as the Lawrencetown Till (LT) (Stea and Kennedy, 1998). Since the shading of the terrain is limited to one direction, the resultant map only highlights features perpendicular to the shading direction and subdues features parallel to it. The application of principal components analysis, which is used to reduce data redundancy and compress multi-channel information into fewer components, has been used on multiple Radatsat images by Paganelli et al. (2003) for structural mapping. When PCA is applied to the 8 shaded relief maps, a new map is constructed where the top three principal components contain over 98% of the information, or the original variance of the 8 maps (Fig. 6). The three components PCA 1,2,3 are projected through the red, green and blue colour guns to form the new composite that highlights all of the topographic features, regardless of their orientation. The advantage of this map, in contrast to the grey-scale images, is that very few areas are in shadow and more features are highlighted (Fig. 6). The composite may be difficult to interpret since it reflects the dominant topographic features of the landscape in the 3 components. The dominant landforms are the North and South Mountains, highlighted in shades of red, separated by the Annapolis Valley in shades of green. The topographic feature which face southeast are in shades of blue (Fig. 6). In addition to the PCA composite the mapped surficial geology deposits of the Lawrencetown Till blanket (TL) have also been superimposed to compare to the terrain roughness. The arrow in figure 6 denotes a change in the glacial landforms visible in the valley. To the west of the arrow, the landforms resemble drumlins associated with an earlier ice movement. To the east of the arrow, streamline landforms are visible in the valley that represents the last movement of ice (Fig. 6). The black box in figure 6 denotes areas where raised beach terraces are visible on the DEM.

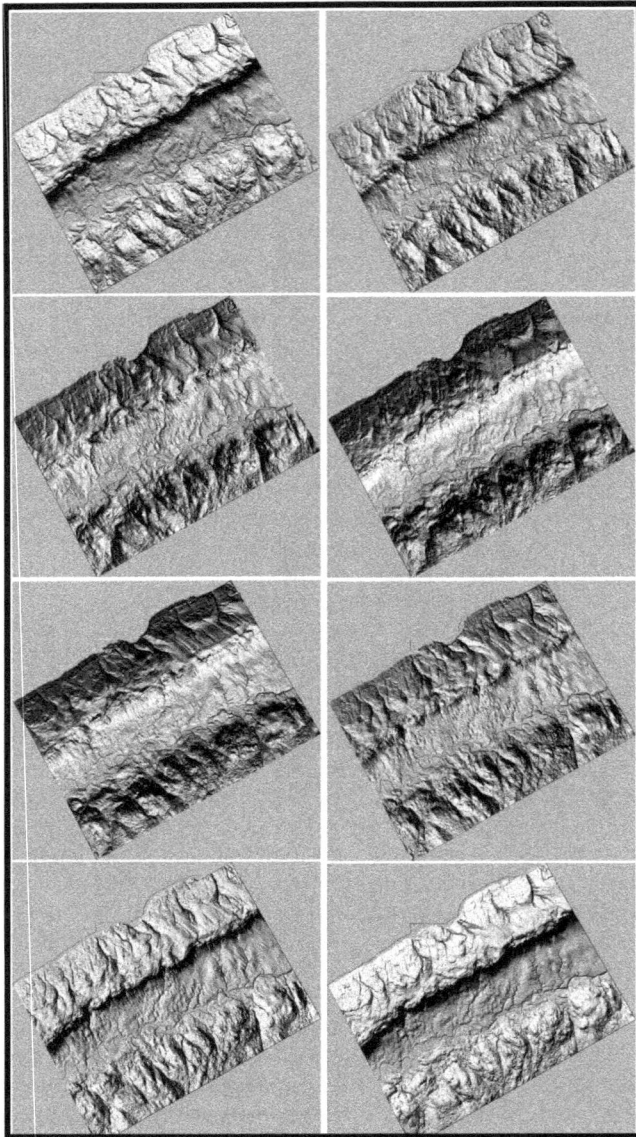

Fig. 5. Shaded relief maps on a lidar DEM for a section of the Annapolis Valley, Nova Scotia. All maps have had a zenith angle of 45 degrees and a vertical exaggeration of 5 times aplied. Top left azimuth of 0 degrees, top right azimuth of 45 degrees, second row left azimuth of 90 degrees, second row right azimuth of 135 degrees, third row left azimuth of 180 degrees, third row right azimuth of 225 degrees, bottom row left azimuth of 270 degrees, and bottom row right azimuth of 315 degrees. Images are approximately 15 km by 10 km. North is at the top of the page.

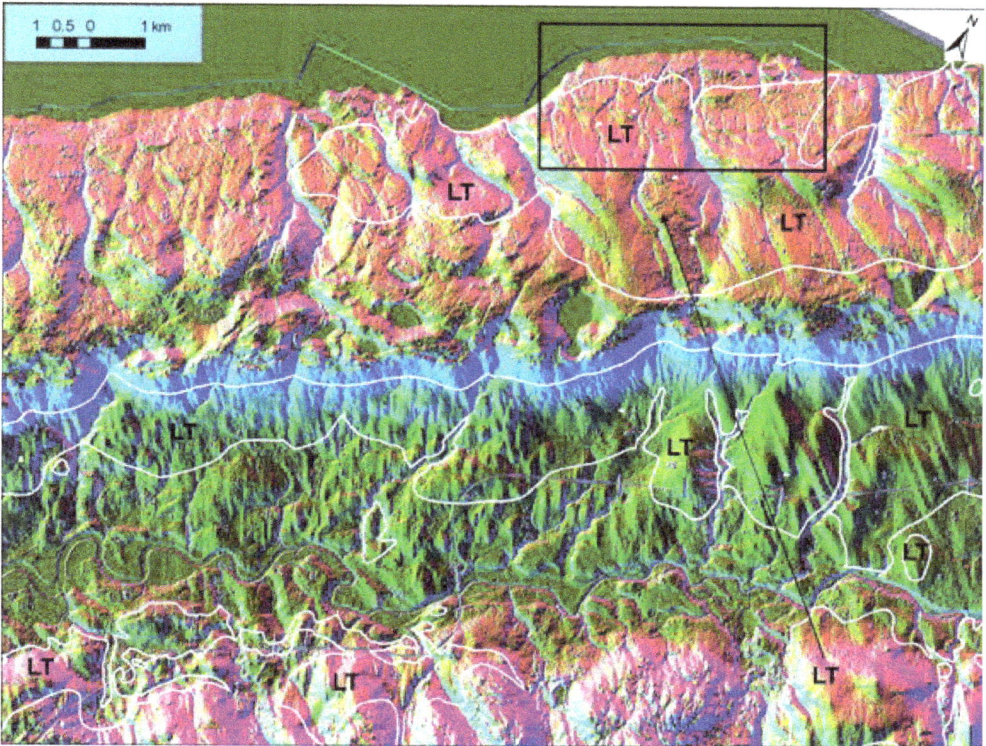

Fig. 6. Principal components 1,2,3 in red, green, blue respectively from the eight shaded relief maps (Fig. 5). The white outline represents the surficial geology boundaries of the Lawrencetown Till blanket (LT).

Another common method to enhance DEM for visualization is to construct a colour shaded relief model (CSR). This has the advantage of the shading enhancing the texture of the topography and the hypsometric colours denoting the absolute elevation of the terrain. An example of a CSR map for this area was constructed by first building a grey scale shaded relief map, with the sun illumination from the northwest (335°) at a zenith angle of 45° with a 5-times vertical exaggeration, then colour was applied to the DEM based on elevation, from below sea-level (hues of blue), to low lying land (green through yellow) to the highest point along the North Mountain (ca 265 m) (red) (Fig. 7). The colourized DEM is then merged with the grey scale shaded relief map in order to provide the texture of the terrain as a result of the shading effect. Since the colours have been applied to the terrain from the lowest elevation corresponding to the shortest visible wavelength (blue) to the highest elevation corresponding to the longest visible wavelength (red), the map appears in 3-D if viewed with Chromadepth™ glasses. Chroma stereoscopy is the technique of using colour to depict depth (Toutin and Rivard, 1995). The glasses are based on a diffraction grating which separates the incident light into different patterns depending on the wavelength. Since the map is coded from low to high elevation by low to high wavelengths of light we see depth when the brain is forced to fuse the multiple image patterns together to form a single image of the terrain. The benefit of this technique is that that map can be viewed and interpreted with or without the 3-D glasses. In comparison to the anaglyph method that requires the red and blue glasses to reveal a 3-D image in black and white. This technique offsets each image in red and blue proportional to the elevation; however, the map only can be easily interpreted when it is viewed with the glasses.

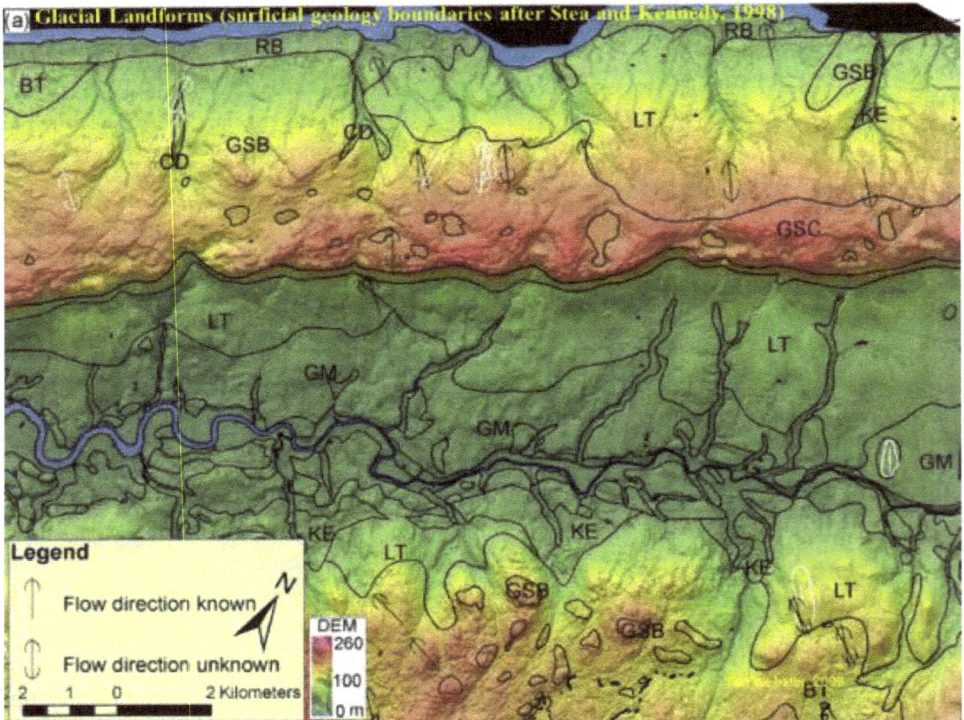

Fig. 7. Colour shaded relief map from lidar DEM. Surficial geological boundaries have been superimposed. The colour scheme is optimized for a Chroma-stereoscopic affect when viewed with 3-D Chromadepth™ glasses.

This technique of merging a grey-scale image with a colour image has been utilized in the past to integrate geophysical data with radar imagery or lidar shaded relief maps (Webster, Murphy and Quinn, 2009). It has also been used to generate "pan sharpened" images utilizing the new optical satellites where a panchromatic band at a 0.5 m resolution is combined with a multispectral (colour) set of bands at a courser resolution such as 2.5 m. The resultant hybrid image has the benefits of the 0.5 m panchromatic detail and the spectral colour information of the courser dataset. In geoscience, this method of data integration is especially useful when datasets that provide complimentary information that can be interpreted when they are integrated. For example, the lidar DEM highlights variations on the surface topography that reflects both bedrock and surficial geology features. Airborne radiometric surveys measure the amount of equivalent uranium, thorium and percent potassium near the surface and have been used in exploration. Since the gamma rays that the

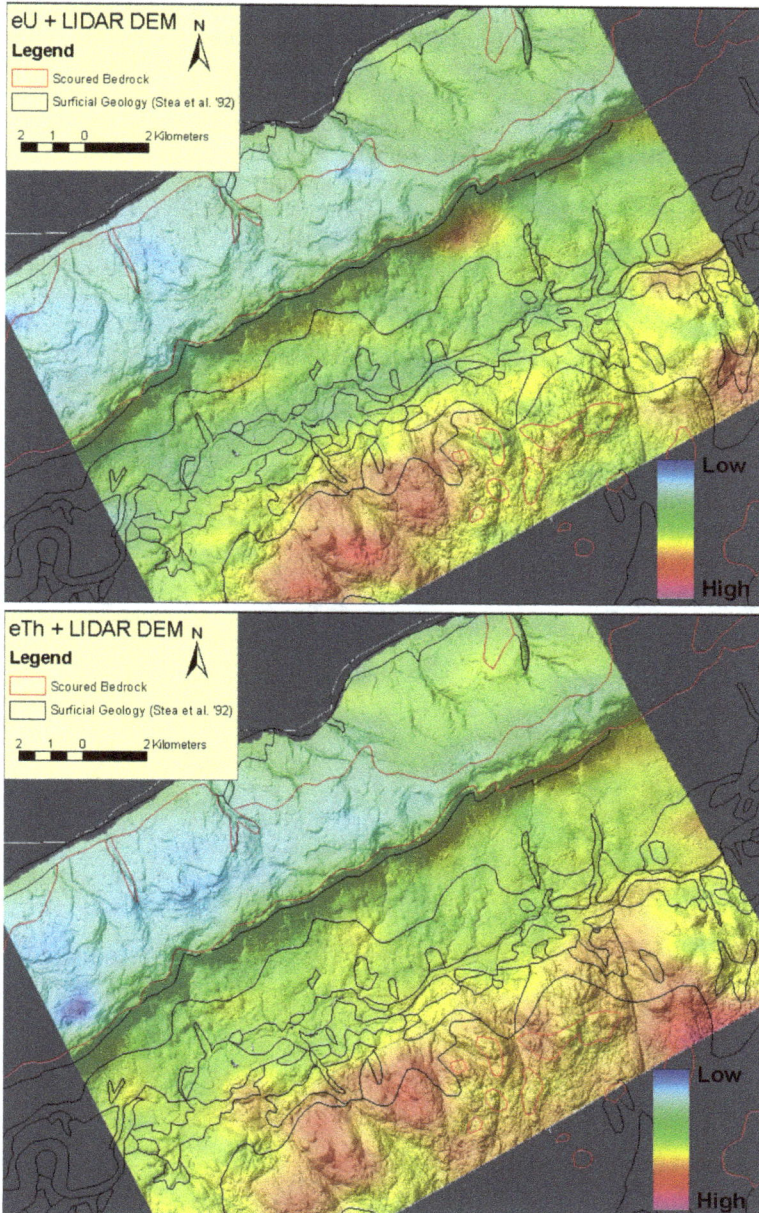

Fig. 8. Radiometric equivalent uranium (top) and thorium (bottom) have been colourized and merged with the lidar shaded relief maps. Low concentrations are colour coded blue through higher concentrations in red. The red areas on the south side of the valley correspond with granite bedrock.

sensor detects do not penetrate vary far through the soil, this geophysical measurement indicates what the concentration of the radioactive isotopes is near the surface. These airborne surveys are often gridded at a course resolution, ca. 250 m where the features appear blurry compared to the detail of lidar maps. The shaded relief lidar has been merged with the colourized equivalent uranium and thorium to produce hydrid maps (Fig. 8). The surficial geology boundaries have been superimposed which indicate the glacial till has a different radiometric signature that the underlying bedrock geology of the North Mountain (Fig. 8). The glacial till contains fragments of the South Mountain Batholith granite which occurs on the south side of the valley (MacDonald and Ham, 1994). The boundary between the scoured glacial bedrock (red outline) and the glacial Lawrencetown Till (Fig. 6-7) is highlighted by the contrast in radiometric element concentrations. The thorium values are anomalously low in the area of a crater within the basalt flow units and may reflect a difference in the chemistry of the basalt in that location (Fig 8 bottom). This technique allows two datasets to be interpreted at the same time and for the courser resolution dataset to be sharpened based on the detail of the grey-scale data. GIS systems allow the user to "fly through' the data or generate perspective views of the terrain and drape other GIS layers, either in the form of imagery or maps on the terrain. This technique further enables us to interpret the terrain and the relationship to lithology or glacial history. The ability to quickly visualize the terrain in a perspective view can often reveal relationships that are not readily visible from the standard top down map view (Fig. 9).

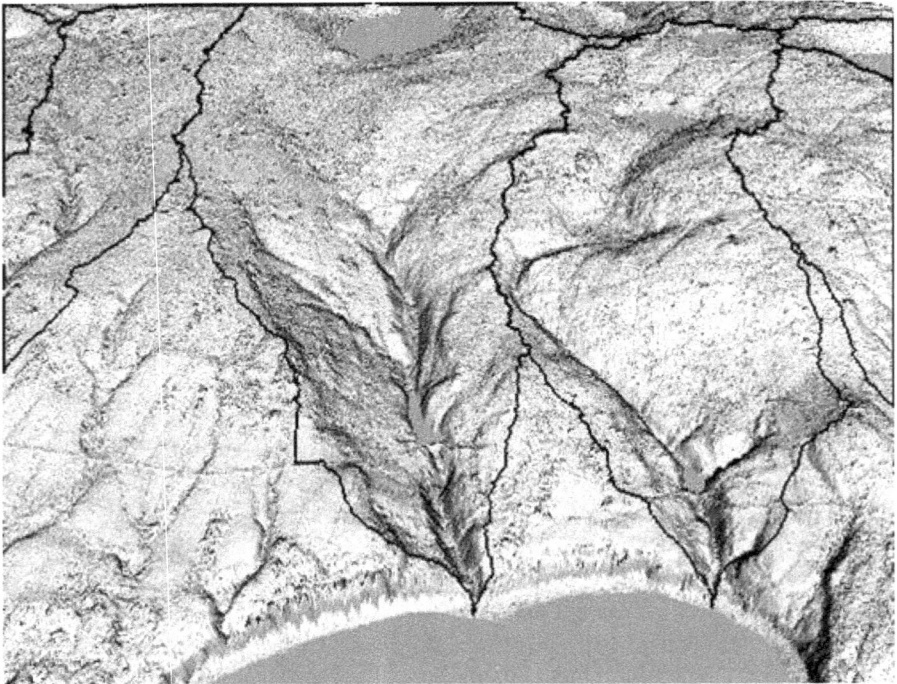

Fig. 9. Perspective view of shaded relief lidar map with the watershed boundaries superimposed.

4. Modern day processes: Watersheds and erosion

Understanding the relationship between stream incision and factors related to fluvial erosion such as rock-uplift, climate, base level changes, and bedrock resistance to erosion (e.g. Stock and Montgomery, 1999; Kirby and Whipple, 2001; Stock et al., 2005) is important for the analysis of landscape evolution (e.g. Kooi and Beaumont, 1996; Dietrich et al., 2003; Pazzaglia, 2003). The availability of high resolution lidar DEMs can facilitate quantitative analysis between incision and watershed morphometrics at sufficiently small scales to allow the examination of isolated influences on stream evolution. Previous studies have considered the relationship between the variations in the resistance of bedrock to erosion (Sklar and Dietrich, 2001) and stream or basin morphometry to the fluvial processes between regions (Belt and Paxton, 2005). However, the variations of bedrock resistance within a region (< 100 km^2) are less constrained, in part due to the scale of studies (Montgomery and Lopez-Blanco, 2003).

Fluvial processes in glaciated terrain are complex because glaciers and streams sequentially may occupy the same valleys but obey different laws of erosion, making the signatures of glacial and fluvial processes difficult to distinguish. Studies applying the stream power law often use the contributing drainage area as a surrogate parameter for stream discharge which, in addition to the local channel slope, controls the stream's ability to incise the underlying bed (e.g. Snyder et al., 2000). However, few studies examine the local hydrological effects of surface and groundwater interaction on discharge (Tague and Grant, 2004). At this scale, factors such as glacial till cover and the fracture density of bedrock can influence infiltration rates and affect peak annual stream discharge.

An example of utilizing a high-resolution lidar DEM to examine metrics of similarly-sized catchments that have been modified by glaciation is presented for the North Mountain within the Fundy Basin. The study area was selected because (i) the catchments are developed on three shallowly dipping volcanic flow units of the Jurassic North Mountain Basalt (NMB) which each have uniform resistance to erosion throughout the study area, (ii) the Bay of Fundy provides a uniform base level for all streams, (iii) there is a clear distinction in till cover thickness over the east and west portions of the study area, and (iv) the age of deglaciation and subsequent fluvial erosion is well documented and uniform throughout. The stream incision depths are related to the variability of the flow unit's resistance to erosion.

The land cover on the North Mountain is influenced by the occurrence of the till cover; farmland (pastures and hayfields) and mixed forest dominate in the east where the till is thickest, whereas the west has mostly mixed forest cover. There are more roads and anthropogenic influences in the east compared to the west where only one paved road occurs along the coast. The coastline varies between gently sloping bedrock platforms and ca. 25 m cliffs that occur in embayments. The streams on the Fundy side of NMB have evenly-spaced mainstems (1.5 km), similar catchment areas (ranging from 2 to 8 km^2) and are all consequent dendritic drainages with stream densities ranging from 0.9 to 2.9 km/km^2. The streambeds are typically 80% bedrock and 20% boulder-covered. Till is present in the streambed of some of the basins, attesting to the youthfulness of these catchments and to the inheritance of some low relief pre-glacial topography. Within the NMB study area, there are similar size basins (2 – 8 km^2) that drain scoured bedrock, and occur in the transition zone with scoured bedrock in their headwaters and glacial till near their outlets, and drain a glacial till blanket covering the basalt. The streams are ephemeral with their peak flows occurring in the spring and fall seasons. Their long profiles are ungraded and have several knick zones.

Watersheds are calculated for the main streams draining into the Bay of Fundy from the lidar DEM based on outlet locations identified on 1:10,000 scale topographic maps (Fig. 10). Most GIS systems can calculate the watershed draining into a stream based on the DEM. The standard D-8 algorithm (Jenson and Dominque, 1988; Costa-Cabral and Burges, 1994) is used to determine down-stream flow direction and sinks (depressions within the DEM treated as errors by the algorithm) are filled in the DEM to allow continuous down stream flow. However, when dealing with DEMs at high-resolution, other considerations must be made. Inspection of the drainage basin boundaries and stream longitudinal profiles indicates that most catchments have sinks. Many of these sinks are adjacent to the raised elevations of a roadbed captured by the high resolution of the lidar. As a culvert could not be represented on the DEM, a "notch" was cut across the roadbed and assigned an elevation of the nearest downstream cell to improve the accuracy of the flow direction algorithm and to prevent excessive erroneous sink filling operations in deriving the catchment basins and stream profiles. This modification improved accuracy of the flow direction algorithm, prevented excessive erroneous sink-filling operations in deriving the catchment basins and stream profiles, and allowed the stream to "pass through the roadbed". The overall result is the generation of a more accurate flow accumulation grid and basin boundary (Fig. 9-10). It

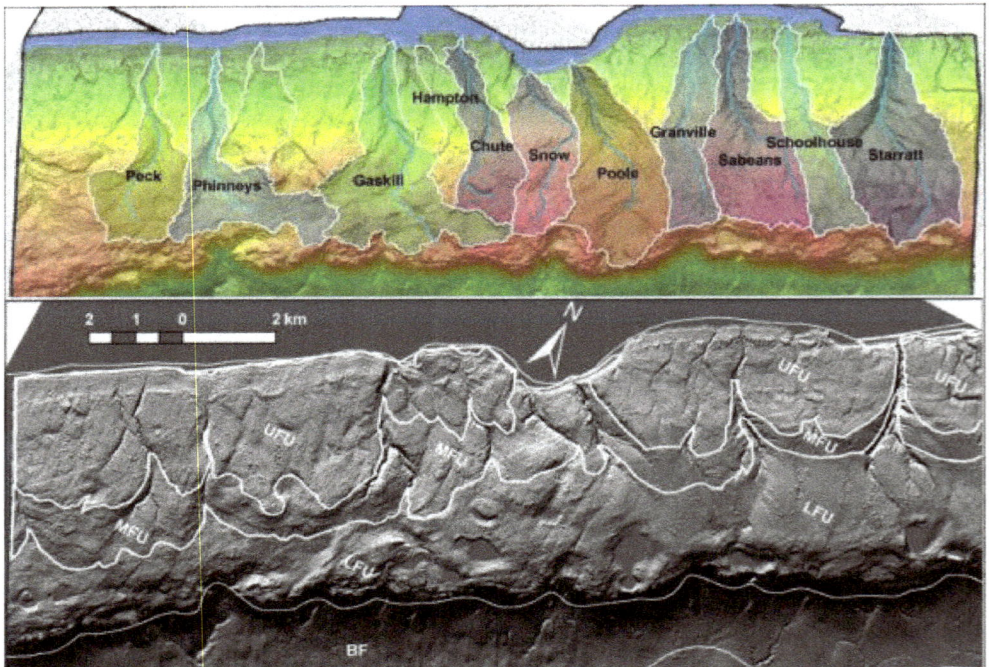

Fig. 10. Top: Lidar DEM with derived watershed boundaries for the streams draining the North Mountain Basalt. Bottom: Basalt flow units for the North Mountain over grey-scale shaded relief lidar DEM. UFU – Upper Flow Unit, MFU – Middle Flow Unit, LFU – Lower Flow Unit.

was determined that the streams from the topographic map and the longitudinal profiles obtained from the original DEM prior to sinks being filled are the most representative based on field observation and used for analysis. The flow units of the NMB have been subdivided into three distinct flow units: the lower flow unit (LFU) consists of a thick (40 - 150 m) massive single flow that is columnar jointed, the middle flow unit (MFU) conformably overlies the LFU, and consists of multiple thin flows that are highly vesicular and amygdaloidal, and the upper flow unit (UFU) conformably overlies the MFU, outcrops along the shore, and consists of 1-2 massive flows (Fig. 10 bottom).

The surface profiles of the drainage divides bordering each basin are averaged and the stream longitudinal profile is subtracted to compute the incision depth along the stream's entire length. The basalt flow units were intersected with the stream longitudinal profiles and the incision depth was summarized for each flow unit (Fig. 11).

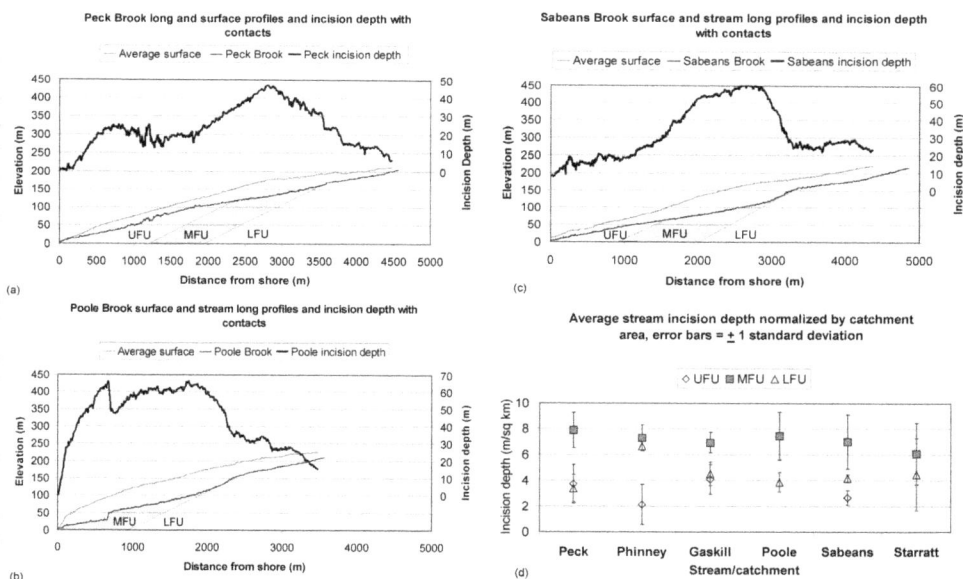

Fig. 11. Stream incision depth diagrams for the main drainage basins along the North Mountain. The surface profiles associated with the drainage divides and the stream long profile are plotted along with the depth of incision (difference between surface and stream profiles). The NMB flow unit (UFU, MFU, LFU) contacts have also been projected to intersection the streambed and related to the depth of incision. (A) Peck Brook profiles and incision. (B) Poole Brook profiles and incision. (C) Sabeans Brook profiles and incision. (D) Average incision depth for each flow unit of the NMB normalized by the drainage area for each basin and error bars indicates ± 1σ.

The stream profiles and incision depths were overlain on the flow unit map of the NMB in order to relate the incision depth to the basalt flow units. The flow units dip approximately 6° to the northwest and have been projected onto the stream profiles (Fig. 11). In general the stream incision depth reaches a maximum within the middle flow unit (MFU). Many knick zones occur either within the MFU or upstream of the contact between the MFU and lower flow unit (LFU). Incision in the upper flow unit (UFU) and LFU is similar in 3 of the 4 basins studied where both units outcrop in the streambed (Fig. 11, D). The average incision depth for the MFU is 45 m compared to 29 and 19 m for the LFU and UFU, respectively. The area percentage of each flow unit per basin and the length percentage of each flow unit per stream suggest that the percentage of flow unit per basin is a better indicator of stream incision depth than the percentage of stream length within a flow unit. The average incision depth is lowest in the catchments where the till cover is thinnest. However, the highest incision depths are associated with the catchments in the transition zone between the thin and thick till blanket areas. The valley cross-sections are used to compute the volume of material removed as described in Mather et al. (2002) for each basin. The elevations associated with the drainage divides were used to construct a paleosurface of the NMB following a similar method to that described by Brocklehurst and Whipple (2002) and Montgomery and Lopez-Blanco (2003). The lidar DEM was then subtracted from this surface in order to quantify the volume of material removed by glacial-fluvial processes and the patterns of erosion for each basin (Fig. 12).

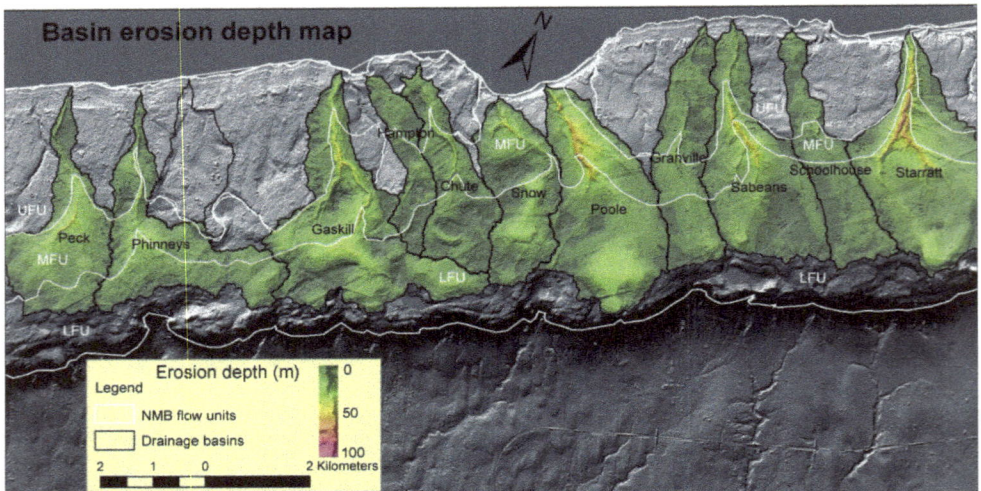

Fig. 12. North Mountain drainage basin erosion depth map with basalt flow unit boundaries. The western basins have incision depth maximums of approximately 50 m and the central and eastern basins have maximum incision depths approaching 100 m.

Erosion rates are calculated from the stream incision depth curves and sediment flux from the erosion depth map assuming erosion began after deglaciation at 12 ka ± 200 yr (1σ) (Stea and Mott, 1998), Table 1.

5. Examples of other lidar DEM geoscience applications

The improved resolution and accuracy under the forest canopy often reveals details that allow traditional geology maps to be improved and contacts between units better defined. Previously, geologists had to rely on sparse outcrop locations along the coast and along

Catchment	Till cover	Volume of sediment removed km³	Maximum sediment flux (km³/ka) assuming erosion started at 12 ka.
Peck	Thin veneer	38.3	3.2
Phinney	Thin veneer	37.4	3.1
Gaskill	Transition zone	81.8	6.8
Poole	Transition zone	91.7	7.6
Sabeans	Thick blanket	47.3	3.9
Starratt	Thick blanket	98.3	8.2

Table 1. Catchments grouped by the amount of till cover and sediment volume removed. Maximum sediment flux per catchment.

stream beds in combination with interpreting aerial photographs that did not penetrate the vegetation canopy. A lidar survey was flown to examine Piping Plover habitat, an endangered shore bird, along the south shore of Nova Scotia (Fig. 13). The area is completely forest covered except along the coast and had been mapped by the Geological Survey of Canada, Open File 1768 (Hope et al., 1988) (Fig. 13).

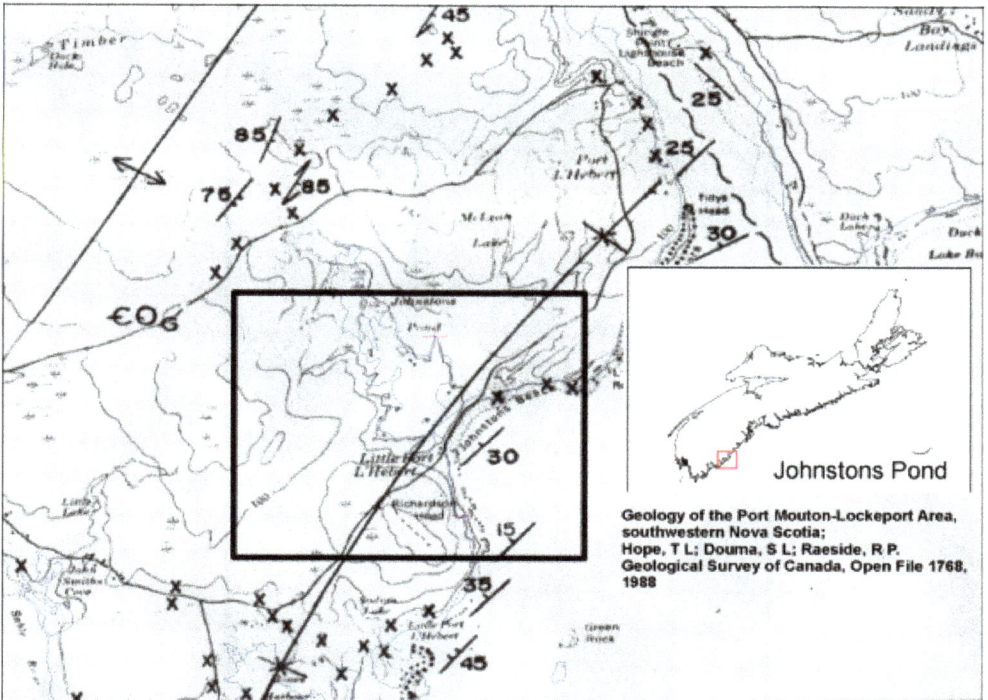

Fig. 13. Geological map of Johnston's Pond area. The black box outlines where the lidar survey was conducted. Note the only geological unit mapped is COg, indicating the Cambro-Ordivician Goldenville formation which is comprised of slates and greywacke. A syncline fold axis passes through the study area.

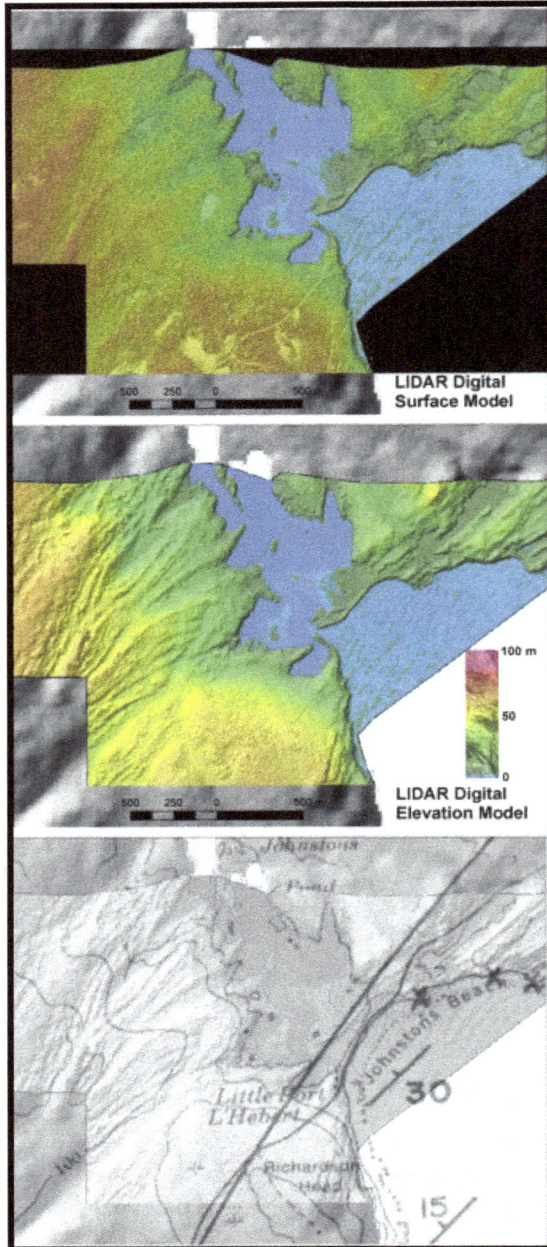

Fig. 14. Lidar surface models. Top: Colour shaded relief of the DSM; Middle: Colour shaded relief of the DEM; Bottom: grey-scale shaded relief of the DEM with the scanned geology (1988).

The lidar survey was conducted during full 'leaf-on' conditions, thus making penetration of the laser pulse to the ground more difficult. The lidar points were classified and surface models constructed, the DSM incorporating all of the lidar returns and the DEM utilizing only the ground points. Colour shaded relief maps of the surface models were constructed and interpreted (Fig. 14). The ability to remove the vegetation points reveals the bedding of the slates and a massive dome structure in the south (Fig. 14). The previous geology indicates that the entire area is made up of sedimentary rocks and is folded into a single syncline which passes directly through the dome structure (Fig. 14, bottom). Based on the visual interpretation of the terrain models, shaded relief and CSR, and a visit to the site for follow up field checks, a new geology map has been derived (Fig. 15). In addition to a new fault being mapped, a granite pluton has also been added to the map. Field evidence to support the occurrence of a fault at this location, where the bedding has been truncated on the CSR DEM, is based on the flat lying sedimentary rocks being tipped vertical in the area proximal to the fault (Fig. 15). A large granite boulder or possible outcrop was found in the field which further supports the interpretation that the topographic dome evident on the lidar is a granite pluton. The variable bed resistance to erosion allows the bedding planes to be traced over large distances even under the forest canopy in the lidar DEM.

Other examples of were the ability to penetrate the forest canopy has assisted geologist in identifying geohazards including sink holes and karst topography is presented next. The Windsor Group represents evaporates, gypsum and salt deposits of Carboniferous age in Nova Scotia. These deposits occur throughout Maritime Canada as sedimentary basins formed on the flanks of the highlands. The area of Oxford, Nova Scotia is used to demonstrate the ability of lidar to map karst topography (Fig. 16). This type of landscape can be a hazard as the bedrock is dissolved by the groundwater and the area can become undermined and local subsidence can occur. The lidar was processed to a DSM and DEM and colour shaded relief maps were constructed (Fig. 16). The bedrock geological boundaries are overlaid to highlight where the Windsor Formation occurs and contain rocks susceptible to the development of karst topography. As can be seen in figure 16, the karst topography crosses under the divided 100 series highway south of the town of Oxford and trends northeast-southwest.

In glaciated terrain, the topography reflects the glacial and fluvial deposits and often masks the bedrock structures. In these areas, the lidar surface models can be used to interpret the unconsolidated sediment deposits and better reconstruct the recent history of the area. If adequate control exists on the locations of bedrock, through outcrop locations or boreholes, a bedrock surface can be constructed and used to derive sediment thickness using the lidar surface model. Lidar was flown along a section of the North Mountain in May 2003 during 'leaf-off' conditions. As mentioned earlier, the North Mountain is underlain by basalt dipping northwest at 6 degrees. Webster et al. (2006) used field checks and the lidar to constrain the individual flow units, especially in areas covered by glacial till. In areas of thick glacial till, the morphology of the flow units is not evident in the lidar because of the smoothed till cover. In this case, planes representing the flow unit boundaries were projected through the DEM and used to define where the flow unit contacts intersected the surface topography (Fig. 17). A series of glacial deposits occur south of Port George along the North Mountain. As a result of the glacial till, the area supports local farms and the land cover is mixed between cleared and forest. The lidar DEM clearly highlights the mound of sediment on the North Mountain (Fig. 18 right). The glacial features evident on the CSR DEM represent a kame deposit with eskers to the south. The kame is formed by the glacier

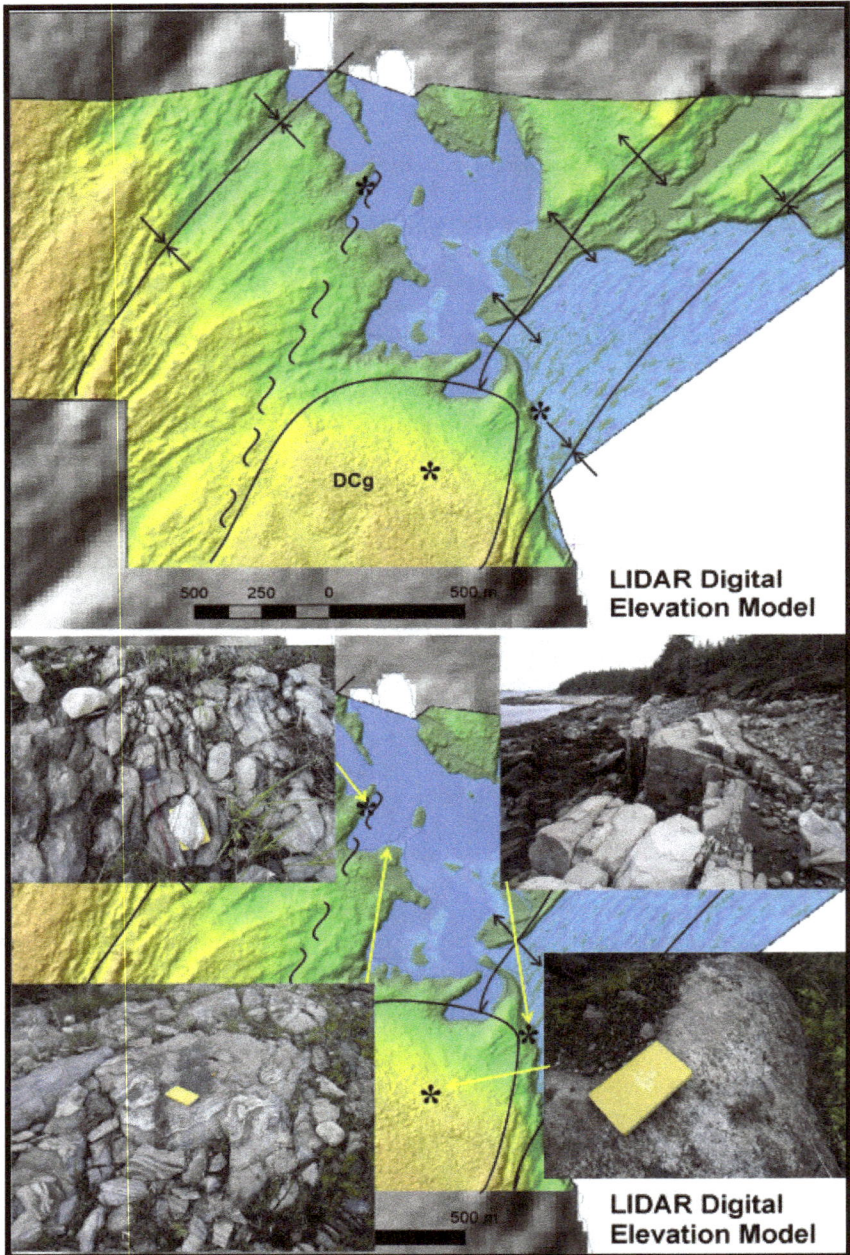

Fig. 15. Top: new interpretation of fold structures and contact between the Goldenville Formation (sedimentary rocks) and granite (DCg large dome to south). Bottom: field photos of rock outcrops that support the interpretation.

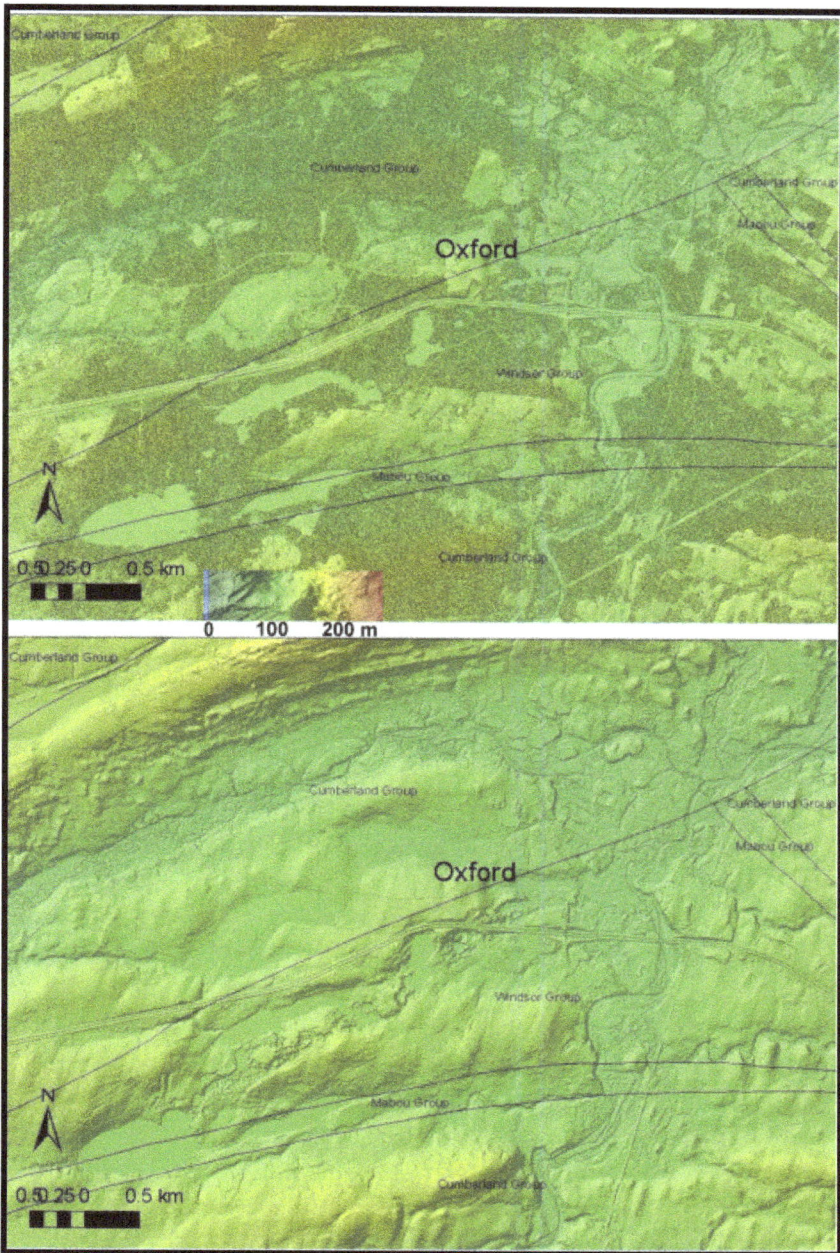

Fig. 16. Top Lidar DSM CSR of Oxford with geological boundaries. Bottom: Lidar DEM CSR with sink holes and karst topography developed below the town of Oxford. Geological boundaries (Keppie, 2000).

remaining stagnant and the trapped sediment dropping out of the ice. The mounds generally do not have well sorted sediment and show no fluvial bedding features. The eskers are the linear ridges running south of the kame and are formed by sediment collected in streams draining the melt-water within the glacier (Fig. 18). The kame and esker systems are comprised more of sand and gravel than the more clay rich glacial till and are a potential aggregate resource.

It appears that the kame and esker deposits are sitting on the bedrock surface or on a very thin glacial till veneer. GIS was used to calculate a surface representing the bedrock, which is dipping at 6 degree northwest (Fig. 19). This surface was constructed by placing a series of points around the perimeter of the kame and esker system and extracting the bedrock elevations from the lidar DEM. The points were then used to construct a TIN and a raster surface was extracted from the TIN using a linear interpolation method (Fig. 19).

Fig. 17. Basalt flow unit contact planes projected through the lidar DEM. The pink plane represents the contact between the Upper Flow Unit (UFU) and the Middle Flow Unit (MFU), the light green plane represents the boundary between the MFU and lower Flow Unit (LFU) and the dark green plane represents the base of the LFU. Adapted from Webster et al. 2006.

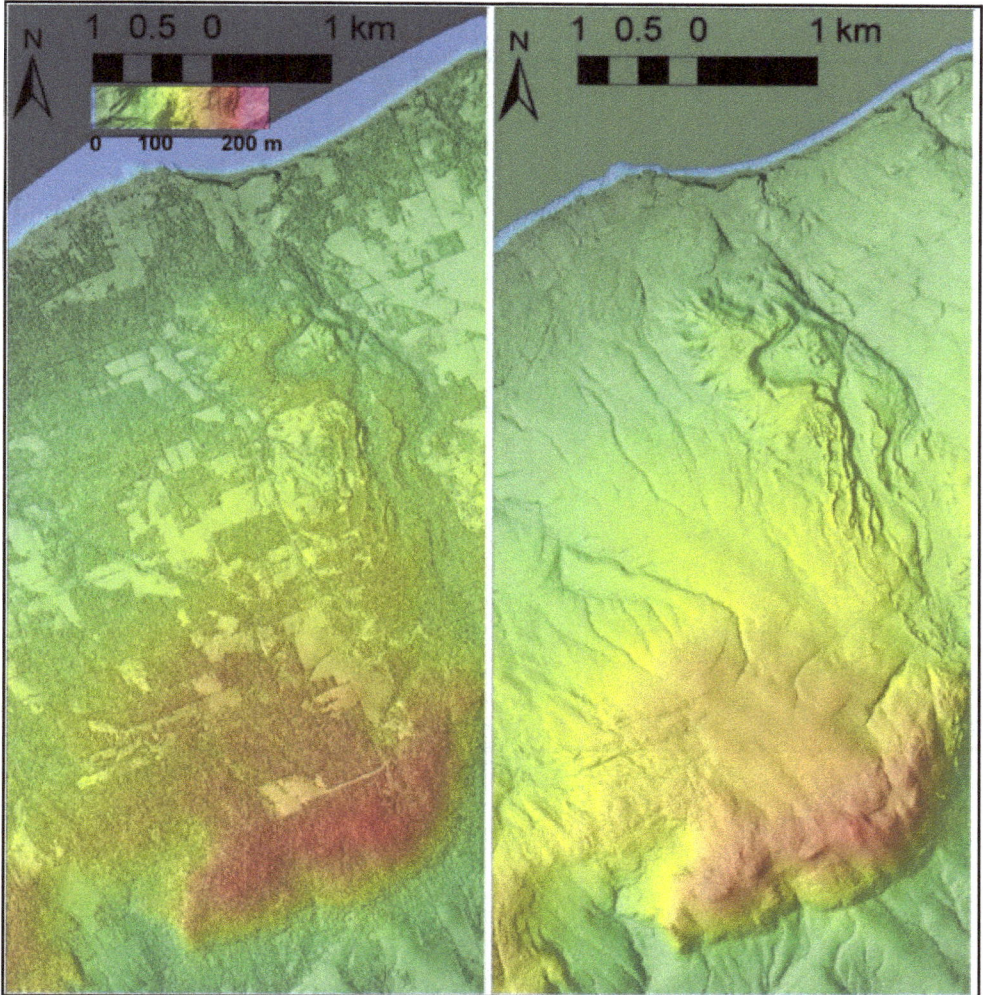

Fig. 18. Port George colour shaded relief lidar surface models. Left: DSM; Right: DEM with the kame and esker glacial deposits clearly visible.

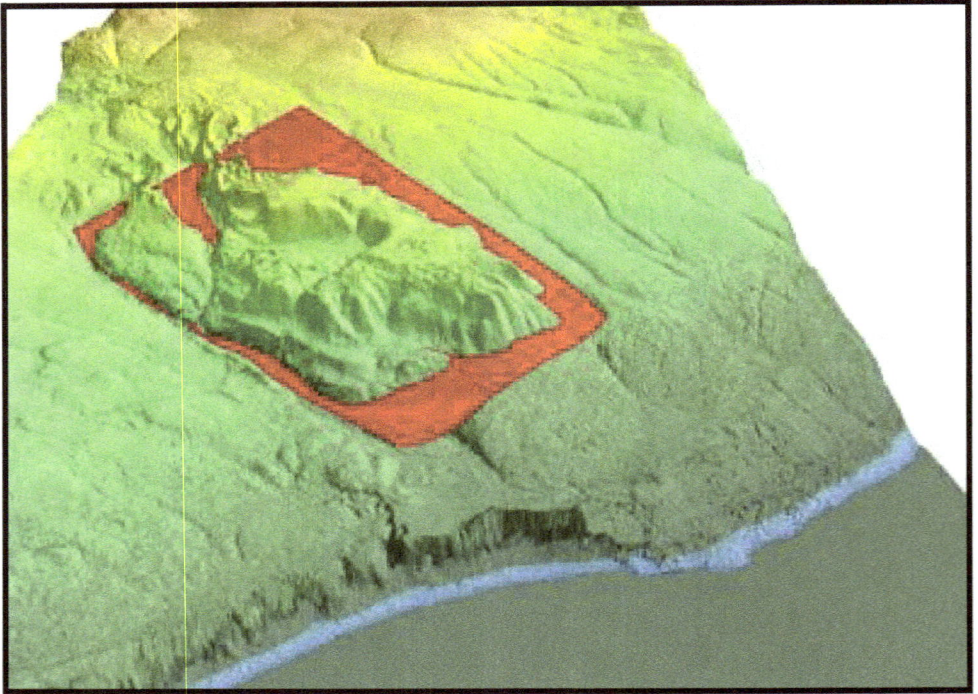

Fig. 19. Perspective view of the lidar DEM along the North Mountain. The red plane represents the bedrock surface that intersects the DEM.

This new surface was used to calculate the thickness of the kame deposit by subtracting the lidar DEM from the bedrock planar surface (Fig. 20). Once the sediment thickness is calculate the volume of sediment can be easily derived. This allows the landowner to assess the potential value of the aggregate resource. As can be seen in figure 20 the thickness is up to 60 m in places and represents a significant amount of material that is available as an aggregate resource.

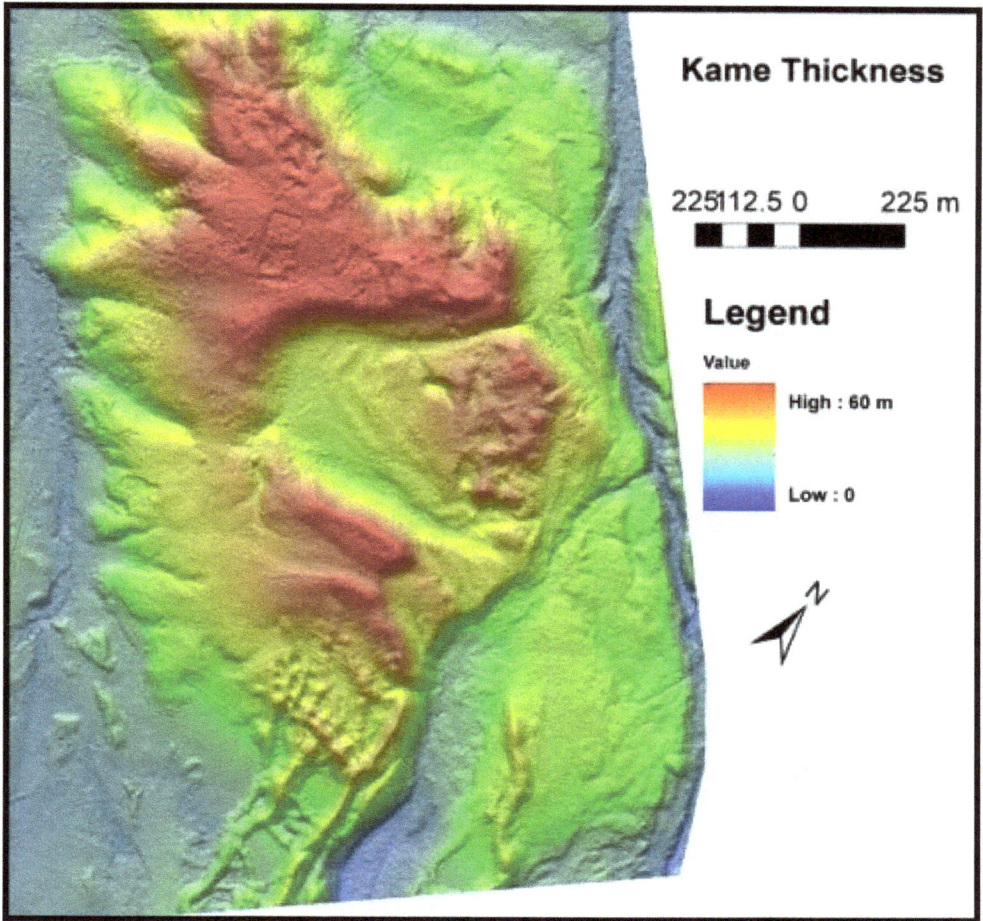

Fig. 20. Thickness of the sediment associated with the kame deposit along the North Mountain near Port George, Nova Scotia.

In some locations along the North Mountain the glacial deposits are thicker than others where ice moved into the Bay of Fundy. In areas of thicker glacial deposits along the coast, the raised beach terraces are more pronounced in the lidar DEM (Webster et al., 2006, A). These terraces represent sea-levels that where 35 m higher than present ca. 12,000 years ago. These higher sea-levels occurred after deglaciation when the melt-water caused the ocean to rise faster than the crust rebounded forming these terraces at the highest elevation.

As noted earlier the influence of glacial till over parts of the North Mountain have smoothed the topography which contrasts between the rough terrain where there is a thin veneer of glacial till and the basalt flows are evident and the smoothed surfaces where the glacial till is thickest (Figs. 6-8). Topography can exhibit a sense of being fractal, which means the measurements we make of the terrain surface are a function of the scale at which we make the measurements (Turcotte, 1992). In other words, the terrain will appear rougher and more undulating as you make observations at larger and larger scales (more detail and smaller areas), as compared to measurements of the terrain taken from data at smaller and smaller scales or courser resolution. Fractal roughness is the difference between the terrain undulations as measured from data at two different scales. To attempt to quantify the difference in roughness between the thin and thick glacial till areas along the North Mountain, a method has been devised that approaches a measure of fractal roughness (Fig. 21).

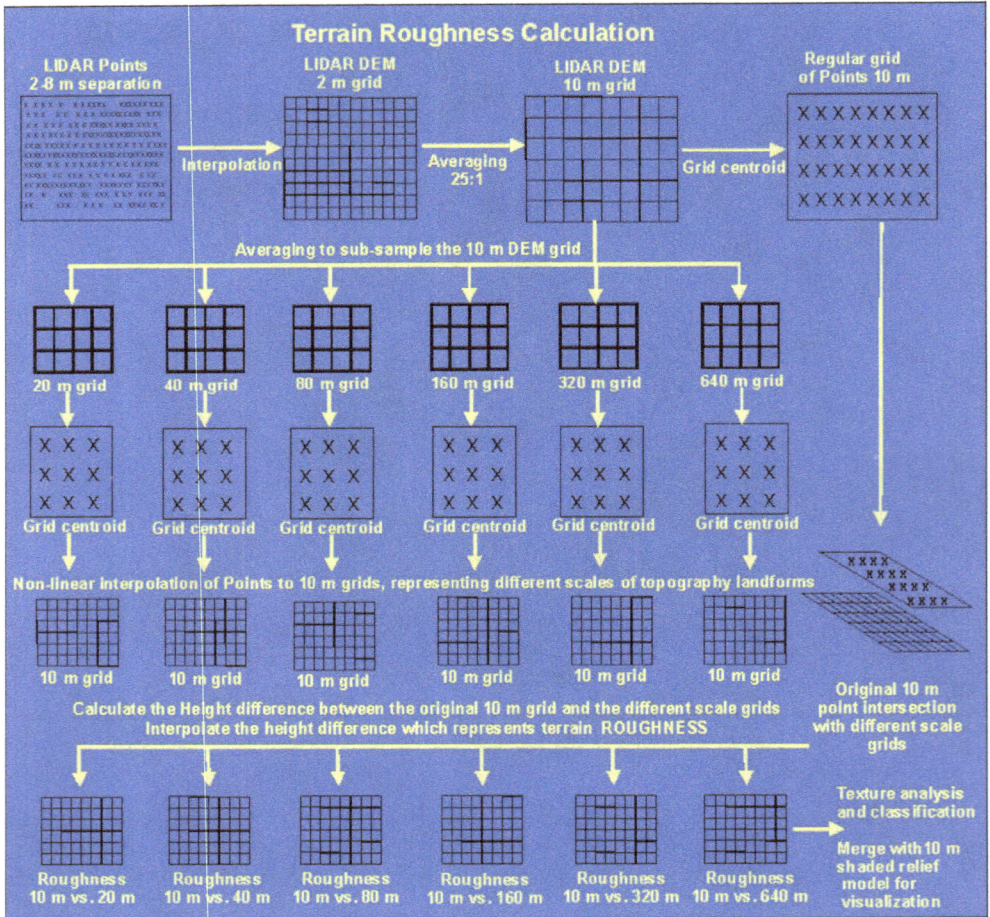

Fig. 21. Schematic of the data processing algorithm to calculate terrain roughness at different scales. Grids are represented by the square grid pattern and points are represented by "X" patterns.

The method involves starting with the irregular point spacing of the classified lidar 'ground' points and interpolating a DEM surface to a regular grid cell size of 2 m. This DEM grid is then averaged to a 10 m grid cell to facilitate processing and reduce the degree of noise of the surface. The 10 m grid is then converted back to points based on the grid centroid locations and assigned the elevation of that cell. The 10 m grid is also sampled up in resolution using an averaging technique to coarser and course DEMs of 20, 40, 80, 160, 320 and 640 m respectively. The choice of what grid cell resolution to resample to is dependent on the scale of the terrain features of interest. In this case, the grid cell resolution was increased by a factor of two each time for the purpose of demonstrating the technique (Fig. 21). These averaged DEMs of variable resolution are then converted to point centroids, where the point spacing would equal the grid cell spacing. These points are then non-linearly interpolated to 10 m grids, where each represents a different scale (decreasing scale with increasing point-grid cell spacing) of the terrain. The original 10 m grid cell points are then used to extract the elevation values from the variable scale grids and the difference between the 10 m scale elevation and the variable scale elevation is calculated. This difference in elevation, from different scale representations of the terrain, is then used to interpolate a new grid which represents the roughness or difference in roughness between the 10 m scale and the coarse scale grids (Fig. 21). The roughness grids have positive and negative values representing valleys and hills respectively on the 10 m DEM that are smoothed over as one moves to coarser and coarser scales of the topography. Profiles were extracted from the different roughness grids for a section of the North Mountain where the glacial till is thin and where it is thick (Fig. 22). The actual roughness profiles for the two

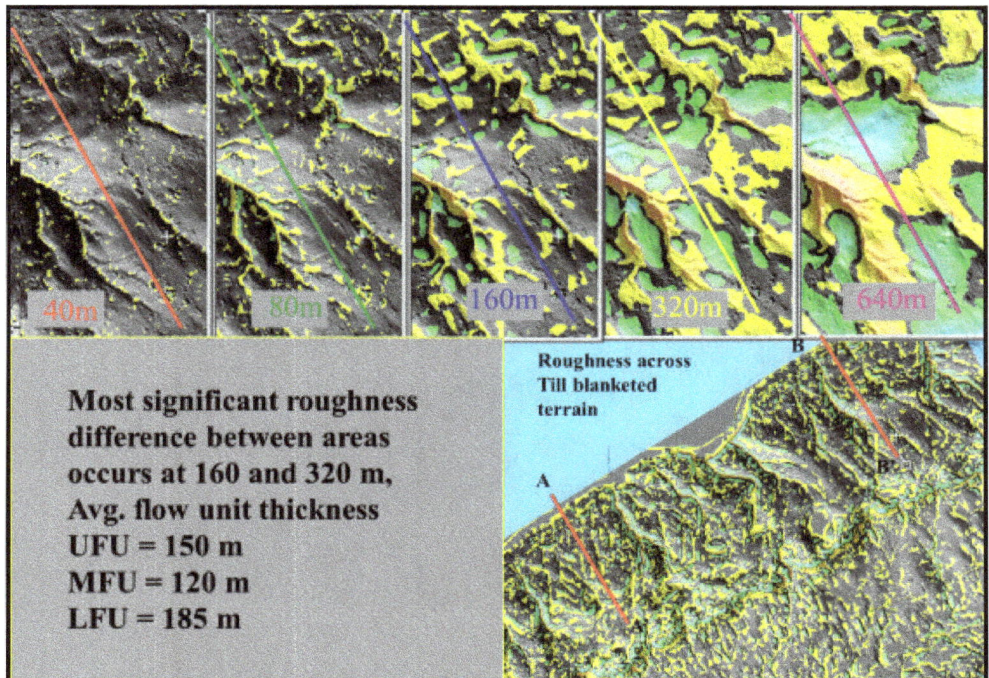

Fig. 22. Example of different terrain roughness grids along the North Mountain. Profile location A-A' is in the area of thin till and rough terrain, while profile B-B' is in the area of thick till and smoother terrain (See Fig. 23 for profiles). The top series of maps shows the location of profile B-B'.

locations are presented in figure 23. As can been seen in this figure the magnitude of the roughness difference between the two sites is significant with profile B-B' being smoother. The differences in roughness at scales of topography closer to the 10 m grid show less differences between the two areas (e.g. 40 and 80 m grid differences) (Fig. 23). The most significant differences between the two profiles occurs once the scale of the DEMs are above 160 m (Blue line Fig. 23). This scale is interpreted to be related to the average volcanic flow unit thicknesses which varies between 150-185 m and are the dominant features that are causing the roughness in the areas of thin glacial till.

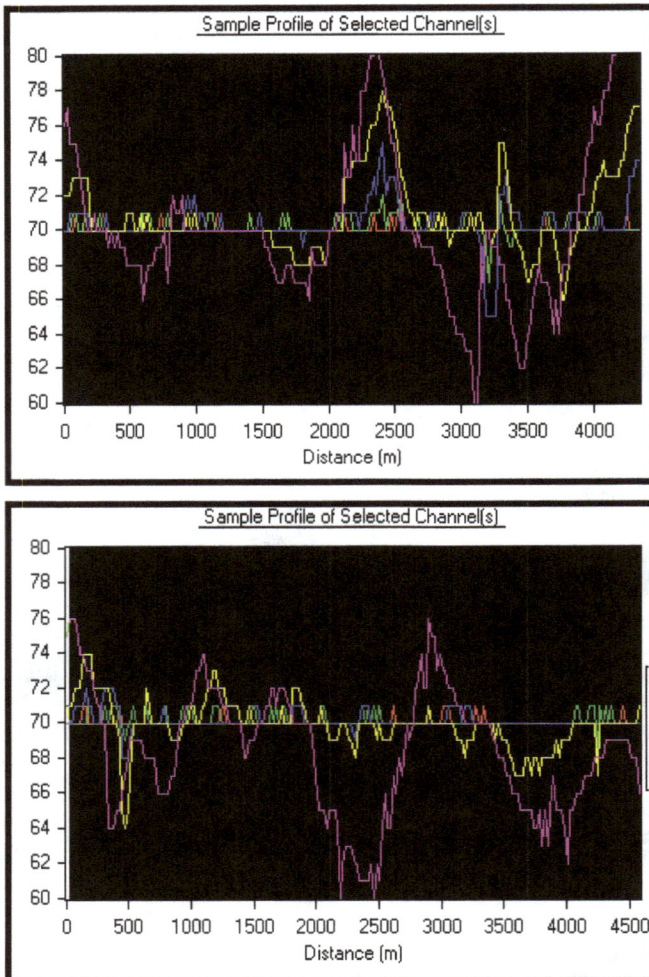

Fig. 23. Profiles of terrain roughness. Top graph is for profile A-A' in Fig. 22 thin glacial till, Top graph is for profile B B' in Fig. 22 thick glacial till. The colour of the profile line corresponds to the line and colour used for the different scales of topography (ie. 40 - red, 80 - green, 160 – blue, 320 - yellow and 640 m – magenta respectively Fig. 22).

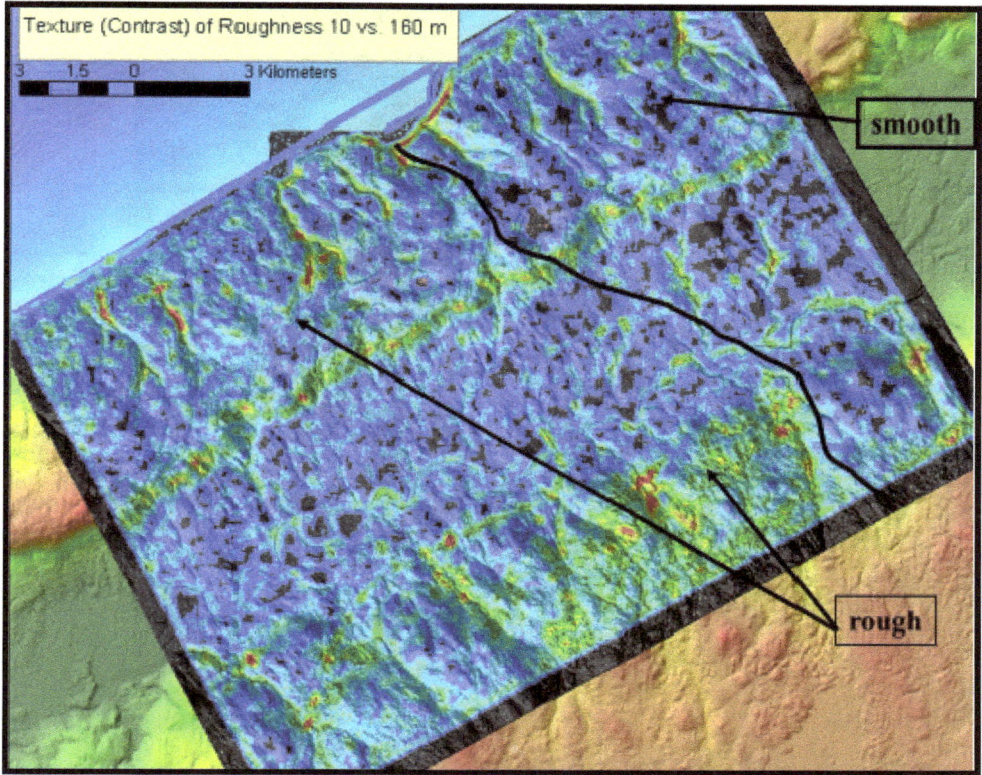

Fig. 24. Texture measure of the roughness grid derived from the difference of the 160 m grid cell terrain and the 10 m grid cell terrain map. Note the patches of dark grey represent a low texture value, or low roughness. The grey boarder around the map is no data.

A further metric was calculated from these difference roughness grids in the form of a texture measurement, which is typically related to the variance or standard deviation of the grid cell values within a moving window. The difference in roughness between the two profile locations begins to be significant at the 160 m scale; this roughness grid was used to calculate the texture in order to determine if the terrain is quantifiably different in these locations (Fig. 24). This map shows a significant contrast between the thick and thin glacial till cover, as marked by the heavy black line across the map. This difference in texture is more pronounced when applied to the 160 m scale roughness map than any of the original lidar DEMs at variable scales.

This approach of quantifying the degree of roughness of the terrain may help in eventually forming a fully automated landscape classification system. For example, if we examine the roughness difference grid at the 320 m scale, we can identify the streamlined landforms and drumlins within the valley floor (Fig. 25). The green areas represent topographic highs and the yellow areas represent topographic lows and the grey background represents smooth terrain (Fig. 25). This approach produces results that can complement other methods where the degree of curvature is estimated for the terrain, ie. concave or convex slopes, which are

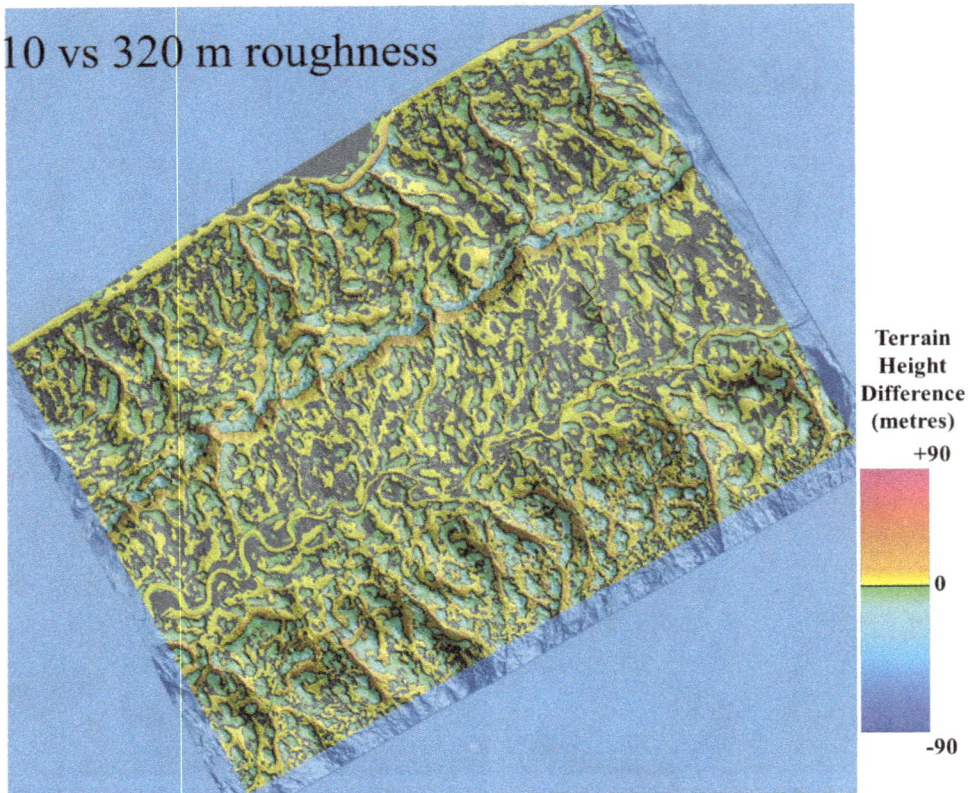

Fig. 25. Example of roughness grid difference between topographic scale 320 m and 10 m. Note the drumlins and streamline glacial landforms are highlighted in the valley floor as well as some of the basalt flow unit boundaries on the North Mountain.

calculated using the second derivative of the DEM. This method can be tailored to the scale of the topographic features of interest and through GIS processing, the features can be automatically extracted and quantified.

Recent advances in laser mapping technology have developed "mobile mapping" systems where laser scanners have been deployed on land and marine vehicles instead of aircraft. One disadvantage of an airborne lidar system is that it does not sample or measure the terrain of steep slopes very well and certainly not of any areas that are covered by an overhang because of the viewing geometry of the system. As a result of this limitation of airborne systems, steeper slopes and cliffs along the coast are not well resolved with airborne lidar DEMs. As with urban buildings and other structures, the cliff face is not resolved to the same level of detail as the rest of the terrain. This limitation affects the ability of airborne lidar to be used for detailed coastal change in these environments. In this case a laser scanner can be setup on the beach and the coastal cliff section can be imaged to capture all of the detail and merged with the airborne lidar to form a complete 3-D surface of the terrain. In urban landscapes, "mobile mapping" systems are being used to capture the sides of the buildings and all of the street furniture (e.g. signs, light poles, fire hydrants etc.). Joggins, Nova Scotia is a world UNESCO Heritage site because of the Carboniferous fossils that occur there in the outcrop exposed along the coast. Unfortunately the cliffs are actively eroding and expected to erode faster as sea-levels rise. An airborne lidar survey was conducted over the area in 2007 and a follow up ground-based lidar survey was conducted in 2009 to obtain details on the cliff face and monitor erosion (Fig. 26). Repeat ground-based surveys are planned in order to measure the change along the cliff face.

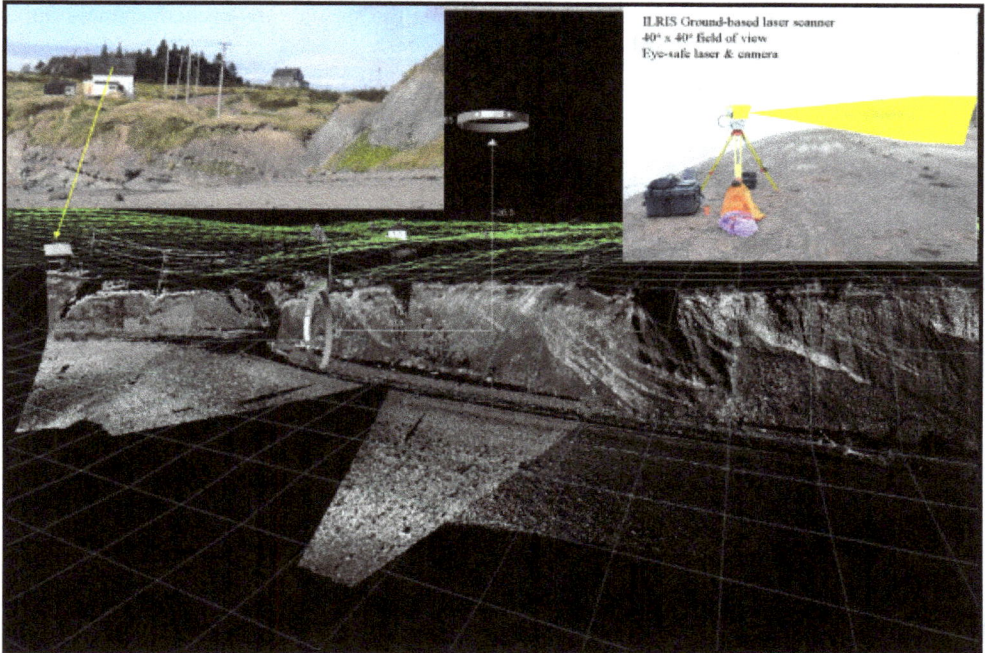

Fig. 26. Point clouds of ground-based lidar scans displayed as grey-scale intensity of the fossil cliffs at Joggins, NS. The wire frame diagram in the background is from an airborne lidar DEM. The top left photo shows the location of a house and power lines at the top of the cliff. The top right photo shows the lidar setup on the beach to image the cliff.

Other coastal areas in Nova Scotia are comprised of glacial till and are even more susceptible to erosion from the sea. Repeat ground-based lidar surveys were conducted at Cape John, Nova Scotia in 2010 and 2011 in order to measure the effects of winter storms on the coastline. Targets are placed within the landscape to be scanned and positioned using survey grade GPS. Once the terrain is scanned, a lidar point cloud can be georeferenced by identifying the targets and their coordinates to transform them into a map projection system so they can be integrated with other spatial data in a GIS. DEMs at 20 cm grid cells representing the bank were constructed from the georeferenced point clouds. Surveys were conducted in Dec. 2010 and Jan. 2011 after a major storm surge event on Dec 21 and 28th 2010 impacted the area. The DEMs from Dec. and Jan. were subtracted to map out the differences in the terrain and calculate the volume of sediment removed during the storm event (Fig. 27). Along a 150 m stretch of the coast, 771 cubic metres of sediment were removed based on the DEM analysis (Fig. 27). The storm surge associated with this event was 1.5 m above the usual water level and a local tide gauge measured the maximum water level to be 2.21 m above CGVD28 or approximate mean sea-level. This does not include breaking waves or wave run-up. A profile of the DEM before (Dec. 16) and after (Jan. 4) the storm indicates the erosion of the bank reached the 4.75 m elevation level. The erosion profile is typical of a coastal section where the tow of the sloped bank has been removed, causing the bank slope to steepen, and some of the material deposited in the near shore. The bank was frozen at the time of the second survey and could not maintain the steep slope and has since slumped during the spring thaw cycle to a stable slope causing the top of the bank to further retreat from the coast.

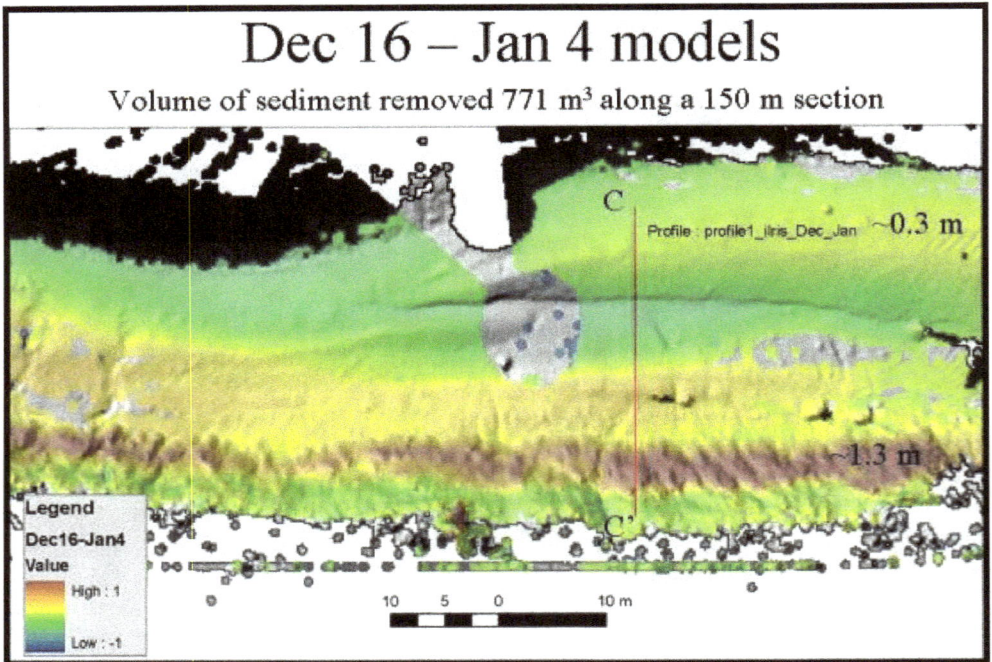

Fig. 27. Difference grid of Dec. and Jan. DEMs derived from ground-based lidar. Note the profile location C-C'.

This approach of utilizing a ground-based lidar allows for detailed analysis of changes of steep slopes that are not easily resolved with airborne techniques. The approach also has the advantage of allowing for quick deployment and is less expensive than an airborne survey. The method is fairly labour intensive and requires targets to be setup and precisely surveyed in order to georeference the scan. The latest "'mobile mapping" systems are equipped with a similar navigation system as the airborne lidars that provide a solution based on the GPS position of the sensor and the angular measurements of an IMU. As a result, the lidar scans are automatically georeferenced in a similar fashion as with the airborne systems, although some urban and cliffed environments present challenges to good GPS satellite reception.

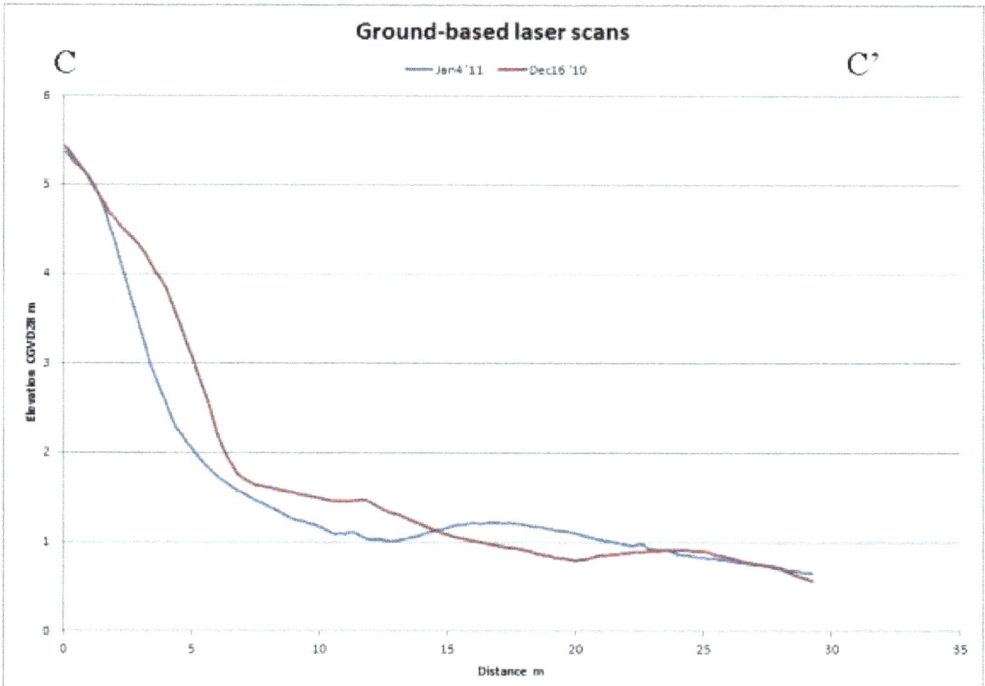

Fig. 28. Profile of coastal glacial till bank at Cape John, NS. before (Dec. 16 - red profile) and after (Jan. 4 – blue profile) a major winter storm event that occurred on Dec. 21, 2010. The upper erosion limit is at the 4.75 m elevation and the bank has been significantly steepened.

6. Conclusions

The objectives of this chapter were be to describe what terrain mapping lidar is and what map products can be derived from it. The ability of lidar to penetrate small openings in the forest canopy and sample the bare earth surface has revolutionized the accuracy and the way DEMs are constructed and used by Geoscientists. This review of lidar included the hardware and software involved in data collection and initial processing at a high level. The lidar point cloud must be processed and the ground points classified. Various surface models can be constructed from the classified point cloud including the Digital Surface Model, incorporating all of the lidar points (ground, buildings and vegetation), and of more importance to the geoscience community the DEM can be constructed by incorporating only the ground points into the model. An additional data product available from a lidar survey is based on the intensity of the reflected lidar pulse. All of the lidar point intensities are used to build the model which essentially resembles a near infrared photograph depending on the wavelength of the lidar laser system, typically 1064 nm. Once the lidar data are processed to map products, various GIS and image processing routines are applied to these data to allow visual and analytical interpretation (shaded relief maps for example).

Since the lidar only provides insights into the surface topography, the concepts of data integration and the generation of hybrid image maps were explained and demonstrated. For example, the grey scale shaded relief lidar was merged with geophysical data in the form of a radiometric survey of equivalent uranium and thorium. Different geophysical datasets can be integrated with the lidar to better understand the relationship between the surface topography and shallow structures and contacts (1st derivative magnetics), to deeper features (e.g. total field magnetics, Bouguer gravity). In glaciated terrains, the topography reflects the surficial deposits of unconsolidated material. The lidar DEM often reveals previously unseen details of the earth's surface that can be used to refine and revise geological maps. This was demonstrated in areas of folded and faulted sedimentary rocks in contact with a granitic pluton. In glaciated terrains, the lidar surface models were used to interpret the glacial and fluvial history of several areas in Nova Scotia. Lidar DEMs were used to measure modern surface processes such as fluvial incision and erosion. GIS and the lidar DEM were used to automatically extract watersheds and longitudinal stream profiles where knick points can be observed and related to erosion rates. Lidar surface models were also used to map geohazards in the form of sink holes and karst topography. In glaciated terrains, the thickness of the unconsolidated material associated with the glacial deposits was calculated using standard GIS techniques. The sediment thickness was calculated by constructing a bedrock surface and subtracting it from the lidar surface model. Other applications of lidar DEMs included research methods to quantify terrain roughness differences in areas of thin glacial till and areas of a thicker till blanket.

Lastly, the chapter concluded with the latest in lidar mapping which includes the use of ground-based scanners and "mobile mapping" which allows the lidar to be mounted on land and marine vehicles. This type of technology can be used to monitor the amount of material removed at open pit mines or gravel quarries. The application of a ground-based lidar was demonstrated to survey a bedrock cliff to establish a baseline of information. Repeat lidar surveys will be used to monitor the rate that the cliff is actively eroding since it is an important fossil heritage site. The impact of storm surge and erosion on a glacial till bank was demonstrated by comparing repeat ground-based lidar surveys. The volume of sediment eroded during a storm event was quantified and the vertical limit of erosion was measured from the lidar derived DEMs of the bank.

Geomatics offers the geoscientist a wide suite of data, tools and techniques to further our understanding of geology and earth surface processes.

7. Acknowledgements

There are several people who have contributed to the research presented in this chapter that I would like to thank. The lidar for the North Mountain was acquired with a grant from the Canada Foundation for Innovation. Some of the research related to the North Mountain was part of my PhD research that was funded through the NSCC and Brendan Murphy of St. Francis Xavier University. Also I would like to acknowledge my PhD supervisors: John Gosse of Dalhousie University, Ian Spooner of Acadia University and Brendan Murphy of St. Francis Xavier University. Lidar for the Oxford and Johnstons Pond areas was flown by AGRG and I would like to thank Bob Maher, Chris Hopkinson, Allyson Fox, David Colville, Ryan Goodale and Doug Stiff for their involvement in those projects. Other research interns and partners who contributed include Angela Templin, and Gordon Dickie and Matt Ferguson of Shaw Resources. The coastal fieldwork team for the repeat surveys of Cape John included Nathan Crowell, Kevin McGuigan and Candace MacDonald. Thanks to Grant Wach and Christian Rafuse for the use of Dalhousie Universities ILRIS laser scanner for the Dec. and Jan. surveys at Cape John. Funding for the repeat surveyes was provided by Will Green of the NS Department of Environment. The Joggins ground surveys were conducted with assistance from Nathan Crowell, Stephanie Rogers, Danik Bourdeau, and Kate Leblanc.

8. References

Belt, K. and Paxton, S.T. 2005. GIS as an aid to visualizing and mapping geology and rock properties in regions of subtle topography. Geological Society of America Bulletin, 117, no. ½: 149-160.

Brock, J.C., Wright, C.W., Sallenger, A.H., Krabill, W.B., Swift, R.N. 2002. Basis and methods of NASA airborne topographic mapper LIDAR surveys for coastal studies. Journal of Coastal Research, 18: 1-13.

Brocklehurst, S.H. and Whipple, K.X. 2004. Hypsometry of Glaciated Landscapes. Earth Surface Processes and Landforms, 29: 907-926.

Costa-Cabral, M.C. and Burges, S.J. 1994. Digital Elevation Model Networks (DEMON): A model of flow over hill slopes for computation of contributing and dispersal areas. Water Resource Research, 30, no. 6: 1681-1692.

Charlton, M.E., Large, A.R., and Fuller, I.C. 2003. Application of airborne LIDAR in River Environments: The River Coquet, Northumberland, UK. Earth Surface Processes and Landforms, 28: 299-306.

ChromaDepth 3-D Glasses. www.Chromatek.com

Dietrich, W.E., Bellugi, D.G., Sklar, L.S., Stock, J.D., Heimsath, A.M., and Roering, J.J. 2003. Geomorphic Transport Laws for Predicting Landscape Form and Dynamics. In Prediction in Geomorphology. Geophysical Monograph 135, pp. 1-30.

Dostal, J., and Dupuy, C. 1984. Geochemistry of the North Mountain Basalts (Nova Scotia, Canada). Chemical Geology, 45: 245-261.

Flood, M. and Gutelius, B. 1997. Commercial implications of Topographic Terrain Mapping Using Scanning Airborne Laser Radar. Photogrammetric Engineering and Remote Sensing, 4: 327-366.

Gomes Pereira, L.M. and Wicherson, R. J. 1999. Suitability of laser data for deriving geographic information - a case study in the context of management of fluvial zones. International Journal of Photogrammetry and Remote Sensing, 54, no. 2-3: 105-114.

Harding, D.L., and Berghoff, G.S. 2000. Fault scarp detection beneath dense vegetation cover: Airborne lidar mapping of the Seatle fault zone, Bainbridge Island, Washington State. In Proceedings of the American Society of Photogrammetry and Remote Sensing Annual Conference, Washington, D.C., pp. 9.

Haugerud, R.A., Harding, D.J., Johnson, S.Y., Harless, J.L., Weaver, C.S., and Sherrod, B.L. 2003. High-resolution lidar topography of the Puget Lowland-A bonanza for earth science. Geological Society of America Today, 13, no. 6: 4-10.

Hope , T.L., Douma, S.L., Raeside, R.P. 1988. Geology of Port Mouton-Lockeport Area, southwestern Nova Scotia. Geological Survey of Canada Open File 1768.

Jenson, S.K., and Dominque, J.O. 1988. Extracting Topographic Structure from Digital Elevation Data for Geographic Information Systems Analysis. Photogrammetric Engineering and Remote Sensing, 54, no. 11: 1593-1600.

Kirby, E. and Whipple, K. 2001. Quantifying differential rock-uplift rates via stream profile analysis. Geology, 29, no. 5: 415-418.

Kooi, H. and Beaumont, C. 1996. Large-scale geomorphologt: classical concepts reconciled and integrated with contemporary ideas via a surface-process model. Journal of Geophysical Research, 101: 3361-3386.

Krabill, W.B., Thomas, R.H., Martin, C.F., Swift, R.N., and Frederick, E.B. 1995. Accuracy of airborne laser altimetry over the Greenland ice sheet. International Journal of Remote Sensing, 16: 1211-1222.

Krabill, W., Abdalati, W., Frederick, E., Manizade, S., Martin, C., Sonntag, J., Swift, R., Thomas, R., Wright, W., and Yungel, J. 2000. Greenland Ice Sheet: high-elevation balance and peripheral thinning, Science 289: 428-430.

MacDonald, M. A. and Ham, J.A. 1994. Geological Map of Bridgetown, South Mountain Batholith Project. Nova Scotia Department of Natural Resources, Halifax N.S. Map 94-08.

Mather, A.E., Stokes, M., and Griffiths, J.S. 2002. Quaternary Landscape Evolution: A Framework for understanding contemporary erosion, southeast Spain. Land Degradation & Development, 13: 89-109.

Maune, D. F. 2001. Digital Elevation Model Technologies and Applications: The DEM Users Manual. Edited by D.F. Maune, American Society of Photogrammetry and Remote Sensing. pp. 1-250.

McKean, J., and Roering, J. 2003. Objective landslide detection and surface morphology mapping using high-resolution airborne laser altimetry. Geomorphology, 1412: 1-21.

Montgomery, D.R., and Lopez-Blanco, J. 2003. Post-Oligocene river incision, southern Sierra Madre Occidental, Mexico. Geomorphology, 55: 235-247.

Paganelli, F., Grunsky, E.C., Richards, J.P., and Pryde, R. 2003. Use of RADARSAT-1 principal component imagery for structural mapping. A case study in the Buffalo Head Hills, northern central Alberta, Canada. Canadian Journal of Remote Sensing, 29, no. 1: 111-140.

Pazzaglia, F.J. 2003. Landscape evolution models. Developments in Quaternary Science. Vol. 1. pp. 247-274.

Sklar, L.S. and Dietrich, W.E. 2001. Sediment and rock strength controls on river incision into bedrock. Geology, 29, no. 12: 1087-1090.

Snyder, N. P., Whipple, K.X., Tucker, G.E., and Merritts, D.J. 2000. Landscape response to tectonic forcing: Digital elevation model analysis of stream profiles in the Mendocino triple junction region, northern California. Geological Society of America Bulletin, 112, no. 8: 1250-1263.

Stea, R.R., and Kennedy, C.M. 1998. Surficial Geology of Bridgetown (NTS sheet 21A/14). Nova Scotia Department of Natural Resources Minerals and Energy Branch, Halifax, N.S. OFM 1998-002.

Stea, R.R. and Mott, R.J. 1998. Deglaciation of Nova Scotia: Stratigraphy and Chronology of Lake Sediment Cores and Buried Organic Sections. Geographie physique et Quaternaire, 50, no. 1: 3-21.

Stock, J.D., Montgomery, D. R, Collins, B.D., Dietrich, W.E., and Sklar, L. 2005. Field measurements on incision rates following bedrock exposure: Implications for process controls on the long profiles of valleys cut by rivers and debris flows. Geological Society of America Bulletin. 117: 174-194.

Stock, J. and Montgomery, D. R. 1999. Geologic constraints on bedrock river incision using the stream power law. Journal of Geophysical Research, 104, no. B3: 4983-4993.

Tague, C. and Grant, G.E. 2004. A geological framework for interpreting the low-flow regimes of Cascade streams, Willamette River basin, Oregon. Water resources Research, 40: W04303.

Toutin, T, Rivard, B. 1995. A New Tool for Depth Perception of Multi-Source Data. Photogrammetric Engineering & Remote Sensing, 61, no. 10: 1209-1211.

Turcotte, D. 1992. Fractals and Chaos in geology and geophysics. Cambridge Press, pp. 199.

Webster, T., Templin, A., Ferguson, M., Dickie, G. 2009. Remote Predictive Mapping of Aggregate Deposits using LiDAR. Canadian Journal of Remote Sensing. Vol. 35, Suppl. 1 (Special Issue on Remote Predictive Mapping), pp. S154-S166.

Webster, T., Murphy, J.B., Quinn, D. 2009. Remote Predictive Mapping of a Potential Volcanic Vent Complex in the Southern Antigonish Highlands using LiDAR, Magnetics, & Field Mapping. Canadian Journal of Remote Sensing. Vol. 35, No. 5, pp. 486-495.

Webster, T.L., Murphy, J.B., Gosse, J.C., and Spooner, I. 2006. The Application of LIDAR-derived DEM analysis for geological mapping: An Example from the Fundy Basin, Nova Scotia, Canada. Canadian Journal of Remote Sensing, Vol. 32. No. 2, pp. 173-193.

Webster, T.L., Murphy, J.B., and Gosse, J.C. 2006A. Mapping Subtle Structures with LIDAR: Flow Units and Phreomagmatic Rootless Cones in the North Mountain Basalt, Nova Scotia. Canadian Journal of Earth Sciences. Vol. 43, pp. 157-176.

Wehr, A., and Lohr, U. 1999. Airborne laser scanning—an introduction and overview, ISPRS Journal of Photogrammetry and Remote Sensing, 54, no. 2-3: 68-82.

Part 4

Environmental Sciences

6

Age Dating of Middle-Distillate Fuels Released to the Subsurface Environment

Gil Oudijk

Triassic Technology, Inc.
USA

1. Introduction

The term "age dating" is defined as: estimating the time frame of a contaminant release to the environment. Because of the high costs of environmental cleanups, age-dating studies have now become an integral part of environmental investigations. Knowledge of the local geology, hydrology and geochemistry are required to perform these studies and, therefore, geologists are commonly involved.

The "middle distillates" include products such as diesel fuel, heating oils, kerosene and jet fuels. Middle-distillate fuels are used throughout the world to power motors, heat residences, fuel jet engines and propel ships, among many other uses. Middle-distillate fuels are commonly stored in aboveground or underground tanks and these tanks are often unprotected and exposed to the elements. Because of corrosion, leaks from storage tanks are a severe environmental problem, especially in locations where groundwater is used for potable supplies. Numerous underground storage tanks (USTs) were installed in North America during the "boom" years following World War II and impacts from leakage are now being found in the subsurface.

An understanding of the problems associated with leaking petroleum USTs has been known since the 1950s (Kehoe, 1960). However, action was not undertaken until the late 1970s and in some places, even much later. For example, the US state of New Jersey did not pass UST regulations until 1986 (State of New Jersey, 1986).

The average non-leaking lifespan of unprotected steel USTs may be as little as 15 years (Robinson et al., 1988). The Canadian province of Nova Scotia requires that USTs older than 25 years be removed (Hankey-Masui, 1998). Thus, numerous leaking USTs existed over the years and many probably continue today. Because of costs, the number of people impacted and the large number of cases, releases of middle-distillate fuels from USTs are a serious problem in North America (Oudijk et al., 1999). In the US states of New Jersey and Maine, several leaks are reported daily to regulatory agencies (Pearson & Oudijk, 1993; McCaskill, 1999). Similar problems exist in Europe (Bennet, 1997).

Remediation costs can be high and cases exist where buildings were removed, razed or structurally supported to complete a cleanup. It is not uncommon for costs to exceed US$500,000 and many cases costing over US$1 million exist. Costs are often borne by insurance policies, although carriers may subrogate and obtain contribution from previous carriers or others responsible. For this reason, carriers and law firms commonly request information on the time frames of releases. Because of costs, many cases are litigated and, consequently, a legally defensible method to age date releases is needed. Kanner (2007) provided the legal criteria needed to defend such methods.

Many methods exist to assess contaminant-release ages, such as UST corrosion models (Morrison, 2000), groundwater flow calculations (Morrison, 2000; Lee et al., 2007), isotope surveys (Oudijk, 2005), tree-ring investigations (Balouet et al., 2007), petroleum-weathering studies (Christensen & Larsen, 1993; Wade, 2001; Douglas et al., 2004; Hurst & Schmidt, 2005; Galperin & Kaplan, 2008; Oudijk, 2009a,b). The most common technique now in use, and normally the least expensive, is the Christensen & Larsen (C&L) method, a procedure employing petroleum-weathering rates. Hurst & Schmidt (2005), with additional data, expanded on the C&L method to date diesel-fuel releases, reporting a best-case precision of ±1.5 years.

2. Fuel composition

2.1 Diesel fuels and heating oils

In North America, diesel fuel and no. 2 heating oil have a similar composition. Diesel fuel may contain some additives during the winter and heating oil often contains a dye for tax purposes. These fuels normally have a density of 0.87 to 0.95 grams per cubic centimeter (g cm^{-3}) at 20°C and are lighter but more viscous than water (Schmidt, 1985; Wang et al., 2003) (Table 1). Heating oils and diesel fuels are composed of hydrocarbons, which are chains or rings of hydrogen and carbon. Hydrocarbons are classified by the number of carbon atoms present. For example, benzene is a C_6 molecule because it contains 6 carbon atoms.

Diesel fuels and no. 2 heating oils are composed predominantly of hydrocarbons in the range of about C_6 to C_{24} (but most of the hydrocarbons are heavier than C_8); sometimes hydrocarbons can be found up to C_{28}. The boiling range is from 150°C to 380°C (Song, 2000; Owen & Coley, 1995)(Figure 1). They contain aromatics (benzene, toluene, o,m,p-xylenes, naphthalenes, phenanthrenes), n-alkanes (such as n-heptadecane), iso-alkanes (such as the isoprenoids: pristane, phytane or norpristane), $cyclo$-alkanes and poly-aromatics plus sulfur-containing compounds such as dibenzothiophenes (Kramer & Hayes, 1987; Potter & Simmons, 1998; Bruya, 2001). The dominant hydrocarbons are the n–alkanes (straight-chain alkanes) and isoprenoids (methyl-substituted "iso-alkanes"). The aromatics include: mono-aromatics (such as benzene and toluene), alkyl-benzenes, naphthalenes, tetralins, biphenyls, acenaphthenes, phenanthrenes, chrysenes and pyrenes (Song, 2000). The predominant poly-aromatic hydrocarbons (PAHs) in no. 2 heating oil and diesel fuel are the napthalenes and phenanthrenes, whereas pyrogenic PAHs, such as chrysene and pyrene, may exist at reduced concentrations.

There can be variations in the composition of diesel fuels. For example, diesel fuels in colder climates tend to contain lighter hydrocarbons to prevent freezing problems (Figure 2).

2.2 Kerosene and jet fuels

Kerosenes are complex mixtures of hydrocarbons generally within a range of C_6 to C_{16} and a boiling range of about 145°C to 300°C (Table 1, Figures 3 & 4). Jet fuels are quite similar in composition to kerosene. The major components of kerosenes are n-alkanes, iso-alkanes and $cyclo$-alkanes. Aromatic hydrocarbons, predominantly alkyl-benzenes and alkyl-naphthalenes, normally comprise less than 25% of the volume (CONCAWE, 1995).

	Kerosene	Jet fuel (Jet-A)	Jet fuel (Jet-B)	Jet fuel (JP-5)	Jet fuel (JP-8)
Color	colorless, multiple dyes	colorless multiple dyes	colorless, multiple dyes	Clear and bright	Clear and bright
Carbon range	C_9 to C_{16}	C_9 to C_{16}	C_9 to C_{16}	C_9 to C_{16}	C_9 to C_{16}
Density	0.81@15°C	0.775 to 0.840 @15°C	0.751 to 0.802 @15°C	0.788 to 0.845 @15°C	0.775 to 0.840 @15°C
Boiling range (°C)	145 to 300	205 to 300	145 to 245	150 to 290	150 to 290
Flash point (°C)	62[b]	38	0	60	38
Kinematic viscosity[a] ($mm^2 s^{-1}$)	1.5 to 2.5 @20°C	8 @-20°C	-	1.28 to 1.60 @40°C	1.05 to 1.58 @40°C
CAS No.	8008-20-6	8008-20-6	8008-20-6	70892-10-3/ 8008-20-6	70892-10-3/ 8008-20-6

	No.2 heating oil/Motor diesel fuel	No. 4 heating oil	No.6 heating oil/Bunker oil
Color	colorless to brown, often dyed red	colorless to brown, can be dyed red	colorless to brown
Carbon range	C_6 to C_{24}	C_{15} to C_{40}	C_{20} to C_{50}
Density	0.87 to 0.95 @ 20°C	0.876 to 0.979 @20°C	0.95 to 1.01 @15°C
Boiling range (°C)	160 to 360	177 to 371	350 to 650
Flash point (°C)	58	>60	>60
Kinematic viscosity ($mm^2 s^{-1}$)	20 to 30 @ 20°C	4 to 50 @20°C	6 to 55 @ 20°C
CAS No.	68476-30-2	68476-31-3	68553-00-4

a One millimetre squared per second ($mm^2 s^{-1}$) equals one centistoke.
b Flash point is for US kerosene. Flash points for European kerosene are normally 40° to 45°C.

Table 1. General characteristics of middle distillate fuels. Sources: Bowden et al., 1988; CONCAWE, 1995;

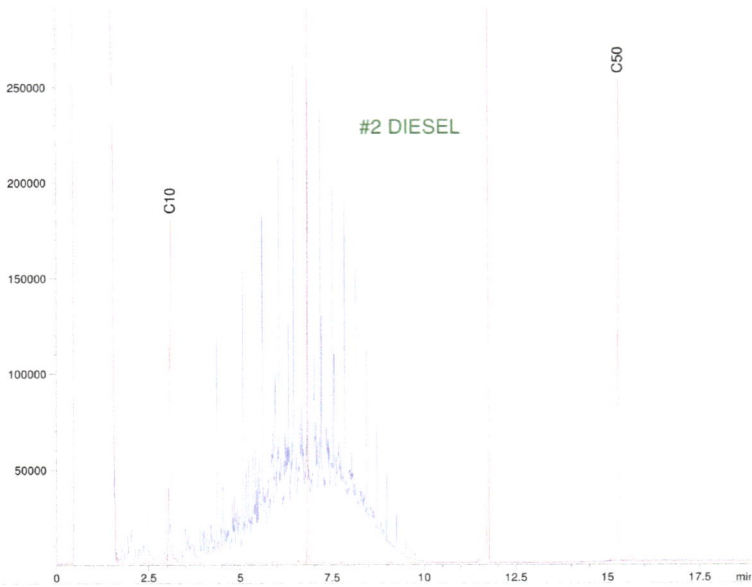

Fig. 1. A GC/FID chromatogram of a motor diesel fuel (2010). Source: Maxxam Analytics (Mississauga, Ontario, Canada).

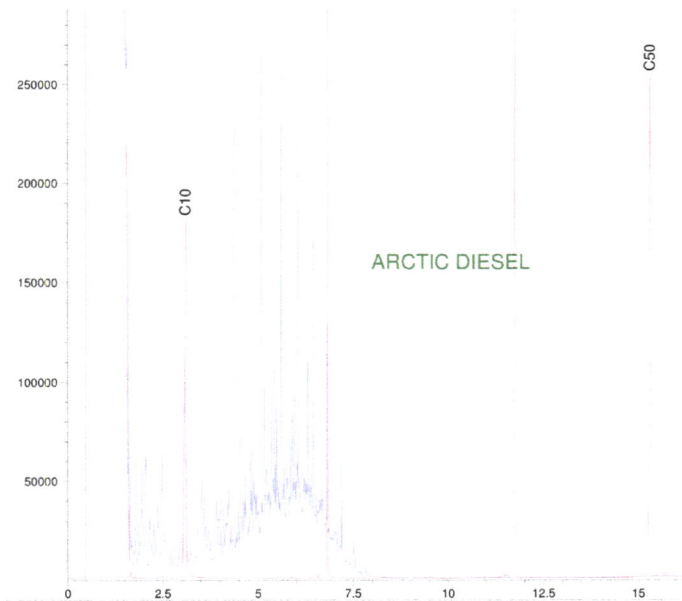

Fig. 2. A GC/FID chromatogram of an Arctic diesel fuel (2010). Source: Maxxam Analytics (Mississauga, Ontario, Canada).

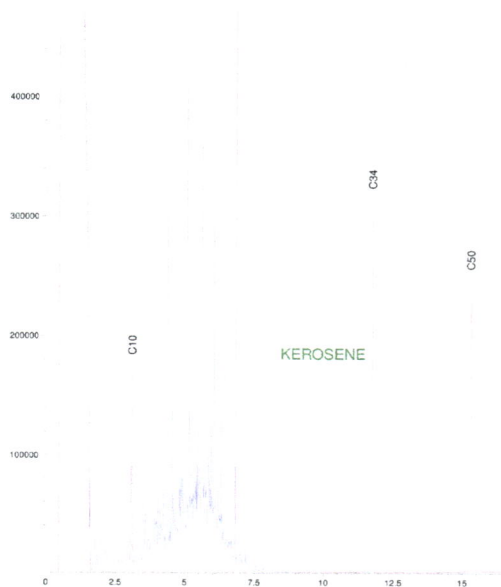

Fig. 3. A GC/FID chromatogram of a kerosene (2010). Source: Maxxam Analytics (Mississauga, Ontario, Canada).

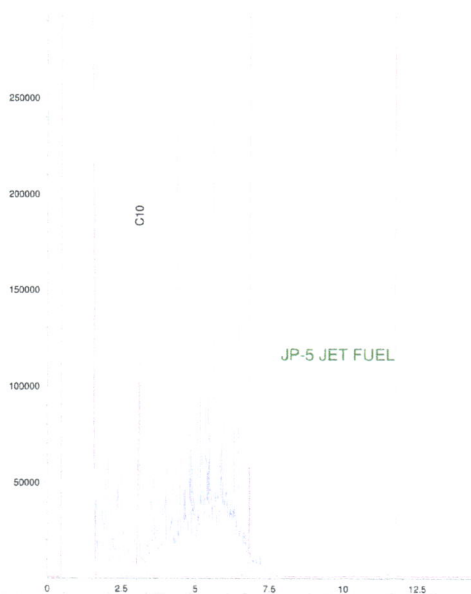

Fig. 4. A GC/FID chromatogram of a JP-5 jet fuel (2010). Source: Maxxam Analytics (Mississauga, Ontario, Canada).

2.3 No. 6 oils and bunker oils

No. 6 oil is a heating fuel, whereas bunker oil fuels ship engines. Both fuels are complex mixtures of hydrocarbons normally within a range of C_{15} to more than C_{30} (CONCAWE, 1998; Stout et al., 2002)(Table 1 & Figure 5). However, these fuels often differ greatly in composition. Furthermore, to prevent freezing problems during the winter, no. 6 oil is often mixed with lighter fuels such as kerosene or no. 2 heating oil.

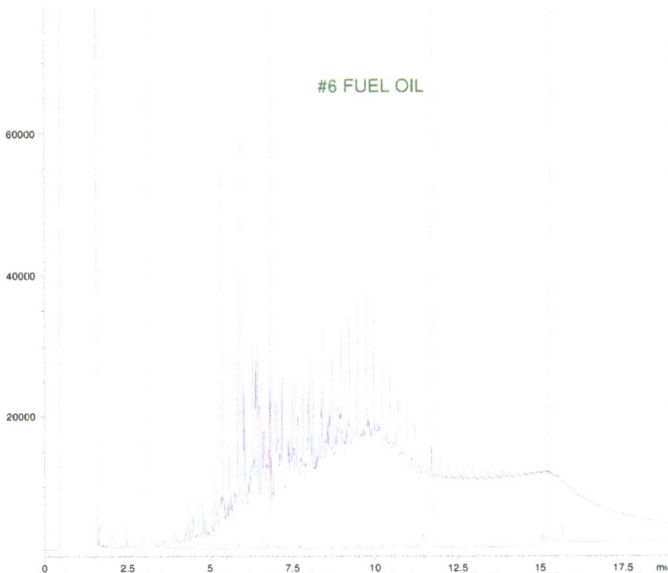

Fig. 5. A GC/FID chromatogram of a no. 6 heating oil (or "no. 6 fuel oil")(2010). Source: Maxxam Analytics (Mississauga, Ontario, Canada).

3. Middle distillate fuels in the subsurface

UST releases are normally slow and often prolonged. Corrosion of steel may be caused by many factors, such as contact with groundwater, ion exchange with clay minerals or stray electrical currents. Holes will begin as pin-sized openings and, with time, expand. Accordingly, petroleum in soil or groundwater is a mixture of ages and the ages will be skewed younger because leakage rates increase with time. The most downgradient portions of a middle-distillate plume are commonly the oldest and most age discrete. Moving closer to the source, for example towards an UST, the oil becomes progressively less age discrete (or less of a mix of ages)(Oudijk et al., 2006). To assess the maximum release age, sampling is needed within these downgradient areas.

4. Assessing a middle-distillate release

Over time, the chemistry of middle distillates placed into an UST may change because the crude-oil source for different refiners can be dissimilar. If the owner changed distributors or the distributor obtained its supply from different refineries, the initial composition of the product will not be identical over the leakage time frame. Accordingly, to properly conduct an age-dating study, it is important to assess if the plume all originates from the same source and if the initial petroleum chemistry differed.

A test to assess different compositions is through isoprenoids ratios. These compounds are relatively resistant to degradation and their ratios may be used to determine if the initial chemistry varied across the plume (Wade, 2005). There are numerous other methods to fingerprint spilled hydrocarbons (Bruce & Schmidt, 1994; Douglas et al., 1996; Galperin & Camp, 2002; Wang & Fingas, 1995a; Wang et al., 2005; Galperin & Kaplan, 2008a). These methods include a comparison of compounds such as PAHs, dibenzothiophenes or bicyclic sesquiterpanes. However, use of pristane/phytane (pr/ph) ratios seems to be the easiest and least expensive.

The pr/ph ratio is dependent on the crude-oil source and may reflect the depositional environment during oil formation (Illich, 1983; ten Haven et al., 1987; Paul et al., 1994; Sun et al., 2004; Peters et al., 2005; Osuji et al., 2009). The refining process, whereby middle distillate are produced from crude oil, normally does not alter pr/ph ratios (Stout & Wang, 2007).

Between 2004 and 2007, pr/ph ratios were calculated from 141 petroleum-saturated soil samples collected from 48 sites in the US state of New Jersey (Oudijk, 2009a). The ratios were then compared to the apparent weathering for each sample, grouped into fresh, moderate, degraded and very-degraded classes (Table 2). Apparent weathering was based on review of chromatograms and dependent on depletion of specific hydrocarbon classes, such as *n*-alkanes, aromatics or *iso*-alkanes (Kaplan et al., 1996; Senn & Johnson, 1987). The data revealed that weathering did not alter pr/ph ratios until significant degradation occurred. Hence, pr/ph values in middle distillates are dependent on their crude-oil source and consequently, they can be a simple and effective fingerprint to assess the origin of relatively unweathered, middle distillates. Other helpful isoprenoid ratios include pristane/norpristane and pristane/farnesane. However, norpristane and farnesane are less resistant to weathering and ratios can be altered in the environment.

Magnitude of petroleum weathering in sample	Number of samples		pr/ph average		Standard deviation
Fresh	41		1.61		0.19
Moderate	39		1.73		0.21
Degraded		49		1.59	0.23
Very degraded	12		1.18		0.40

Table 2. Mean and standard deviation for pristane/phytane values from soil samples (contaminated with no. 2 heating oil or motor diesel fuel) collected in the US states of New Jersey, Pennsylvania and New York, 2002-2007. Source: Oudijk (2009a).

5. Factors influencing petroleum weathering

Petroleum degradation in soil is predominantly controlled by (Stout et al., 2002b):
- evaporation, occurring when petroleum is in contact with air, causing constituents to volatilize;

- dissolution, occurring when petroleum is in contact with water, causing constituents to dissolve, and
- biodegradation, the digestion of petroleum constituents by microbes.

Researchers found that normally, but not always, the predominant weathering process in the subsurface is biodegradation (de Jonge et al., 1997; Kaplan, 2003). With a subsurface leak, evaporation is often not a factor and the remaining processes dominate. Because hydrocarbons in middle distillates are relatively insoluble in water, especially the heavier ones, biodegradation often predominates over dissolution (Christensen & Larsen, 1993).

Many microbes can use hydrocarbons as a sole energy source in their metabolism (Zobell, 1946). Energy is obtained through transfer of electrons between donors such as organic carbon, although some reduced forms of nitrogen, iron and sulfur also play a role. Dissolved oxygen (O_2) produces the most energy per mole of organic carbon oxidized than any other commonly-available electron acceptor and it is preferred by subsurface microbes (McMahon & Chapelle, 2008).

Atlas & Bartha (1992) concluded that, in one environment, spilled petroleum could persist almost indefinitely, whereas under other conditions, the same hydrocarbons might be completely removed within a few hours or days. Therefore, each environmental setting is specific and significant differences could exist in the rates and types of biodegradation. In general, biodegradation depends on:

- presence of microbes with the metabolic capacity to degrade the petroleum;
- recalcitrance of compounds (in the middle-distillate mixture);
- growth and activity factors, such as temperature, nutrients, electron acceptors and pH, influencing the microbial-population dynamics; and
- bioavailability (de Jonge et al., 1997). Bioavailability is the amount of contaminant present in the soil or groundwater that is available for uptake by microbes (Harmsen et al., 2005).

Microbes with the potential to degrade hydrocarbons in soil and ground water include bacteria, fungi, and yeasts, although bacteria are normally the most plentiful, followed by fungi (Markovetz et al., 1968; Leahy & Colwell, 1990). The byproducts of microbial degradation of, for example n-alkanes, are normally alcohols, aldehydes, and then fatty acids and possibly ketones of similar chain length plus water, CO_2 or CH_4 (Klug & Markovetz, 1967; Atlas & Bartha, 1992). Dashti et al. (2008) found that bacteria prefer n-alkanes, whereas fungi prefer the oxidized byproducts; however, the same consortium of microbes could degrade both the original alkanes and the degradation byproducts.

For biodegradation to occur, electron acceptors, such as O_2 and NO_3^-, and nutrients, such as NH_4^+ and PO_4^{3-}, are needed. Aromatics can be mineralized in the absence of O_2 under denitrifying, iron-reducing, methanogenic and/or sulfate-reducing conditions, whereas n-alkanes can mineralize under sulfate-reducing or denitrifying conditions (Bregnard et al., 1996; Ehrenreich et al., 2000). However, biodegradation rates under anaerobic conditions may be slow. Under aerobic conditions, the n-alkanes and mono-aromatics are the first hydrocarbons to be depleted. They are usually followed by alkyl-benzenes, alkyl-naphthalenes, alkyl-cyclo-hexanes and then the isoprenoids, thiophenes and PAHs (Cerniglia, 1984; Singer & Finnerty, 1984; Hostettler & Kvenvolden, 2002). The lighter hydrocarbons in each series are commonly removed earliest. For example, the naphthalenes often degrade in series of methylnaphthalene to dimethylnaphthalene to trimethylnaphthalene (Garrett et al., 2003). According to Kaplan et al. (1996), the most resistant isoprenoid is pristane, although our field data do not confirm this conclusion (Oudijk, 2009a)(Table 2).

In uncontaminated soils, hydrocarbon-degrading bacteria often constitute less than 1% of the microbial community. In polluted soils, hydrocarbon-degrading bacteria are often 10% of the community and, in some cases, may comprise 100% (Atlas, 1981; Atlas & Bartha, 1992). Furthermore, previously polluted environments, although remediated, tend to contain elevated percentages of hydrocarbon-degrading microbes (Leahy & Colwell, 1990). Therefore, microbial weathering may be accelerated in urban soils in comparison to pristine soils.

Environmental factors that may influence petroleum biodegradation, but are not always investigated or quantified in environmental investigations, include (Atlas, 1981; Leahy & Colwell, 1990):

5.1 Hydrocarbon physical state

The physical state of the spilled hydrocarbons, such as separate, dissolved, vapor or adsorbed phases, impacts biodegradation rates. Dissolved or vapor phases are often more susceptible to weathering processes.

The addition of large quantities of separate phase can suppress or completely stop bacterial growth. However, some researchers found that addition of large quantities to previously polluted environments increases bacterial growth (Colwell, 1978). At a previously pristine site, Hostettler & Kvenvolden (2002) found unweathered separate-phase crude oil almost 20 years after a spill. De Jonge et al. (1997) found that n-alkane biodegradation decreased significantly when petroleum concentrations exceeded 4,000 milligrams per kilogram (mg/kg). Lapinskiene et al. (2005) found that diesel concentrations in excess of 30,000 mg/kg were generally toxic to microbes in an aerated soil. Swindell & Reid (2006) found that, at total diesel concentrations of 20 mg/kg, 200 mg/kg and 2,000 mg/kg, phenanthrene was biologically removed from soil. However, at 20,000 mg/kg, phenanthrene removal was retarded.

The degree of separate-phase spreading will also influence biodegradation. A thick pool of separate phase will biodegrade slower compared to a pool that spread across the water table (Atlas & Bartha, 1992). The thinner pool will have more surface area in contact with ground water or the unsaturated zone, allowing increased dissolution and volatilization and hence more biological activity.

Concentrations of hydrocarbons in dissolved or vapor phases can have a strong influence on the petroleum-degrading microbes. Highly elevated dissolved petroleum concentrations may also limit biological alteration (Atlas, 1981).

5.2 Soil and groundwater chemistry

Factors such as pH or salinity can reduce biological activity. For example, elevated salt concentrations can prevent microbes from consuming petroleum. However, in some instances, salt-tolerant microbes exist and can accelerate biodegradation (Atlas, 1981; Diaz et al., 2002). The optimum pH for microbial activity is normally about 8 and soil water or groundwater commonly exhibits lower pH values (Atlas & Bartha, 1992). In some instances, soil with pH values greater than 6 can exhibit accelerated biodegradation. However, biodegradation often produces acids, allowing the pH to lower, often to as low as 3 or 4. A pH value in this range can often inhibit microbial degradation (Zaidi & Imam, 1999). The presence of certain elements in the soil or groundwater, in particular heavy metals, is toxic to certain microbes and can reduce or prevent biodegradation.

5.2.1 Redox conditions

Under aerobic conditions, n-alkanes commonly degrade readily, whereas isoprenoids are generally recalcitrant. Bouchard et al. (2008) found that, based on isotopic studies, biological degradation of n-alkanes in aerobic, unsaturated sand was dependent on chain length with smaller molecules degrading quicker.   Isoprenoids, such as pristane, can weather under anaerobic conditions (Bregnard et al., 1997), whereas light n-alkanes may become recalcitrant compared to heavier n-alkanes (Hostettler & Kvenvolden, 2002; Siddique et al., 2006; Hostettler et al., 2008). In particular, Bregnard et al. (1997) found that pristane can weather under nitrate-reducing conditions. Hostettler & Kvenvolden (2002) found that under anaerobic conditions the degradation order is the same compared to aerobic conditions: n-alkanes are removed first followed by alkyl-*cyclo*-hexanes and *iso*-alkanes. However, anaerobic conditions can cause the order to reverse within each homologous series. Heavier n-alkanes may be removed first and the same is true for alkyl-*cyclo*-hexanes. Other researchers finding similar reversals include Setti et al. (1995)(and references therein). However, Davidova et al. (2005) did not find a reversal in the degradation order, at least under sulfate-reducing conditions, and Stout & Uhler (2006) and Galperin & Kaplan (2008b) contend that reversals are caused by other means. Also, n-alkane degradation up to C_{28} was observed under sulfate-reducing conditions (Caldwell et al., 1998). Therefore, use of n-alkane/isoprenoid ratios, as a measure of weathering under anoxic or sub-anoxic conditions, may be problematic.

Under nitrate-reducing or methanogenic conditions, nitrogen gas (N_2) or methane (CH_4) can form through degradation of aromatics. If the gas accumulates, it can limit groundwater flow and retard biological processes (Reinhard et al., 2000).

Fungi degrade long-chain n-alkanes (n-nonane to n-octadecane) in preference to shorter-chain varieties (Merdinger & Merdinger, 1970; Teh & Lee, 1973). Because fungi are dependent on oxygen for growth, depletion of long-chain n-alkanes may be indicative of fungi, instead of low redox. However, Jovancicevic et al. (2003) found that an accumulation of heavier, even-numbered n-alkanes, such as n-C_{16} and n-C_{18}, may occur during biodegradation because of the presence of algae.

5.2.2 Temperature

Near-ground-surface temperatures fluctuate greatly, whereas underground temperatures remain somewhat constant. Biological alteration of spilled petroleum generally increases with temperature. Furthermore, volatilization of lighter n-alkanes at colder temperatures may decrease.

Atlas (1981) found that degradation was an order of magnitude greater at 25°C compared with 5°C, whereas Sexstone et al. (1978) found diesel contamination in Arctic soils 28 years after a spill. Ludzack & Kinhead (1956) found that motor oil rapidly oxidized at 20°C, but not at 5°C. Margesin & Schinner (2001) found that diesel degradation at a cold, high-altitude location occurred mostly during the summer and at a reduced rate. Man (1998) found that n-alkane depletion was similar regardless of temperature if the range was between 10°C and 22°C. Bonroy et al. (2007) found that heating-oil biodegradation rates in shallow soil almost doubled during the summer months compared to the winter.

Ground cover can impact the temperature of surface soils and consequently the temperature of percolating rainwater (Huang et al., 2008). Paved surfaces, such as asphalt or concrete, retain heat, whereas grass-covered or forested areas cool quicker during summer months. Increased temperature will decrease petroleum viscosity, allowing increased spreading, additional surface area in contact with groundwater, and enhanced biodegradation (Atlas & Bartha, 1992).

5.2.3 Contact with water

Many constituents of middle distillates exhibit low aqueous solubilities. Aromatics are more soluble than aliphatics of the same carbon number, whereas *cyclo*-alkanes tend to be slightly more soluble than *n*-alkanes (Bobra, 1992). Two compounds often used to represent petroleum weathering are the *n*-C_{17} alkane (*n*-heptadecane: $C_{17}H_{36}$) and pristane (2,6,10,14-tetramethylpentadecane: $C_{19}H_{40}$)(or "*n*-C_{17}/pr"). Bregnard et al. (1997) reported that pristane's aqueous solubility is less than 0.1 microgram per litre (μg/l), whereas Ritter (2003) found that solubility differences (in petroleum) between *n*-C_{17}, *n*-C_{18}, pristane and phytane are small. Middleditch et al. (1978) reported *n*-heptadecane concentrations in seawater ranging from 2 to 747 μg/l. Leahy & Colwell (1990) report that microbial degradation of long-chain *n*-alkanes ($\geq C_{12}$) occurs at rates that exceed the rates of hydrocarbon dissolution.

LaFargue & Barker (1988) found that *n*-alkanes lighter than C_{14} in crude oils were susceptible to dissolution, whereas the heavier *n*-alkanes were not. Isoprenoids heavier than C_{16} were not susceptible to dissolution, whereas the C_{13} through C_{15} isoprenoids were somewhat vulnerable.

For a given carbon number, ring formation, unsaturation, and branching cause an increase in aqueous solubility. Therefore, one could expect that when dissolution occurs, aromatics of a given carbon number would decrease first, followed by *cyclo*-alkanes, *iso*-alkanes and *n*-alkanes (Palmer, 1991).

Dissolution of hydrocarbons into groundwater or soil water may be impacted by:

* the surface area of hydrocarbons in contact with water, also known as the oil-water ratio. A higher ratio may impart greater dissolution; accordingly, geologic materials with a greater porosity may allow greater dissolution (Bobra, 1992);
* ambient groundwater chemistry and, in particular, temperature, pH and oxidation-reduction potential (ORP). The aqueous solubility of hydrocarbons often increases with temperature; however, the relationship between variables such as pH or ORP and solubility is often compound specific and possibly site-specific;
* the magnitude of precipitation and recharge. Recharge commonly increases dissolution, and
* the groundwater migration rate. Slow-moving groundwater will lessen transfer of hydrocarbons to a dissolved state, whereas the opposite occurs with rapidly migrating groundwater (Fried et al., 1979). In column experiments, Miller et al. (1990) found that the rate of mass transfer between a toluene separate phase and the aqueous phase was directly related to the groundwater migration rate.

As a result of mass transfer, dissolution and biodegradation are coupled processes because contact with water stimulates biological activity. Addition of petroleum to groundwater or soil water can allow indigenous bacteria to multiply and preferentially attack *n*-alkanes (Solevic et al.,2003). Therefore, contact with groundwater may cause dissolution of lighter *n*-alkanes and isoprenoids and induce microbial degradation of lighter and heavier *n*-alkanes and isoprenoids. Degradation can also begin inside an UST if sufficient water infiltration occurs (Gaylarde et al., 1999).

A rapidly fluctuating water table will foster emulsification and can enhance biological activity because of greater contact between the separate phase and water. Therefore, production of an emulsification can increase biodegradation rates (Atlas & Bartha, 1992).

5.2.4 Light

The rate of photochemical reactions is directly proportional to the number of photons absorbed by a chemical. Nearness to the Equator or an increase in altitude will accelerate the reactions (Sukol et al., 1988). Photodecomposition is not a significant process in the subsurface, although immediately adjacent to the ground surface, it may be important.

5.2.5 Oxygen and nutrients

Aerobic microbes need electron acceptors and nutrients to degrade petroleum. Lack of oxygen and nutrients may limit biological activity. Even though anaerobic microbes exist, anaerobic degradation is normally slower. For example, Bonin & Betrand (2000) found lowering oxygen contents could stop n-heptadecane mineralization. Numerous researchers found that oxygen availability is the most important factor in petroleum degradation (Raymond et al., 1976; Song et al., 1990). Factors affecting oxygen availability in soil include (Atlas & Bartha, 1992):

- *Drainage*: in water-logged soils, oxygen diffusion can be slow and bacterial movement restricted;
- *Soil texture*: coarse-grained soils have higher permeabilities and oxygen can be quickly replenished. Furthermore, coarser textures allow greater contact area between water and petroleum, increasing dissolution. However, for reasons stated earlier, medium-grained soils may exhibit the most biodegradation potential;
- *Proximity to the ground surface*: in laboratory column experiments, degradation was 3 to 5 times greater at the top versus the base (Atlas, 1981). This observation is related to proximity to greater oxygen abundance, temperature and recharge. Biological degradation can vary significantly over short distances in the horizontal and vertical directions. Variations will be dependent on nutrient and oxygen content and microbial diversity of geologic layers (Maila et al., 2005), and
- *Quantity of hydrocarbons*: Areas saturated with hydrocarbons may exhaust oxygen faster than it can be resupplied. Oxidation of 1 litre (L) of hydrocarbons can exhaust the dissolved oxygen in close to 400,000 L of water (Atlas & Bartha, 1992). Furthermore, large quantities of separate phase may decrease soil permeability with respect to water.

5.2.6 Bacteriocides

For biodegradation to occur, toxic concentrations of bacteriocides must not exist. Bacteriocides are elements or compounds toxic to bacteria. For example, H_2S may be toxic to some microbes (Prince & Walters, 2007). Under sulfate-reducing conditions, H_2S may form through biodegradation of aromatics.

5.3 Soil composition: Chemistry, lithology and texture

Coarser-grained soils permit freer movement of liquids such as soil gas, soil water and groundwater, allowing replenishment of oxygen, nutrients and microbes. Pore diameters of less than 3 micrometres are an obstacle to bacteria, thereby limiting biodegradation (Aichberger et al., 2006). Zibiske & Risser (1986) found that medium-grained soil might have the most biodegradation potential: a combination of sufficient permeability and soil-surface area is the cause for increased biological activity. Increased surface area allows attachment of a greater number of microbes.

One cause for the persistence of spilled petroleum in the subsurface is a concept known as burial (Owens et al., 2008). If petroleum migrates into an enclosed area, for example, a sand layer sandwiched between clay, replenishment of nutrients and oxygen may be limited and petroleum could last for many years or decades.

5.3.1 Soil chemistry

The chemical composition of soil will impact conditions such as pH, redox and cation/anion exchange capacities (McVay et al., 2004). For example, soil derived from or overlying carbonate-type rocks will tend to exhibit higher pH values, whereas sandier soil (derived from sandstones, quartzites, etc.) will be less buffered and impacted more readily by acid rain. Higher organic carbon content tends to induce more biological activity in the soil. The organic carbon content commonly lessens in older soil and is often high in glacial sediments (Jobbágy & Jackson, 2000).

5.3.2 Soil moisture

Soils lacking moisture normally exhibit decreased biodegradation rates. The lack of moisture prevents influx of oxygen and nutrients and reduces contact between microbes and spilled petroleum. Waterlogged soils may retard biological processes. Laboratory studies performed by Schroll et al. (2006) showed a linear relationship between soil moisture and pesticide biodegradation. Bekins et al. (2005) reported on a crude-oil release where the shallowest soil samples exhibited the least petroleum degradation. The lack of degradation was attributed to reduced moisture within the shallow soil.

5.4 Petroleum chemistry

The chemical composition of petroleum products can influence weathering rates. Distillates derived from certain crudes can weather at varying rates, despite similar compositions (Atlas, 1981). Eganhouse et al. (1996) reports that certain petroleum constituents may inhibit degradation of others. For example, degradation rates of heavier n-alkanes may increase once lighter n-alkanes are removed.

Contaminant mixtures also impact biodegradation. In one study, iso-alkanes degraded individually, but when introduced with other hydrocarbons, degradation proceeded slowly. This finding suggests a competition effect (Kampbell & Wilson, 1991). However, there is evidence to the contrary, suggesting that degradation for some compounds is more rapid when in a mixture (Smith, 1990).

5.5 Distance from source

Distance from the source of the release will impact petroleum weathering. Because of the effects of source-area sequestration, increased surface area, and decreased contaminant mass, peripheral portions of the middle-distillate plume often weather at a faster rate than the core area (Parsons, 2003). It is unlikely that petroleum will weather at a uniform rate across the plume (Landon & Hult, 1991).

5.6 Hydrologic conditions

In areas with fluctuating water tables, separate phase can become engulfed by groundwater, forming an emulsion and enhancing biodegradation. Bekins et al. (2005) reports that in areas of significant recharge, enhanced degradation can occur because of increased contact with nutrient-rich water. At sites exhibiting rapid groundwater migration rates, mass transfer to the aqueous phase may increase, thereby enhancing hydrocarbon degradation.

5.7 Vegetation

Nearby plants and associated microbes can metabolize petroleum and convert it to harmless byproducts through a process known as phytoremediation. Microbial populations can be 5 to 100 times greater in the vicinity of roots, an area called the rhizosphere (Frick et al., 1999; Kechavarzi et al., 2007). McPherson et al. (2007) found that diesel removal in soil can be up to 40% greater when poplar trees exist. Hence, heavily vegetated areas may increase weathering of spilled petroleum.

Increased vegetation will also increase the number and density of roots in the subsurface. Because of transpiration, increased vegetation will lessen recharge and possibly decrease petroleum dissolution.

6. Sequence of biodegradation

The *n*-alkanes and aromatics (benzene, toluene, ethylbenzene and *o, m, p*-xylenes) are commonly the first compounds to be removed through biological processes (Chapelle, 2001). The *n*-alkanes are more readily converted to long-chain fatty acids (for subsequent beta-oxidation) compared to unsaturated or branched-chain hydrocarbons.

Because it has the highest solubility, benzene is commonly the first mono-aromatic to be depleted from a middle-distillate separate phase (Kaplan et al., 1996). However, Barker et al. (1987) found benzene to be the most persistent aromatic in ground water. Depletion is then normally followed by alkyl-benzenes and alkyl-naphthalenes. Alkyl-naphthalenes appear more resistant than alkyl-benzenes. Furthermore, homologues with longer alkyl chains will be more resistant to biodegradation (Kaplan et al., 1996). For example, a C_1-naphthalene (such as 1-methylnaphthalene) is normally less resistant than a C_4-naphthalene (such as diethylnaphthalene). Alkyl-*cyclo*-hexanes are commonly more resistant than *n*-alkanes and alkyl-benzenes and may be found in the environment much later in the life of a spill. In general, compound classes in order of decreasing susceptibility to biodegradation are *n*-alkanes > *iso*-alkanes (except isoprenoids) > low-molecular-weight aromatics > *cyclo*-alkanes (Leahy & Colwell, 1990).

Kaplan et al. (1997) found that weathering of petroleum products could be divided into seven progressive stages, which we term the *Kaplan Stages*. Similar weathering stages have been presented by Philp & Lewis (1987), Peters et al. (2005), Zytner et al. (2006) and Prince & Walters (2007). The *Kaplan Stages* are depicted on Table 4. Biodegradation including and beyond Stage 5 indicates substantial alteration and normally implies residence times greater than 20 years (Kaplan, 2003; Peters et., 2005).

7. Christensen & Larsen method

Microbes preferentially digest some hydrocarbons, leaving behind a biomarker (Christensen & Larsen, 1993). A biomarker is an organic compound that can be structurally related to its precursor molecule, which occurs as a natural product in a plant, animal, bacteria, spore, fungi or petroleum (Philp & Lewis, 1987). Biomarkers are often resistant to degradation. For example, the isoprenoids: pristane, phytane, norpristane and farnesane, are resistant to microbial alteration, and their relative concentrations compared to n-alkanes, can be used as a proxy for weathering (Schaeffer et al, 1979). Therefore, ratios, such as n-C_{17} alkane to pristane (n-C_{17}/pr) or n-C_{18} alkane to phytane (n-C_{18}/ph) have been used as a measure of biodegradation. These n-alkanes and isoprenoids have similar solubilities and partitioning coefficients and the absence of n-alkanes is a result of biological activity and not transport or sorption (Bregnard et al., 1996).

Biodegradation of n-alkanes with molecular weights of up to n-C_{44} is known (Atlas, 1981). However, under aggressive conditions, isoprenoids may be susceptible to microbial oxidation; farnesane and norpristane are the most vulnerable (Pirnik et al., 1974; Pirnik, 1977; Nakajima et al., 1985).

The Christensen & Larsen (C&L) study reported a linear correlation between the n-C_{17}/pr ratio and the diesel-fuel age in soil from numerous spills where release dates were known. The n-C_{17}/pr ratio has been used as a measure of biodegradation for several decades (Atlas, 1981; Swannell et al., 1996), especially with marine spills. Christensen & Larsen (1993) report that statistical analysis of the correlation between the n-C_{17}/pr ratio and known spill ages can provide an age estimate to ± 2 years at a 95% confidence level, with some slight variability for releases <5 and >20 years old. Kaplan et al. (1996) provided an equation to calculate the C&L age where,

$$T(year) = -8.4(n\text{-}C_{17}/pr) + 19.8$$

According to Christensen & Larsen (1993), their method may be valid if several conditions are met:

- samples are collected from below an impervious cover such as asphalt or concrete;
- samples are obtained from at least 1 m below the ground surface;
- samples are acquired from at least 1 m above the water table;
- petroleum concentrations in the samples are at least 100 mg/kg, and
- the release is sudden.

Christensen & Larsen (1993) do not define a sudden release, but it can be assumed that a discharge lasting 1 year or less is implied. Most UST releases are slow and prolonged.

The C&L method dealt solely with contaminated soil samples. It did not apply to groundwater or separate-phase samples.

There has been much discussion on the validity of the C&L method (Alimi, 2002; Kaplan, 2002; Stout et al., 2002a; 2002b; Wade, 2002; Galperin & Kaplan, 2008c). Several claim that the method is invalid (Bruya, 2001; Smith et al., 2001; Shepperd & Crawford, 2003; Zemo, 2007). For example, Hostettler & Kvenvolden (2002) found weathered products (crude oils and distillates) with n-C_{17}/pr ratios in excess of 3.0. Stout & Douglas (2007) presented a case study where the C&L method failed to accurately predict the age of a known and sudden release of diesel fuel. However, several recent studies conclude that the method is viable, although with limitations; for example, more than one sample is recommended and knowledge of the original n-C_{17}/pr ratio is needed (Wade, 2001; Hurst, 2003; Hurst & Schmidt, 2005; Oudijk et al., 2006; Hurst & Schmidt, 2007; Oudijk, 2007; Hurst & Schmidt, 2008). Galperin & Kaplan (2008d) recently provided a model based on different initial n-C_{17}/pr values.

As discussed earlier, de Jonge et al. (1997) found that biodegradation rates decreased significantly when petroleum concentrations exceeded 4,000 mg/kg. Accordingly, one might argue that a window exists, only between 100 mg/kg and 4,000 mg/kg, where the C&L method might be valid.

To assess the validity of the assumption the C&L, nine samples of heating oil and motor diesel were collected from residential tanks and commercial service stations in the northeast United States in 2007. The samples were analyzed with a GC/FID to evaluate n-C_{17}/pr ratios. Furthermore, a literature review was conducted to establish n-C_{17}/pr ratios in middle distillates and crude oils (Palacas et al., 1982; Collins et al., 1994; Buruss & Ryder, 1998; Porter & Simmons, 1998; Wang et al., 2003; Chung et al., 2004; Environment Canada, 2004; Hurst & Schmidt, 2005; Blanco et al., 2006; Hwang et al., 2006; Stout et al., 2006; Røberg et al., 2007).

Christensen & Larsen (1993) claim that n-C_{17}/pr ratios for fresh diesel fuel range from around 2.0 to 2.4 (based on Figure 4 of their article). Based on 11 samples, they obtained an average n-C_{17}/pr value of 1.98 with a standard deviation (σ) of 0.83. Hurst & Schmidt (2005) conducted a search of n-C_{17}/pr ratios in fresh distillates and crude oil and found a mean value of 2.3±0.7. However, our samples revealed n-C_{17}/pr ratios ranging from only 0.95 to 1.54 with a mean of 1.15 and σ of 0.18 (Table 3). There are several potential reasons for the discrepancy between our findings and the others:

- n-C_{17}/pr ratios were previously around 2.0, but more recently lowered to the 0.95-to-1.54 range because of changes in crude-oil sources;
- lower n-C_{17}/pr ratios are an artifact of only northeast-US refineries, and
- C&L reveal a mean value of around 2.0, but data are highly variable. Assuming the cited σ value, a 95% confidence interval would be between 1.15 and 2.81.

Type	Town	State	n-C_{17}/pr	pr/ph
Heating oil	Frenchtown	New Jersey	1.25	1.87
Heating oil	North Bellemore	New York	1.10	1.69
Heating oil	Toms River	New Jersey	1.54	1.46
Diesel fuel	Morrisville	Pennsylvania	1.05	1.89
Diesel fuel	Millstone	New Jersey	1.11	1.74
Diesel fuel	North Brunswick	New Jersey	1.29	1.56
Diesel fuel	South Plainfield	New Jersey	0.95	1.62
Diesel fuel (1)	Trenton	New Jersey	1.00	1.66
Diesel fuel (2)	Trenton	New Jersey	1.04	1.85
		Average:	1.15	1.70
		Standard deviation:	0.18	

NOTES: Laboratory analyses performed by Precision Testing Labs, Inc., Toms River, New Jersey. Based on Hurst and Schmidt (2005), the origin of these heating oils and diesel fuels may be Venezuelan and Canadian crude oils, which have average n-C_{17}/pr ratios of 1.4 and 1.0, respectively. Because much of New Jersey's heating oil originates from the Hess Corporation refinery in Port Reading, New Jersey, and Hess obtains crude oil from Petroleo de Venezuela, SA (PDVSA), this conclusion seems probable. The Venezuelan crude oil is fairly immature and exhibits low n-C_{17}/pr values. Furthermore, as of 2008, much of the United States' East Coast crude oil comes from the oil sands of Alberta, Canada (Oudijk, 2009a), which also exhibit much low n-C_{17}/pr values.

Table 3. n-C_{17}/pristane (n-C_{17}/pr) and pristane/phytane (pr/ph) ratios in samples of fresh no. 2 heating oil and motor diesel fuel collected in the US states of New Jersey, Pennsylvania and New York in 2007. Source: Oudijk (2009a).

Stage	Description
1.	Abundant n-alkanes, dye still present
2.	Light-end n-alkanes removed (such as n-C_8 in diesel fuels)
3.	Middle-range n-alkanes (n-C_9 through n-C_{14} in diesel fuels), benzene, toluene removed
4.	More than 90% of n-alkanes removed
5.	Alkyl-$cyclo$-hexanes & a kyl-benzenes removed
6.	Isoprenoids, C_1-naphthalenes, benzothiophenes and alkyl-benzo-thiophenes removed, C_2-naphthalenes selectively reduced
7.	Phenanthrenes, dibenzothiophenes and other PAHs reduced

NOTE: In many North American localities, a dye is placed into no. 2 heating oil to distinguish it from motor diesel fuel for taxing purposes. Accordingly, fresh motor diesel fuel would not have this dye.

Table 4. Stages of biodegradation of no. 2 heating oil or motor diesel fuel, known as the Kaplan Stages. Based in part on Kaplan et al. (1997) and Peters et al. (2005).

Our literature review showed that n-C_{17}/pr ratios for crude oil worldwide range from <1.0 to about 7.0. The n-C_{17}/pr ratio in diesel fuel or heating oil would not be significantly different from its crude source, although Stout & Wang (2007) report that if the fuel is blended with cracked components during refining, n-C_{17}/pr ratios may be altered.

Based on the crude-oil data and our findings, C&L ages for today's fresh diesel fuel are unreliable. Therefore, it is unlikely that n-C_{17}/pr ratios can presently assist in age-dating studies, especially if litigation ensues. Because original n-C_{17}/pr ratios have changed, the C&L method may no longer be appropriate for age dating, at a minimum in North America, and a new method is needed.

8. Age-dating methodology

Significant laboratory studies and/or field investigations have not been performed to determine specific weathering rates of spilled middle distillates. Furthermore, Chapelle & Lovely (1990) report that laboratory studies tend to overestimate biodegradation rates. Field studies with known spill time frames are not plentiful. Therefore, specific data on subsurface weathering rates are generally not available. To obtain such data may be an extremely cumbersome endeavor because of the numerous variables involved. Studies of this type would need to address all the different geological, hydrological and biological conditions, which are numerous.

Previous age-dating methods for spilled middle distillates have been based, for the most part, on the chemistry of the petroleum. These methods have, in general, used weathering or biodegradation rates as a proxy for age. Because weathering at and within each spill site could be different, such a method can be problematic. Cherry et al. (1984) found that "Because the proportion of each [microbial] species present at any point in space and time is environmentally dependent, predictions of actual organic transformation pathways and rates are all but impossible (p. 57)". In their study of a crude-oil spill, Bekins et al. (2005) concluded that ". . . techniques for dating the time of a spill on the basis of the degree of degradation may yield very different results. . . (p. 140)". Accordingly, the use of only degradation rates for age dating is not sound and a technique is needed that considers many parameters, such as weathering, geology, site history and the numerous site-specific environmental factors.

Because a mix of historical and scientific data will be used for our age estimates, each with possibly a large error range, a purely quantitative method, such as the equation used by Kaplan et al. (1997) (equation 1), is not practical. For that reason, a semi-quantitative method is proposed. This technique is based on an evaluation of five major factors and 15+ sub-factors, some of which are used to select a site-specific, weathering-potential regime (Atlas, 1981; Atlas & Bartha, 1992; Providenti et al., 1993) (Tables 4 and 5).

With the technique described here, five site-specific weathering-potential regimes are proposed to describe each release site (Table 6). The regimes are: very weak, weak, moderate, aggressive and very aggressive, and they are based on site-specific environmental factors. To obtain the age-date range, the weathering regimes are compared through a matrix to the Kaplan Stages, as described in Oudijk (2009a) and Table 7.

Environmental conditions	Chemical conditions
Soil permeability & effective porosity	*n*-alkane depletion
Water-table depth/hydrologic setting	*n*-alkane distribution
Soil cover	aromatic content/depletion
Organic-matter content of soil	unresolved complex mixture (UCM)
Groundwater and soil-water	size and location
salinity & pH	Carbon range
Lithology/chemical composition of	Ratio of heavy versus light
geologic formation	hydrocarbons
Sulfide content of soil & groundwater	Depletion of:
Vegetation	*iso*-alkanes
Dissolved oxygen content of	PAHs
ground water/oxidation- reduction potential	alkyl-*cyclo*-hexanes
(aerobic v. anaerobic or oxic v. anoxic)	alkyl-benzenes
Moisture content of soil	thiophenes
Temperature	
Indigenous microbial community	
Co-presence of other contaminants	
Presence of electron acceptors	
Presence of emulsifying agents	
Distance from source	

Table 5. Examples of environmental factors impacting the weathering of middle-distillate fuels and resulting chemical responses

Regime	Examples and description
Very aggressive	Water-logged soils; areas prone to flooding; surficial or very shallow releases; moderate to high permeability and effective porosity; stray electrical currents, heavily vegetated; nutrient-rich soils and/or groundwater, moderate pH (7–8); high DO content (at or near saturation) and ORP in groundwater; lack of soil cover; high recharge rates, and history of environmental pollution (acclimated microbes).
Aggressive	High water table; highly-permeable and porous soils; lack of soil cover; shallow release; moderate pH (7–8); high oxygen content in soil and groundwater (>5 mg/L); high organic-matter content; stray electrical currents; high salinity content of groundwater; high soil sulfide content; high soil moisture; heavily vegetated, and urban environment (not pristine).
Moderate	Moderate water content in soil; moderate depth to groundwater; moderate permeability and porosity; moderate pH (5–9); moderate oxygen content in soil and groundwater (>2 mg/L); lack of stray electrical currents, and moderate vegetation.
Weak	Low moisture content in soils; water-logged soil (as per biodegradation); very high or very low pH; low oxygen content in soil and groundwater (<1 mg/L); deep water table; no stray electrical currents; low organic-matter content in soil; non-existent or sparse vegetation, and pristine environment.
Very weak	Extremely cold, hot or harsh environment; extremely high or extremely low pH; total lack of oxygen; pristine; no soil moisture, and sterilized environment (elevated bacteriocides).

Table 6. Site-specific weathering-potential regimes. Source: Oudijk (2009a)

9. Assessing petroleum weathering with chromatograms

Petroleum weathering may be assessed through collection of soil or separate-phase samples and laboratory analysis with a gas chromatograph (GC) equipped with flame-ionization (GC/FID) or mass spectrometry (GC/MS) detectors (Senn & Johnson, 1987).

To assess the magnitude of weathering in each sample, either the peak height or area for the n-alkanes and iso-alkanes (in particular, the isoprenoids) must be calculated. Calculation of the peak areas is preferred; however, peak heights are acceptable if there is a linear relationship between heights and areas (Wade, 2001). There are two methods to calculate the peak height: either directly from the base of the chromatogram, or from the base of the UCM. The UCM method is preferred (Hostettler et al., 1999).

Assessment of petroleum weathering is needed to determine into which *Kaplan Stage* a sample is placed. There are several factors to consider:

- *Compound depletion*: Specific compounds are more resistant to biodegradation and their presence or depletion can be used to assess weathering;
- *Carbon range*: No. 2 heating oil and diesel fuel are normally within a range of C_9 through C_{24}. Lighter hydrocarbons (less than C_9, but not the mono-aromatics) may be evidence of the presence or mixture with gasoline or kerosene. A heavier fraction may be evidence of increased weathering (Wang & Fingas, 1995b). In addition, heavier constituents (greater than C_{24}) may be evidence of a mixture with no. 6, lubricating or motor oils;
- *The n-alkane distribution*: The n-alkanes in middle distillates, such as diesel fuel, heating oils or kerosene, normally show an even distribution, evidenced by a bell-shaped

Weathering regime:	Very aggressive	Aggres-sive	Moderate	Weak	Very Weak
Fresh fuel	0	0	0	0	0
Kaplan Stages:					
1. Abundant n-alkanes	<0.25	0-2	0-4	0-8	0-10
2. Light n-alkanes removed, benzene & toluene removed	<0.5	2-4	4-8	8-16	10-20
3. Middle-range n-alkanes removed, ethylbenzene & xylenes removed	<1	4-6	8-12	16-24	20-30
4. More than 50% of the n−alkanes removed	<2	6-8	12-16	24-32	30-40
5. More than 90% of n-alkanes removed, alkyl-benzenes and alkyl-*cyclo*-hexanes begin to degrade	<3	8-10	16-20	32-40	40-50
6. All n-alkanes removed, alkyl-benzenes	<4	10-12	20-24	40-48	50-60
7. Isoprenoid removal significant	<5	>12	>24	>48	>60

NOTE: The age ranges cited above must be compared to site-specific information, such as underground storage tank (UST) age, UST condition and the extent of contamination, to assess their accuracy The age ranges provided in this table should be used solely as a guide. Accordingly, additional information is needed to estimate the actual age as described in the text herein. In some situations, however, these age ranges may not apply and should not be used at all. Such situations, for which an age estimate cannot be done with the method described herein, include but are not limited to the following: multiple releases as well as changes of environmental conditions since the release has occurred. Such conditions should be carefully evaluated and excluded before applying this age-dating method.

Table 7. Matrix of *Kaplan Stages* and weathering-potential regimes providing potential age ranges in years for a release of a middle-distillate fuel. Source: Oudijk (2009b).

envelope. In diesel and no. 2 heating oils, the envelope reaches a maximum at C_{14} to C_{17} (Kaplan et al., 1996). In kerosene and jet fuels, the maximum is normally between C_{10} and C_{12}. An uneven or jagged distribution is often evidence of weathering (Figure 6a through 6c);

- *Unresolved complex mixture (UCM)*: The UCM is the hump at the base of a GC/FID trace (Figures 6a through 6c) and a mixture of complex *cyclo*- and *iso*-alkanes that are unresolvable through gas chromatography (McGovern, 1999). UCMs are a typical appearance on chromatograms for crude oil and crude-oil distillates (Frysinger et al., 2003). The UCM normally increases in relative height and width as biodegradation proceeds (Wang & Fingas, 1995b). The presence of multiple UCMs is commonly evidence that more than one distillate is present, for example, a mixture of no. 2 and no. 6 heating oil, and

- *Heavy versus light n-alkanes.* Under aerobic conditions, lighter n-alkanes are normally removed quicker compared to the heavier n-alkanes (Mohantya & Mukherji, 2008). A comparison of heavy n-alkanes, such as n-C_{20} through n-C_{22}, versus lighter n-alkanes, such as n-C_8 through n-C_{10}, can demonstrate the magnitude of evaporation. Lighter n-alkanes are often more volatile (Wang & Fingas, 1995b). Experiments have shown that

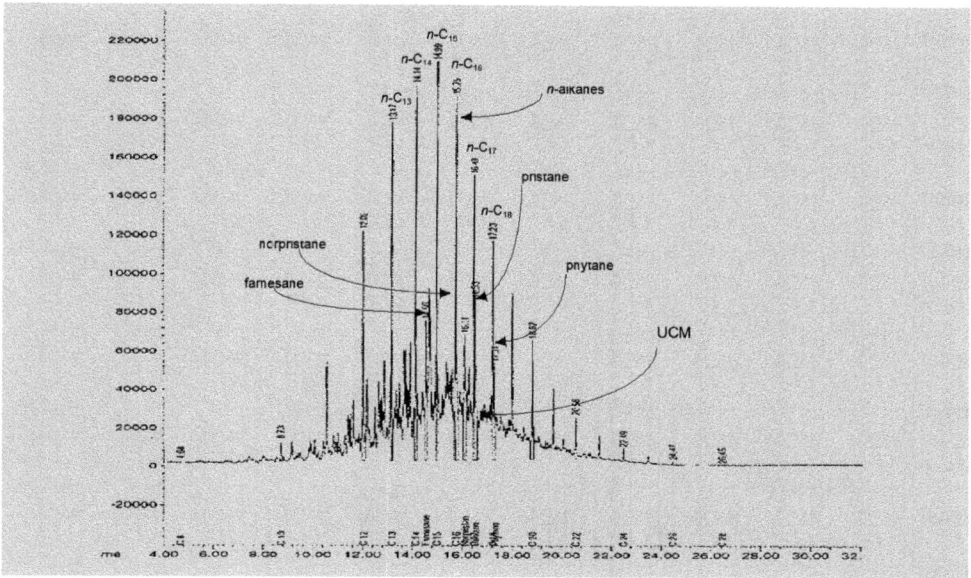

Fig. 6a. GC/FID chromatogram for a 2007 fresh motor diesel fuel from New Jersey (USA) showing the n-alkane peak envelope and the unresolved complex mixture (UCM). Source: Precision Testing Labs, Inc., Toms River, New Jersey (USA).

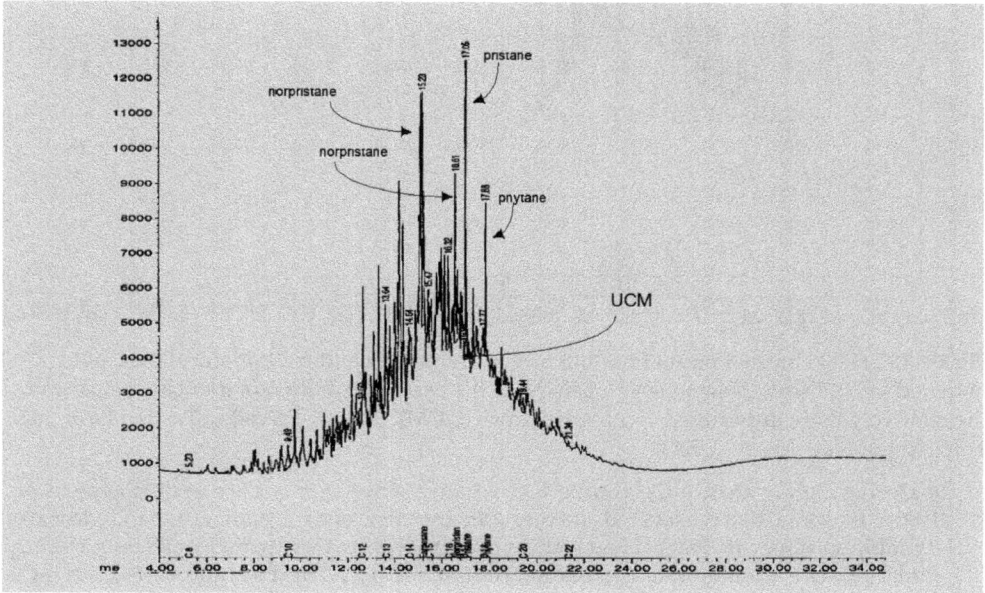

Fig. 6b. GC/FID chromatogram for a weathered motor diesel fuel obtained in 2007 from New Jersey (USA). Source: Precision Testing Labs, Inc., Toms River, New Jersey (USA).

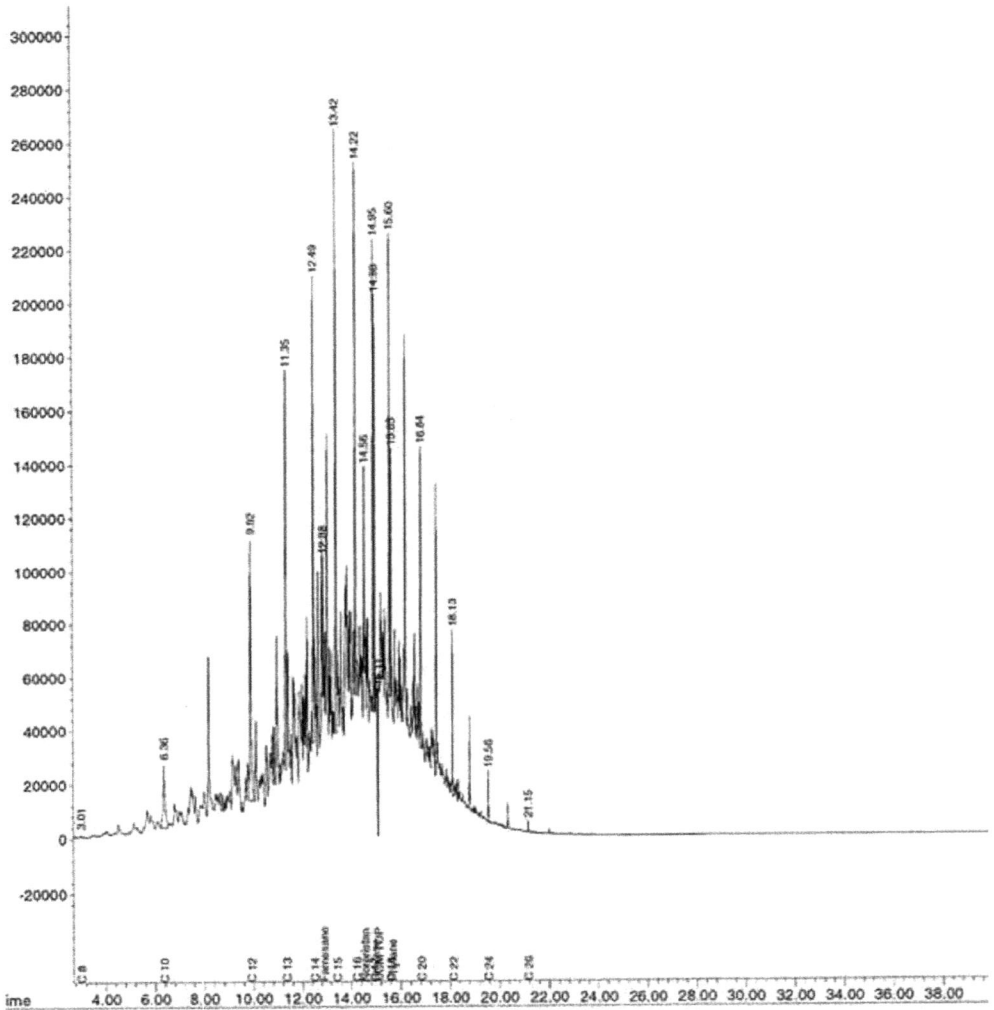

Fig. 6c. GC/FID chromatogram for a mixture of weathered and fresh motor diesel fuel obtained in 2011 from New Jersey (USA). Note the even distribution of n-alkane peaks and the relatively large unresolved complex mixture (UCM). Source: Precision Testing Labs, Inc., Toms River, New Jersey (USA).

n-alkanes lighter than n-C_{11} can be lost within 9 days in a surface spill (Payne et al., 1991). In some cases, elevated salinity can increase evaporation rates and decrease dissolution (Oyewo, 1988). However, hydrocarbons heavier than C_{14} are only slightly impacted by evaporation or dissolution (Blumer et al., 1970). The ratio of n-C_{10} to n-C_{20} (n-C_{10}/n-C_{20}) is normally 0.5 to 1.5 in an unweathered diesel fuel, whereas evaporated diesel fuel often exhibits lower n-C_{10}/n-C_{20} values. Therefore, n-C_{10}/n-C_{20} values can help to assess the magnitude of evaporation and dissolution. Furthermore, a formula, similar to the weathering index (WI*) suggested by Wang & Fingas (1995b), used to assess weathering for diesel fuels or no. 2 heating oil, is:

$$WI^* = (n\text{-}C_8 + n\text{-}C_{10} + n\text{-}C_{12} + n\text{-}C_{14})/(n\text{-}C_{16} + n\text{-}C_{18} + n\text{-}C_{20} + n\text{-}C_{22})$$

Under aerobic conditions, lower WI* values are indicative of weathering, whereas higher values are evidence of less degradation.

GC/FID traces can show evidence of a mixture of fresh and weathered middle distillates. Evidence for this phenomenon is normally an enlarged UCM overlain with an envelope of evenly distributed n-alkanes. A mixture of highly weathered and fresh product is often evidence of two (or more) releases, although it does not necessarily reveal more than one source. Furthermore, a subsurface release could be superimposed by a surficial spill, overfill or piping failure (Figure _).

The above methods can be employed to assess the weathering characteristics of middle-distillate fuels such as kerosene, the jet fuels, diesel fuels (such as motor diesel or railroad diesel), heating oils (such as no. 2, no. 4 and no. 6) and bunker oil. In some cases, such as with no. 6 oil/bunker oil, quantifying the n-alkane/isoprenoid ratios may be difficult because of low concentration of the marker compounds. Furthermore, evaluation of ratios such as $n\text{-}C_{10}/n\text{-}C_{20}$ in the heavier oils may not be possible.

10. Site-specific environmental and non-environmental factors

A proper age-dating study will include information on the following five factors and several associated sub-factors: 1. site history; 2. on-site environmental conditions; 3. extent and magnitude of known impact; 4. condition (and age) of the UST (or other types of sources), and 5. other conditions. The environmental factors can be used to select a site-specific, weathering-potential regime (Table 5).

10.1 Site history

Historical factors include: 1) first date when petroleum was observed in the environment or when problems first began, such as a malfunctioning furnace; 2) age of the UST (or other types of sources, such as aboveground storage tanks or pipelines). Quite commonly, the UST age is not known and it must be assumed that it is the same as the site, service station or residence (although not always correct), and 3) known or calculated petroleum quantity in the subsurface.

It is assumed that the petroleum age will be less than the UST age, but older than the date of its first environmental appearance. Furthermore, trouble with the furnace, because of water or lack of oil, may be a clue surrounding the onset of UST failure in a residential case. The average lifespan of an UST may be as low as 15 years, whereas the commonly used "rule-of-thumb" is 25 years. The 15- to 25-year average lifespan should be kept in mind when estimating a release age. However, we have seen USTs develop leakage within days (because of improper installation) and USTs older than 75 years in close-to-perfect condition.

A calculation of the petroleum quantity in the subsurface may be helpful, although this result is often fraught with error. The calculation may be performed by: 1) computing the petroleum quantity through soil-sampling results and separate-phase-thickness measurements in wells, or 2) comparing the amount of fuel delivered versus average usage (with the use of "degree days" to estimate fuel usage). The calculated value may then be divided by an estimated leakage rate to obtain the time frame. A minor leakage rate could be 0.01 L/day, whereas a high rate might be greater than 0.5 L/day.

10.2 Environmental conditions

Environmental factors impacting age-dating evaluations include: 1) depth to groundwater, 2) lithology and texture of geologic materials; 3) geochemical conditions of soil and groundwater, such as: a) pH; b) salinity; c) redox potential, and d) dissolved oxygen content; 4) biological conditions; 5) overlying soil cover, and 6) other factors.

10.2.1 Depth to groundwater

Petroleum in soil samples collected beneath the water table may experience increased dissolution, in particular, lighter n-alkanes. The hydrologic locality must also be considered. For example, in recharge zones, the water table may fluctuate several meters seasonally. Therefore, soil-sampling locations may have previously been within groundwater. To assess this problem, governmental agencies often have nearby observation wells with water-level data spanning decades and these data should always be consulted.

The amount of recharge is dependent on soil cover, vertical permeability and topographic location (such as within a hill or valley). Soil samples collected beneath a building may be subjected to less water contact. However, petroleum in samples collected beneath covers such as grass or bare ground may experience increased dissolution and, consequently, additional weathering.

10.2.2 Lithology and texture of geologic materials

Soil texture will impact drainage and permeability. Poor drainage and low permeability prevent oxygen and nutrient replenishment. Fine-grained soils, such as clays, more commonly exhibit anaerobic conditions. Soils exhibiting high cation-exchange capacities, such as silts or clays, or contain large organic-carbon contents, may adsorb petroleum readily. Adsorption decreases with increasing temperature and soil moisture and adsorbed petroleum is less available to microbes (Providenti et al., 1993). Hence, biological activity may be subdued and weathering minimized in fine-grained soils; however, large organic carbon contents could also induce greater microbial activity.

Inadequate soil hydration depresses microbial metabolism and movement. Furthermore, lack of moisture decreases nutrient replenishment (Providenti et al., 1993). However, waterlogged soils can also limit oxygen concentrations. Accordingly, soil-moisture extremes may decrease biological activity and weathering.

10.2.3 Geochemical and biochemical conditions of soil and groundwater

Subsurface geochemical conditions impact petroleum weathering. Extreme pH limits biological activity, whereas redox dictates if aerobes or anaerobes exist. Both microbes use middle distillates as substrates, but aerobic degradation is often quicker. The dissolved oxygen content and oxidation-reduction potential (ORP) of groundwater (if the water table is shallow) can help to identify these conditions.

10.2.4 Biological conditions

Microbes are a part of the geochemical framework of saturated and unsaturated zones. Populations may increase in response to petroleum releases and alter geochemical conditions.

Certain bacteria may increase or decrease mineralization rates of hydrocarbons. Furthermore, microbes can act as emulsifying agents, suspending petroleum in the aqueous phase and enhancing dissolution and biodegradation (Zajic et al., 1974). Additionally, vegetation impacts weathering, especially in the rhizosphere where microbial action is plentiful.

10.2.5 Overlying soil cover
The type and extent of soil cover will impact infiltration and potentially the magnitude of petroleum dissolution. This factor could have an impact on n-alkane depletion. Christensen & Larsen (1993) recommend that samples be collected at least 1 m below the ground surface. Petroleum located at shallow depths may be subject to increased volatilization or photodecomposition. Furthermore, shallow locations face greater temperature changes and increased weathering.

10.2.6 UST-related factors
In addition to affecting petroleum weathering, environmental conditions will also impact UST corrosion. These environmental factors include:
1. soil exhibiting low resistivity increases ion exchange between steel and soil minerals. Highly-corrosive soil is often composed of clay with elevated sulfide concentrations and low pH;
2. coarse-grained and/or angular-grained soil. The periodic filling of the UST can cause angular stones to puncture the UST;
3. increased recharge, such as an adjacent roof leader or a location within a low area, will increase UST contact with water, and
4. periodic or constant immersion in water, in particular acidic groundwater.

Anthropogenic conditions that can impact UST corrosion include:
- quality of the UST materials, such as the steel thickness. Based on our field observations, North American USTs installed prior to the 1970s are often constructed of a thicker gage steel and, therefore, less susceptible to corrosion. In coastal areas, where the water table is shallow, aboveground tanks are often used underground. These tanks are commonly constructed with thin-gage steel and more susceptible to corrosion;
- UST size. Larger diameter USTs will corrode at an accelerated rate, possibly because of the greater load on the steel (Holt, 1997);
- improper installation procedures, such as use of jagged backfill;
- lack of cathodic protection;
- use of dissimilar materials in UST construction, such as a mix of galvanized- and stainless-steel, and
- stray electrical currents from nearby buried power lines.

Many of these factors also apply to aboveground tanks (ASTs). Furthermore, ASTs are particularly susceptible to lightning strikes, which can cause immediate failure.

10.3 Extent and magnitude of impact
The extent and magnitude of impact can be measured by: 1) area of impacted soil and/or groundwater and the quantity of separate-phase-saturated soil; 2) vertical extent of impact, and 3) extent and thickness of separate phase on the water table.

It is assumed that a middle-distillate plume with a significant distance will also exhibit a significant age. If sufficient data are available, it may be possible to calculate the migration rate and back-calculate the time frame.

The volume of petroleum in the environment will have an impact on weathering. Samples collected within a large pool of separate phase may not exhibit any weathering many years after a release. Accordingly, samples collected within a highly contaminated location may not provide productive evidence.

10.4 UST condition
The UST condition can be assessed by: the number, size and location of corrosion holes. Corrosion often takes considerable time to develop. Metallurgists are commonly consulted to provide opinions on the UST release ages. However, soil conditions should be evaluated to determine their connectivity.

Stray electrical currents may have a significant impact on unprotected steel USTs. In particular, central air-conditioning units, which normally run on underground 220-volt currents, are often the culprits with newer buildings or homes. Unfortunately, USTs are commonly installed adjacent to these units and leakage can often initiate within 5 years.

10.5 Other considerations
There may be additional site-specific factors in addition to the five listed. For example, USTs are often abandoned in-place, possibly by a previous owner. The abandonment date might represent when a previous owner suspected leakage. Ecosystem responses may also need to be evaluated. For example, contamination may induce stressed vegetation or impacts to water bodies. The time needed to produce such impacts may be significant. Investigators need to evaluate these factors and determine if they are sufficiently important to consider in the matrix.

11. Recommended sampling and laboratory analyses

Oudijk et al. (2006) and Oudijk (2009b) provided guidelines for age-date sampling. Samples can be either impacted soil or separate phase, although soil samples are preferred. Samples as distant as possible from the source are needed to assess the maximum release age; however, a sufficient quantity of petroleum must be present to perform laboratory analyses. A hydrocarbon concentration of greater than 1,000 mg/kg is recommended.

An important decision is the sampling locations. It is assumed that locations distant from the source represent older ages. However, downgradient locations may be more susceptible to excessive weathering. Furthermore, petroleum in samples collected from within separate-phase pools may not weather as much as locations proximate or outside the pool. Accordingly, an understanding of sampling locations with respect to accumulations of separate phase is needed. As discussed by Wade (2001) and Oudijk et al. (2006), age dating based on one sample is unwise.

Field analyses of groundwater samples are needed (unless the water table is deep and inaccessible). The analyses should include pH, dissolved oxygen (DO), ORP, specific conductance, temperature and salinity. The samples should be laboratory analyzed by GC/FID. We have found that analyses for n-alkanes by GC/MS are often inaccurate. The GC/FID analyses must be conducted so that sufficient separation exists between peaks. For example, the n-C_{17} alkane and pristane elute very close to each other. To enhance peak separation, a run time of about 40 minutes is recommended. In some cases, some of the samples should be analyzed for aromatics such as benzene, toluene, ethylbenzene and o, m, p-

xylenes and base/neutral extractable compounds (B/Ns), also targeting C_1- and C_2-naphthalenes, alkyl-benzenes, alkyl-*cyclo*-hexanes and dibenzothiophenes. It is further recommended that a sample of fresh fuel be collected from each site for comparison purposes. However, it is possible that the fresh oil may be significantly different from the spilled oil.

12. Evaluating the age range

To evaluate the age, a Kaplan Stage is selected for each sample, whether soil or separate phase. Based on known environmental conditions, a weathering-potential regime is then chosen for the site. A matrix, comparing the Kaplan Stages to the weathering-potential regimes, is provided detailing potential release ages (Table 7). Compared to Kaplan et al. (1997), the Kaplan Stages on Table 7 were modified to include additional parameters. These potential age ranges are based on:

- under very aggressive conditions, n-alkanes can be completely removed in less than 5 years (Hurst, 2003). For example, in marine environments, which are very aggressive, n-alkanes can be removed in a matter of days (Colwell, 1978). In the 2002 *Prestige* tanker spill off the coast of Spain, the n-C_{17}/pr ratios in nearby sediments were cut in half after less than one year (Blanco et al., 2006). With marine spills, processes in addition to biodegradation, such as volatilization and dissolution, may cause the n-alkane depletion. Conversely, under these same aggressive conditions, PAHs may still last decades (DeLaune et al., 1990);
- under very-weak conditions, such as a Arctic environments, n-alkanes can persist in soil for decades (Sexstone et al., 1978a; Collins et al., 1994);
- high concentrations of nutrients and oxygen, indicative of aggressive environments, can allow complete middle-distillate degradation in soil within less than one decade (Bregnard et al., 1996). However, removal or lessening of oxygen and nutrients can effectively retard hydrocarbon removal (Bonin & Bertrand, 2000). Unless extreme conditions exist, such as permafrost or drought, complete or near-complete removal of hydrocarbons is normally accomplished within 20 to 30 years;
- benzene and toluene often biodegrade and dissolve quicker than ethylbenzene and *o,m,p*-xylenes (although not always)(Kaplan et al., 1996). Based on our field observations, these aromatics are removed rapidly at most sites, and under moderate weathering conditions, *iso*-alkanes, such as pristane and phytane, and the alkyl-naphthalenes are commonly the predominant compounds after about 20 years (Caredda et al., 2007).

The matrix was constructed with the following assumptions:

- the very-aggressive column represents a marine-spill situation. The n-alkanes degrade quickly in such environments. There are reports that n-alkanes may persist for two or three years (Colwell, 1978), but they will, in general, be gone within 4 years (and often much earlier). With regard to n-alkane depletion, time frames of weeks or months are more common than years (de Souza & Triguis, 2004). Accordingly, <4 years can be placed into the matrix for Stage 6 in an aggressive environment.

the very-weak column represents spills in an Arctic or Antarctic climate. The n-alkanes degrade slowly here (Sexstone et al., 1978a; Collins et al., 1994). Sexstone et al. (1978b) reported that biodegradation in tundra soil could be slow "with no major preferential utilization of classes of hydrocarbons during the period of exposure". The longer-chain n-alkanes (>C_{10}) are commonly solid at temperatures less than 10° C (Whyte et al., 1998). Kershaw & Kershaw (1986) found significant depletion at surface locations from a 35-year old spill in the Canadian Northwest Territories, but with depth, up to 80% of the oil persisted. Collins et al. (1994) found only marginal depletion of n-alkanes in subsurface soils from a 12-year-old crude-oil spill in a permafrost region of Alaska. Gore et al. (1999) and Kerry (1993) also found minimal subsurface biodegradation in a similar Antarctic environment. A very weak environment is where microbes are dormant, cannot come in contact with hydrocarbons or have been removed because of toxicity. Accordingly, >60 years can be placed into the matrix for Stage 6 in a very-weak environment;

- under moderate conditions, n-alkanes are normally removed from subsurface soils in about 20 years (Christensen & Larsen, 1993; Kaplan, 2003). Accordingly, 20 to 24 years can be placed into the matrix for Stage 6 in a moderate environment;

- degradation follows a clear sequential pattern as depicted by Kaplan et al. (1996) and Table 4. This sequential pattern is normally the case. However, there are cases where different compounds of the same class degrade at significantly different rates. For example, Olson et al. (1999) found that components within the aliphatic fraction of diesel fuel degraded at different rates, although the aliphatic fraction, as a whole, degraded quicker than the aromatic fraction;

- removal of n-alkanes tends to be linear, instead of exponential (Christensen & Larsen, 1993; Hurst & Schmidt, 2005; Galperin & Kaplan, 2008c). Therefore, age ranges are extrapolated in a linear manner from Stage 6 to Stage 1 (zero-order kinetics). Chapelle (2001) explains that there is often a lag time between introduction of a contaminant to a soil or groundwater system and acclimation of microbes. However, in the first days or weeks after release and acclimation, once that acclimation has occurred, biodegradation rates may be significant. Díez et al. (2007) found that biodegradation rates of heavy oils can be slow, even in a marine environment. Accordingly, it can be assumed that as oil ages and becomes more viscous, the rate of biological consumption decreases. Colwell (1978) found that rates could initially be logarithmic for marine spills and later linear. Bonroy et al. (2007) reports that biodegradation rates will vary seasonally. Walker et al. (1976) found that the degradation rate of the alkane fraction of a crude oil was linear, whereas the rate for aromatics varied. Ostendorf et al. (2007) found that n-alkane degradation rates in unsaturated soil followed zero-order kinetics, whereas aromatics followed first-order. Bjorklof et al. (2008) found that petroleum degradation rates can be linear, but they are mass-transfer dependent. Therefore, rates will more likely be linear in a permeable soil where the petroleum can dissolve more easily. It is assumed that rates averaged across the years are linear, although it is understood that this assumption may not apply everywhere, and

- age ranges can then be extrapolated between the very-aggressive and very-weak environments and the moderate environments.

The matrix should only be used to provide potential ages and should not be the sole factor for an age-date opinion. The matrix is a method to lead the investigator towards the correct age, but it is not the final say.

13. Critiquing the matrix age range

Once a potential age range is obtained from the matrix, it must be compared to several factors to assess its reliability. Other factors that may impact the age are:

- age of the UST;

- time frames of known or possible pollution events, such as appearance of petroleum in waterways, furnace failures, excessive oil usage or abandonments. This process may entail a review of fuel bills to determine when excessive usage began;
- state of the UST, such as the number and sizes of corrosion holes and pitting;
- extent and magnitude of petroleum-impacted soil and groundwater, and
- miscellaneous factors such as high-voltage underground electrical lines.

With the date of petroleum appearance, such as when oil began seeping into a creek, the upper and lower ages can be constrained. For example, if it is known that an UST was installed in 1980 and the furnace began to take in water in 2004, we can conclude that UST failure occurred between 1980 and 2004.

If heating oil was released, the history of the heating system should be considered. The dates of repairs and service calls should be reviewed. The fuel company is often called when a furnace fails and causes for any failures should be determined. The date of any on-site work that included excavating should also be identified. Gardening is a common cause for severing feed and return lines.

Reports should be obtained on the UST condition. If one pin-sized hole is present, it is probable that the leak began within the last 1 to 2 years. However, if 25 coin-sized holes are found, it is more likely that the leak began 10 years ago or later.

Calculations of plume extent are important in assessing the reasonableness of estimated age ranges. A large and extensive plume is more indicative of an older release. However, care is needed in making such conclusions because aggressive biodegradation can greatly reduce plume sizes. If monitoring wells are available, groundwater migration rates may be calculated and time frames for plume movement assessed and compared to geochemical data.

Lastly, underground power lines should be assessed. Residential USTs are often adjacent to air-conditioning units and associated power lines can quickly instigate leakage. At a site in New Jersey, a power line produced a sequence of holes on the top and bottom of an underlying UST. The power line had only existed for 2 years.

Table 7 provides an age range, for example, of 4 to 8 years. This range could be interpreted as 6 years ±2 years. It is stressed that the values on Table 7 are only potential ranges and might contain significant error, if they are not verified through alternative means. Each release is different and on-site petroleum weathering may not necessarily fall within the confines of this technique.

An additional geochemical method to critique the age-date estimate is to establish the sulphur content of the oil. In many countries, the government limits sulphur contents in on-road diesel fuel (Table 8). By determining the sulphur content, the age of the diesel fuel can often be constrained.

14. Range of error

Error is the difference between an observed or calculated value and the true value. Usually, we do not know the true value, but can approximate error ranges from earlier experiments or theoretical predictions (Bevington & Robinson, 1992). The purpose of calculating error ranges is to evaluate the confidence of our results.

All geochemical data collected to assess some type of natural phenomenon contain error caused by sampling and analysis (Miesch, 1967). Subsurface modeling is plagued by uncertainties that are both epistemic (reducible through observations) and aleatory (irreducible because of inherent stochasticity) (Srinivasan et al., 2007). The amount of error

	Maximum sulphur concentration allowable (mg/l)	Regulated as of
Australia	10	2009
Canada	15	1997
	500	1994
China	2,000	2002
European Union	50	2005
New Zealand	10	2009
Singapore	50	2005
Taiwan	50	2007
USA	15	2007[a]
	500	1993

[a] Mandated year was 2006 in California and 2010 in Alaska.

Table 8. Maximum allowable sulphur concentrations allowable in motor diesel fuel in selected countries

often cannot be found in textbooks or with a formula. It is commonly based on a 'rule-of-thumb' or experience. Nevertheless, some calculations can be performed to provide insight into potential error ranges.

To assess error ranges in our calculations, we can use the C&L method as an example. C&L uses the $n-C_{17}$/pristane ratio as its basis and, to calculate the ratio, an analysis with a gas chromatograph is needed. This analysis generally has a precision of about 5%. Furthermore, the method assumes that $n-C_{17}$/pristane ratios for unweathered oil range from 2.0 to 2.2. Hence, the starting point has an error of about 10%. Therefore, errors are summed and the resulting error is about 15%. This exercise assumes that the C&L method perfectly reflects weathering processes in the subsurface. This, we know to be untrue. Consequently, we can assume that the method's error is at least 20% or ±2 years

In their article, Christensen & Larsen (1993) cite an error range of ±1 year or about 10%. Hurst & Schmidt (2005), employing a similar method, cite a minimum error of 15% or ±1.5 years. These cited error ranges seem optimistic and there are several references in the scientific literature arguing that the range is greater (Stout et al., 2002; Oudijk, 2007).

Because the age-dating method described herein is semi-quantitative, it may be difficult to assign an error range. Furthermore, there have not yet been experiments to test these techniques and evaluate what these error ranges may be. Nevertheless, one should assume that the range, under the best of scenarios, may be ±2 years and, under many circumstances, it may only provide a 5-year age window. In other circumstances, the method may only be able to constrain the age, wherein an example of an opinion might be 'less than 5 years' or 'greater than 10 years.'

One important limitation to this method is the presence of multiple, overlapping releases. It may be possible to distinguish the releases, but applying an age to each may be exceedingly difficult.

The limitations of the methods described herein need to be fully understood by its users and should be explained whenever an age-dating opinion is provided. As stated by Morgan & Bull (2007, p. 56),

"Fundamental differences exist between forensic geoscience and its sister disciplines, fundamental enough to make the unwary geoscientist succumb to philosophical and practical pitfalls which will not only endanger the outline of their report, but may well indeed provide false-negative or false-positive results leading to contrary or inaccurate conclusions. In the law, such outcomes have devastating and untenable consequences".

The need to apply age-dating methods cautiously cannot be understated. It must be noted that the methods described herein may not be applicable to all spill sites. It is certainly possibly that site-specific circumstances could exist to preclude these methods from consideration. The investigator will need to carefully consider each site individually and the possibility exists that other methods, such as tree-ring studies, isotopic analyses or groundwater migration rate calculations, may be more applicable or financially viable.

15. Conclusions

Because of costs associated with cleanups, many cases come to litigation. Age dating is often central to the litigation. For a plaintiff or defendant to be successful, a legally defensible age-dating method is needed. The C&L method, which is solely dependent on chemistry, has been strongly attacked inside and outside the courtroom. Hence, a revised method is needed and proposed here.

Each pollution site has specific characteristics that must be evaluated. Investigators need a thorough understanding of the geologic and hydrologic conditions, in addition to the nature of the release. Investigators also need to combine knowledge of chemistry, microbiology, and site-specific history to provide an opinion on the release time frame. Because scientific processes associated with releases are too complex to model through a simple formula, a qualitative or semi-quantitative technique is needed. The methods described herein are an effort to develop such an approach.

16. References

Aichberger, H., Loibner, A. P., Celis, R., Braun, R., Ottner, F. & Rost, H. 2006. Assessment of factors governing biodegradability of PAHs in three soils aged under field conditions. *Soil Sediment Contam.* 15:73–85.

Alimi, H. 2002. Invited commentary of the Christensen and Larsen technique. *Environ. Forensics* 3:5.

Atlas, R. M. 1981. Microbial degradation of petroleum hydrocarbons: An environmental perspective. *Microbiol. Rev.* 45:180–209.

Atlas, R. M. & Bartha, R. 1992. Hydrocarbon biodegradation and oil spill bioremediation. *Adv. Microbial Ecol.* 12:287–315.

Balouet, J-C., Oudijk, G., Smith, K. T., Petrisor, I., Grudd, H. & Stocklassa, B. 2007. Applied dendroecology and environmental forensics. Characterizing and age dating environmental releases: Fundamentals and case studies. *Environ. Forensics* 8:1–17.

Barker, J. F., Patrick, G. C. & Major, D. 1987. Natural attenuation of aromatic hydrocarbons in a shallow sand aquifer. *Ground Water Monitor Remed.* 7:64–71.

Bekins, B. A., Hostettler, F. D., Herkelrath, W. N., Delin, G. N., Warren, E. & Essaid, H. I. 2005. Progression of methanogenic degradation of crude oil in the subsurface. *Environ. Geosc.* 12:139–152.

Bennet, S. M. 1997. Groundwater contamination from leaking home heating oil systems. *J. Environ. Hydrology* 5:1–6.

Bevington, P. R. & Robinson, D. K. 1992. *Data Reduction and Error Analysis for the Physical Sciences, 2nd edition*: New York, NY: McGraw Hill, Inc.

Bjorklöf, K., Salminen, J., Sainio, P. & Jørgensen, K. 2008. Degradation rates of aged petroleum hydrocarbons are likely to be mass transfer dependent in the field. *Environ. Geochem. Health* 30:101–107.

Blanco, C. G., Prego, R., Azpiroz, M. D. G. & Fernandez-Dominguez, I. 2006. Characterization of hydrocarbons in sediments from Laxe Ria and their relationship with the Prestige oil spill (NW Iberian Peninsula): *Ciencias Marina* 32:429–437.

Blumer, M., Sass, J., Souza, G., Sanders, H. Grassle, F. & Hampson, G. 1970. *The West Falmouth Oil Spill: Persistence of the Pollution Eight Months After the Accident*. Woods Hole Oceanographic Institution Reference No. 70-44.

Bobra, M. 1992. Solubility Behaviour of Petroleum Oils in Water. Ottawa, ON: Environment Canada, Environmental Protection Directorate.

Bonin, P. & Betrand, J. C. 2000. Influence of oxygen supply on heptadecane mineralization by Pseudomonas nautical. *Chemosphere* 41:1321– 1326.

Bonroy, J., van de Steene, J., van de Velde, R., van Eetvelde, G., Verplancke, H., Verfaillie, F. & Boucneau, G. 2007. Monitored natural attenuation of domestic fuel oil in the unsaturated zone. In *In Situ and On-Site Bioremediation – 2007*. Proceedings of the Ninth International In Situ and On-Site Bioremediation Symposium, eds., Gavaskar, A. R. & Silver, C. F. Columbus, OH: Battelle Press.

Bouchard,D., Hunkeler, D. & Hohener, P. 2008. Carbon isotope fractionation during aerobic biodegradation of n-alkanes and aromatic compounds in unsaturated sand. *Org. Geochem.* 39:23–33.

Bowden, J. N., Westbrook, S. R. & LePera, M. E. 1988. A survey of JP-8 and JP-5 properties. U. S. Army Belvoir Research, Development and Engineering Center, Contract No. DAAK70-87-C-0043.

Bregnard, T. P-A., Hohener, P., Haner, A. & Zeyer, J. 1996. Degradation of weathered diesel fuel by microorganisms from a contaminated aquifer in aerobic and anaerobic microcosms. *Environ. Toxicol. Chem.* 15:299–307.

Bregnard, T. P-A., Häner, A., Höhener, P., & Zeyer, J. 1997. Anaerobic degradation of pristane in nitrate-reducing microcosms and enrichment cultures. *Appl. Environ. Microbiol.* 63:2077–2081.

Bruce, L. G. & Schmidt, G. W. 1994. Hydrocarbon fingerprinting for application in forensic geology: Review with case studies. *AAPG Bulletin* 78:1692–1710.

Bruya, J. 2001. Chemical fingerprinting. In *Practical Environmental Forensics*, eds. J. P. Sullivan, F. J. Agardy & R. K. Traub, New York, NY: John Wiley & Sons.

Buruss, R. C. & Ryder, R. T. 1998. Composition of crude oil and natural gas produced from 10 wells in the Lower Silurian "Clinton" sands, Trumbull County, Ohio. U.S. Geological Survey Open-File Report 98–799.

Caldwell, M. E., Garrett, R. M., Prince, R. C. & Suflita, J. M. 1998. Anaerobic biodegradation of long-chain n-alkanes under sulfate-reducing conditions. *Environ. Sci. Technol.* 32:2191–2195.

Caredda, P., Pintus, M., Ruggeri, C., Viola, A., Tamburni, E., Suardi, E., Franzetti, A. and Bestetti, G. 2007. Laboratory tests and bioremediation of a site chronically contaminated by diesel. In *In Situ and On-Site Bioremediation – 2007*. Proceedings of the Ninth International In Situ and On-Site Bioremediation Symposium, eds., Gavaskar, A. R., and Silver, C. F. Columbus, OH: Battelle Press.

Cerniglia, C. E. 1984. Microbial transformation of aromatic hydrocarbons. In *Petroleum Microbiology*, ed. R. M. Atlas. New York, NY: MacMillan Publishing Company.

Chapelle, F. H. 2001. *Ground-water microbiology and geochemistry, Second Edition*. New York: John Wiley & Sons.

Chapelle, F. H. & Lovely, D. R. 1990. Rates of microbial metabolism in deep coastal plain aquifers. *Appl. Environ. Microbiol.* 56:1865–1874.

Cherry, J. A., Gillham, R. W. & Barker, J. F. 1984. Contaminants in groundwater: Chemical processes. In *Groundwater Contamination (Studies in Geophysics)*. Washington, DC: National Academy Press, 46– 64.

Christensen, L. B. & Larsen, T. H. 1993. Method for determining the age of diesel spills in the soil. *Ground Water Monitor Remed.* 13: 142– 149.

Chung, J. W., Lee, W. S., Yoon, J. Y. & Kim, H. G. 2004. The study for identification of waterborne spilled oil by fast gas chromatography [in Korean]. *J. Korean Soc. Marine Environ. Eng.* 7:122–130.

Collins, C. M., Racine, C. H. & Walsh, M. E. 1994. The physical, chemical, and biological effects of crude oil spills after 15 years on a Black Spruce forest, interior Alaska. *Arctic* 47:164–175.

Colwell, R. R. 1978. Toxic effects of pollutants on microorganisms. In *Principles of Ecotoxicology*, ed. G. C. Butler, New York, NY: John Wiley & Sons.

CONCAWE. 1995. Kerosines/jet fuels . CONCAWE product dossier no. 94/106, Brussels.

CONCAWE. 1998. Heavy fuel oils. CONCAWE product dossier no. 98/109, Brussels.

deCourcy Hinds, M. 1979. Homeowners buying larger oil tanks. *New York Times*, July 5. http://www.nytimes.com.

Dashti, N., Al-Awadhi, H., Khanafer, M., Abdelghany, S. & Radwan, S. 2008. Potential of hexadecane-utilizing soil-microorganisms for growth on hexadecanol, hexadecanal and hexadecanoic acid as sole sources of carbon and energy. *Chemosphere* 70:475–479.

Davidova, I. A., Gieg, L. M., Nanny, M., Kropp, K. G. & Sulfita, J. M. 2005. Stable isotopic studies of n-alkane metabolism by a sulfate-reducing bacterial enrichment culture. *Appl. Environ. Microbiol.* 71:8174–8182.

DeLaune, R. D., Gambrell, R. P., Pardue, J. H. & Patrick, W. H. 1990. Fate of petroleum hydrocarbons and toxic organics in Louisiana coastal environments. *Estuaries* 13:72–80.

Diaz, M. P., Boyd, K. G., Grigson, S. J. W. & Burgess, J. G. 2002. Biodegradation of crude oil across a wide range of salinities by an extremely halotolerant bacterial consortium MPD-M, immobilized onto polypropylene fibers. *Biotech. Bioeng.* 79:145– 153.

Diez, S., Jover, E., Bayona, J. M. & Albaigés, J. 2007. Prestige oil spill. III. Fate of a heavy oil in the marine environment. *Environ. Sci. Technol.* 41:3075–3082.

Douglas, G. S., Bence, A. E., Prince, R. C., McMillen, S. J. & Butler. E. L. 1996. Environmental stability of selected petroleum hydrocarbons source and weathering ratios. *Environ. Sci. Technol.* 30:2332– 2339.

Douglas, G. S., Stout, S. A., Uhler, A. D., McCarthy, K. J. & Emsbo-Mattingly, S. D. 2007. Advantages of quantitative chemical fingerprinting in oil spill source identification. In *Oil Spill Environmental Forensics*, eds. Z. Wang & S. A. Stout, San Diego, CA: Academic Press, 267– 299.

Douglas, G. S., Ziegler, S., Pinzone, C., Hardenstine, J. & McCarthy, K. 2004. The application of the Federal On Road Diesel Fuel Sulfur Reduction Act of 1993 to the age dating of diesel fuels: A case study (Abs.). In *The Annual International Conference on Soil, Sediment & Water*. Amherst, MA: Association of Environmental Health and Soils.

Eganhouse, R. P., Dorsey, T. F., Phinney, C. S. & Westcott, A. M. 1996. Processes affecting the fate of monoaromatic hydrocarbons in an aquifer contaminated by crude oil. *Environ. Sci. Technol.* 30:3304–3312.

Ehrenreich, P., Behrends, A. & Harder, J. 2000. Anaerobic oxidation of alkanes by newly isolated denitrifying bacteria. *Arch. Microbiol.* 173:58–64.

Environment Canada. 2004. Oil composition and properties for oil spill modeling. USEPA 3D-6152-NAFX.

Frick, C. M., Farrell, R. E. & Germida, J. J. 1999. Assessment of phytoremediation as an in-situ technique for cleaning oil-contaminated sites: Report prepared for the Petroleum Technology Alliance of Canada by the Department of Soil Science, University of Saskatchewan, Saskatoon, SK. Saskatoon, SK: University of Saskatchewan.

Fried, J. J., Muntzer, P. & Zilliox, L. 1979. Ground-water pollution by transfer of oil hydrocarbons: *Ground Water* 17:586–594.

Frysinger, G. S., Gaines, R. B., Xu, L. & Reddy, C. M. 2003. Resolving the unresolved complex mixture in petroleum-contaminated sediments. *Environ. Sci. Technol.* 37:1653–1662.

Galperin, Y. & Camp, H. 2002. Petroleum product identification in environmental samples: Distribution patterns in fuel-specific homologous series. *Soil Sediment Contam.* April/May: 27–29.

Galperin, Y. & Kaplan, I. R. 2008a. Zero-order kinetics model for the Christensen-Larsen method for fugitive fuel age estimates. *Ground Water Monit. Remed.* 28 (2): 94-97.

Galperin, Y. & Kaplan, I. R. 2008b. Forensic environmental geochemistry in dispute resolution— Case history 2: Differentiating sources of diesel fuel in a plume at a fueling station. *Environ. Forensics* 9:55–62.

Galperin, Y. & Kaplan, I. R. 2008c. Comments on the reported unusual progression of petroleum hydrocarbon distribution patterns during environmental weathering: *Environ. Forensics* 9:117–120.

Galperin, Y. & Kaplan, I. R. 2008d. Age significance of nC_{17}/pr ratios in forensic investigations of refined product and crude oil releases: Discussion. *Environ. Geosc.* 1585–86.

Galperin, Y. & Kaplan, I. R. 2008e. Zero-order kinetics model for the Christensen-Larsen method for fugitive fuel age estimates. *Ground Water Monitoring Rev.* 28:94–97.

Garrett, R. M., Rothernburger, S. J. & Prince, R. C. 2003. Biodegradation of fuel oil under laboratory and Arctic marine conditions. *Spill Sci. Technol. Bull.* 8 (3): 297-302.

Gaylarde, C. C., Bento, F. M. & Kelly, J. 1999. Microbial contamination of stored hydrocarbon fuels and its control. *Revista Microbiol.* 30:1–10.

Gore, D. B., Revill, A. T. & Guille, D. 1999. Petroleum hydrocarbons ten years after spillage at a helipad in Bunger Hills, East Antarctica. *Antarctic Science* 11:427–429.

Hankey-Masui, K. M. 1998. Bioremediation of Nova Scotia tills contaminated with weathered heating oil. Master's thesis, Dalhousie University, Halifax, Nova Scotia, Canada.

Harmsen, J., Rulkens, W. & Eijsackers, H. 2005. Bioavailability: concept for understanding or tool for predicting? *Land Contam. Reclam.* 13 (2): 161-171.

Harper, R. G. 2000. Comparisons of independent petroleum supply statistics. Petroleum Supply Monthly. Washington, DC: Energy Information Administration www.eia.doe.gov.

ten Haven, H. L., de Leeuw, J. W., Rullkotter, J. & Sinninghe Damste´, J. S. 1987. Restricted utility of the pristane/phytane ratio as a palaeoenvironmental indicator: *Nature* 330:641–643.

Holt, G. S. 1997. Risk assessment program of underground storage tank systems. In IIR Corrosion Conference. Johannesburg, South Africa: South African (SA) Oil Industry Corrosion Control Group.

Hostettler, F. D. & Kvenvolden, K. A. 2002. Alkylcyclohexanes in environmental geochemistry. *Environ. Forensics* 3:293–301.

Hostettler, F. D., Pereira, W. E., Kvenvolden, K. A., van Geen, A., Luoma, S. N., Fuller, C. C. & Anima, R. 1999. A record of hydrocarbon input to San Francisco Bay as traced by biomarker profiles in surface sediment and sediment cores. *Marine Chem.* 64: 115-127.

Hostettler, F. D., Bekins, B. A., Rostad, C. E. & Herkelrath, W. N. 2008. Response to commentary on observed methanogenic biodegradation progressions. *Environ. Forensics* 9:121–126.

Huang, L., Li, J., Zhao, D., and Zhu, J. 2008. A fieldwork study on the diurnal changes of urban microclimate in four types of ground cover and urban heat island of Nanjing, China. *Building and Environment* 43:7–17.

Hurst, R. W. 2003. Invited commentary on Dr. Isaac Kaplan's paper "Age dating of environmental organic residues". *Environ. Forensics* 4:145–152.

Hurst, R. W. & Schmidt, G. W. 2005. Age significance of nC17/pr ratios in forensic investigations of refined product and crude oil releases. *Environ. Geosc.* 12:177–192.

Hurst, R. W. & Schmidt, G. W. 2007. Age significance of nC17/pr ratios in forensic investigations of refined product and crude oil releases: Reply. *Environ. Geosci.* 14:111–112.

Hurst, R. W. & Schmidt, G. W. 2008. Age significance of nC17/pr ratios in forensic investigations of refined product and crude oil releases: Reply. *Environ. Geosc.* 15:85–86.

Hwang, E-Y., Park, J-S., Kim, J-D. & Namkoong, W. 2006. Effects of aeration mode on the composting of diesel-contaminated soil. *Ind. Eng. Chem. Res.* 12:694–701.

Illich, H. A. 1983. Pristane, phytane, and lower molecular weight isoprenoid distributions in oils. *AAPG Bulletin* 67:385–393.

Jobbágy, E. G. & Jackson, R. B. 2000. The vertical distribution of soil organic carbon and its relation to climate and vegetation. *Ecol. Applications* 10 (2): 423-436.

de Jonge, H., Freijer, J. I., Verstraten, J. M., Westerveld, J. & van der Wielen, F. W. M. 1997. Relation between bioavailability and fuel oil hydrocarbon composition in contaminated soils. *Environ. Sci. Technol.* 31:771–775.

Jovancicevic, B., Polic´, P., Vrvic´, M., Scheeder, G., Teschner, M. & Wehner, H. 2003. Transformations of n-alkanes from petroleum pollutants in alluvial groundwaters. *Environ. Chem. Letters* 1:73–81.

Kampbell, D. H. & Wilson, J.T. 1991. Bioventing to treat fuel spills from underground storage tanks. *J. Haz. Materials* 28:75–80.

Kanner, A. 2007. Meeting the burden for admissibility of environmental forensic testimony. *Environ. Forensics* 8:19–23.

Kaplan, I. R. 2002. Invited commentary of the Christensen and Larsen technique. *Environ. Forensics* 3:7.

Kaplan, I. R. 2003. Age dating of environmental organic residues. *Environ. Forensics* 4:95–141.

Kaplan, I. R., Galperin, Y., Alimi, H., Lee, R-P. & Lu, S-T. 1996. Patterns of chemical changes during environmental alteration of hydrocarbon fuels. *Ground Water Monitoring Rev.* 16:113–125.

Kaplan, I. R., Galperin, Y., Lu, S-H. & Lee, R. P. 1997. Forensic environmental geochemistry: Differentiation of fuel-types, their sources and release time. *Org. Geochem.* 27:289–317.

Kechavarzi, C., Petterson, K., Leeds-Harrison, P., Ritchie, L. & Ledin, S. 2007. Root establishment of perennial ryegrass (L. perenne) in diesel contaminated subsurface soil layers. *Environ. Pollution* 145: 68–74.

Kehoe, R. A. 1960. Impacts of pollution on health. In *Proceedings of the National Conference on Water Pollution.* Washington, DC: US Department of Health, Education and Welfare, Public Health Service, 60–84.

Kerry, E. 1993. Bioremediation of experimental petroleum spills on mineral soils in the Vestfold Hills, Antarctica. *Polar Biology* 13:163–170.

Kershaw, G. P. & Kershaw, L. J. 1986. Ecological characteristics of 35-year-old crude-oil spills in tundra plant communities of the Mackenzie Mountains, N.W.T. *Can. J. Botany* 64:2935– 2947.

Klug, M. J. & Markovetz, A. J. 1967. Degradation of hydrocarbons by members of the Genus Candida. *J. Bacteriology* 93:1847–1852.

Kramer, W. H. & Hayes, T. J. 1987. Water soluble phase of number 2 fuel oil: Results of a laboratory mixing experiment. New Jersey Geological Survey Technical Memorandum 87:3.

LaFargue, E. & Barker, C. 1988. Effect of water washing on crude oil compositions. *AAPG Bulletin* 72:263–276.

Landon, M. K. & Hult, M.F. 1991. Evolution of physical properties and composition of a crude oil spill, in Mallard, G. E. & Aronson, D.A. (Eds.) *U.S. Geological Survey Toxic*

Substances Hydrology Program. Proceedings of the Technical Meeting, Monterey, California. U.S. Geological Survey Water Resources Investigations Report 91-4034, 641-645.

Lapinskiene, A., Martinkus,P. & Rebzdaite,V. 2005. Eco-toxicological studies of diesel and biodiesel fuels in aerated soils. *Environ. Pollution* 142:432–437.

Leahy, J. G. & Colwell, R. R. 1990. Microbial degradation of hydrocarbons in the environment. *Microbiol. Rev.* 54:305–315.

Lee, Y-K., Ryu, H-W., Lee, J-Y., Han, J-G., Chang, K-H. & Lee, J-H. 2007. The occurrence time estimation of oil contaminated source considering hydrogeological features. *Geosc. J.* 11:95–103.

Ludzack, F. L. & Kinkead, D. 1956. Persistence of oily wastes in polluted water under aerobic conditions. *Ind. Eng. Chem. Res.* 48:263–267.

McGovern, E. 1999. Bioremediation of Unresolved Complex Mixtures in Marine Oil Spills: Final Report. Dublin, Ireland: Marine Institute Fisheries Research Centre:. EU Project B$-3040/95/928/jnb/C4.

McMahon, P. B. & Chapelle, F. H. 2008. Redox processes and water quality of selected principal aquifer systems. *Ground Water* 46:259–271.

McPherson, A., Fleming, I., Farrell, R. & Headley, J. 2007. Monitoring phytoremediation of petroleum hydrocarbon-contaminated soils in a closed and controlled environment. In *In Situ and On-Site Bioremediation – 2007. Proceedings of the Ninth International In Situ and On-Site Bioremediation Symposium*, eds., Gavaskar, A. R., and Silver, C. F. Columbus, OH: Battelle Press.

Maila, M. P., Randima, P., Surridge, K., Drønen, K. & Cloete, T. E. 2005. Evaluation of microbial diversity of different soil layers at a contaminated diesel site. *Int. Biodeterior. Biodegrad.* 55:39–44.

Man, A. G. 1998. Temperature effect on bioremediation of diesel fuel contaminated Winnipeg Clay. Master's thesis, University of Manitoba, Winnipeg, Manitoba, Canada.

Margesin, R. & Schinner, F. 2001. Bioremediation (natural attenuation and biostimulation) of diesel-oil-contaminated soil in an alpine glacier skiing area. *Appl. Environ. Microbiol.* 67:3127–3133.

Markovetz, A. J., Cazon, J. & Allen, J. E. 1968. Assimilation of alkanes and alkenes by fungi. *Appl. Microbiol.* 16:487–489.

McCaskill, W. D. 1999. A report from Maine on the trials and tribulations of leaking aboveground home heating oil tanks. *LUSTLine Bull.* 33:20–23.

McVay, K. A., Radcliffe, D. E., West, L. T. & Cabrera, M. L. 2004. Anion exchange in saprolite. *Vadose Zone J.* 3:668–675

Merdinger, E. & Merdinger, R. P. 1970. Utilization of n-alkanes by Pullularia pullulans. *Appl. Microbiol.* 20:651–652.

Middleditch, B. S., Basile, B. & Chang, E. S. 1978. Discharge of alkanes during offshore oil production in the Buccaneer Oilfield. *Bull. Environ. Contam. Toxicol.* 20:59–65.

Miesch, A. T. 1967. Theory of error in geochemical data: U. S. Geological Survey Professional Paper 574–A.

Miller, C. T., Poirier-McNeill, M. M. & Mayer, A. S. 1990. Dissolution of trapped nonaqueous phase liquids: Mass transfer characteristics. *Water Resour. Res.* 26(11): 2783-2796.

Mohantya, G. & Mukherji, S. 2008. Biodegradation rate of diesel range *n*-alkanes by bacterial cultures *Exiguobacterium aurantiacum* and *Burkholderia cepacia. Int. Biodet. Biodegrad.* 61 (3): 240-250.

Morgan, R. M. & Bull, P. A. 2007. The philosophy, nature and practice of forensic sediment analysis. *Prog. Phys. Geog.* 31: 43–58.

Morrison, R. D. 2000. Application of forensic techniques for age dating and source identification in environmental litigation. *Environ. Forensics* 1:131–153.

Nakajima, K., Sato, A., Takahara, Y. & Iida, T. 1985. Microbial oxidation of isoprenoid alkanes, phytane, norpristane and farnesane. *Agri. Biol. Chem.* 49:1993–2002.

Olsen, J. J., Mills, G. L., Herbert, B. E., and Morris, P. J. 1999. Biodegradation rates of separated diesel components. *Environ. Toxicol. Chem.* 18:2448–2453.

Ostendorf, D. W., Schoenberg, T. H., Hinlein, E. S. & Long, S. S. 2007. Monod kinetics for aerobic biodegradation of petroleum hydrocarbons in unsaturated soil microcosms. *Environ. Sci. Technol.* 41:2343–2349.

Osuji, L. C., Ogali, R. E. & Kalu, A. U. 2009. The use of pristane and phytane biomarkers: A rethink of Cognoscenti. *Scientia Africana* 8 (2): 42-52.

Oudijk, G. 2005. The use of atmospheric contaminants to estimate the minimum age of environmental releases impacting groundwater. *Environ. Forensics* 6:345–354.

Oudijk, G. 2007. Age significance of nC17/pr ratios in forensic investigations of refined product and crude oil releases: Discussion. *Environ. Geosc.* 14:110–111.

Oudijk, G. 2009a. Age dating heating-oil releases. Part 1. Heating-oil composition and subsurface weathering. *Environ. Forensics* 10(2): 107-119.

Oudijk, G. 2009b. Age dating heating-oil releases, Part 2: Assessing weathering and release time frames through chemistry, geology and site history. *Environ. Forensics* 10(2): 120-131.

Oudijk, G., Duffy, B. & Ochs, L. D. 1999. Aerobic bioremediation of residential petroleum releases. In *In Situ and On-Site Bioremediation of Petroleum Hydrocarbons and Other Organic Compounds*, eds. Alleman, B. C. & Leeson, A., 271– 276.

Oudijk, G., Obolensky, M. & Polidoro, K. 2006. The use of the Christensen-Larsen model to age date residential heating oil releases: Conditions, limitations, and recommended practices. *Environ. Claims J.* 18:257–273.

Owens, E. H., Taylor, E. & Humphrey, B. 2008. The persistence and character of stranded oil on coarse-sediment beaches. *Marine Poll. Bull.* 56:14–26.

Owen, K. & Coley, T. 1995. *Automotive Fuels Reference Book, Second Edition*. Warrendale, PA: Society of Automotive Engineers.

Oyewo, E. O. 1988. *Preliminary Studies on the Initial Weathering of Oil on Water*. Nigerian Institute for Oceanography and Marine Research Technical Paper No. 26.

Palacas, J. G., Snyder, R. P., Baysinger, J. P. & Threlkeld, C. N. 1982. Geochemical analysis of potash mine seep oils, collapsed breccia pipe oil shows and selected crude oils, Eddy County, New Mexico. U. S. Geological Survey Open-File Report 82–421.

Palmer, S. E. 1991. Effect of Biodegradation and Water Washing on Crude Oil Composition: Chapter 4: *Petroleum Generation and Migration*. AAPG Special Volumes, Volume TR: Source and Migration Processes and Evaluation Techniques, 47–54.

Parsons Engineering Science. 2003. *Final, Light nonaqueous-phase liquid weathering at various fuel release sites, 2003 Update.* U. S. Air Force Center for Environmental Excellence Science and Engineering Division, Brooks City-Base San Antonio, Texas.

Payne, J. R., McNabb, G. D. & Clayton, J. R. 1991. Oil-weathering behavior in Arctic environments. *Polar Research* 10: 631–662.

Pearson, G. & Oudijk, G. 1993. Investigation and remediation of petroleum product releases from residential storage tanks. *Ground Water Monitoring Rev.* 13:124–128.

Peters, K. E., Walters, C. C. & Moldowan, J. M. 2005. *The Biomarker Guide: Volume 1, Biomarkers and Isotopes in the Environment and Human History.* Cambridge, UK: Cambridge University Press.

Philp, R. P. & Lewis, C. A. 1987. Organic geochemistry of biomarkers. *Ann. Rev. Earth Plan. Sci.* 15:363–395.

Pirnik, M. P. 1977. Microbial oxidation of methyl branched alkanes. *CRC Critical Rev. Microbiol.* 5:413–422.

Pirnik, M. P., Atlas, R. M. & Bartha, R. 1974. Hydrocarbon metabolism by Brevibacterium erythrogenes: Normal and branched alkanes. *J. Bacteriol.* 119:868–878.

Potter, T. L. & Simmons, K. E. 1998. Composition of Petroleum Mixtures, vol. 2. Amherst, MA: Amherst Scientific Publishers.

Prince, R. C. & Walters, C. C. 2007. Biodegradation of oil hydrocarbons and its implications for source identification. In *Oil Spill Environmental Forensics*, eds. Z. Wang & S. A. Stout. San Diego, CA: Academic Press, 349–379.

Providenti, M. A., Lee, H. & Trevors, J. T. 1993. Selected factors limiting microbial degradation of recalcitrant compounds. *J. Ind. Microbiol. Biotechnol.* 12:379–395.

Raymond, R. L., Hudson, J. O. & Jamison, V. W. 1976. Oil degradation in soil. *Appl. Environ. Microbiol.* 31:522–535.

Reinhard, M., Hopkins, G., Cunningham, J. & Lebron, C. A., 2000. Enhanced in situ anaerobic bioremediation of fuel-contaminated ground water: Naval Facilities Engineering Command, Washington, DC. Contract Report CR 00–005-ENV.

Ritter, U. 2003. Solubility of petroleum compounds in kerogen— implications for petroleum expulsion. *Org. Geochem.* 34:319– 326.

Røberg, S., Stormo. S. K. & Landfald, B. 2007. Persistence and biodegradation of kerosene in high arctic intertidal sediment. *Marine Environ. Res.* 64:417–428.

Robinson, J. E., Scott, D. W., Knocke, W. R. & Conn, W. D. 1988. Underground storage tank disposal: Alternatives, economics, and environmental costs. Virginia Polytechnic Institute and State University, Center for Environmental and Hazardous Materials Studies, Bulletin 160.

Schaeffer, T., Cantwell, S. G., Brown, J. L., Watt, D. S. & Fall, R. R. 1979. Microbial growth on hydrocarbons: Terminal branching inhibits biodegradation. *Appl. Environ. Microbiol.* 38 (4): 742-746.

Schmidt, P. F. 1985. *Fuel Oil Manual, 4th edition*: New York, NY: Industrial Press, Inc.

Schroll, R., Becher, H. H. & Dorfler, U. 2006. Quantifying the effect of soil moisture on the aerobic microbial mineralization of selected pesticides in different soil. *Environ. Sci. Technol.* 40:3305– 3312

Senn, R. B. & Johnson, M. S. 1987. Interpretation of gas chromatographic data in subsurface hydrocarbon investigations. *Ground Water Monitoring Rev.* 7:58–63.

Setti, L., Pifferi, P. G. & Lanzarini, G. 1995. Surface tension as a limiting factor for aerobic n-alkane biodegradation. *J. Chem. Technol. Biotechnol.* 64:41–48.

Sexstone, A., Everett, K., Jenkins, T. & Atlas, R. M. 1978a. Fate of crude and refined oils in North Slope soils. *Arctic* 31:339–347.

Sexstone, A., Gustin, A. & Atlas, R. M. 1978b. Long term interactions of microorganisms and Prudhoe Bay crude oil in tundra soils at Barrow, Alaska. *Arctic* 31:348–354.

Shepperd, D. A. & Crawford, T. R. 2003. Age-dating diesel fuel: Facts and fallacies: *Proceedings of the American Academy of Forensic Sciences* IX:101–102.

Siddique, T., Fedorak, P. M. & Foght, J. M. 2006. Biodegradation of short-chain n-alkanes in oil sands tailings under methanogenic conditions. *Environ. Sci. Technol.* 40:5459–5464.

Singer, M. E. & Finnerty, W. R. 1984. Microbial metabolism of straight- chain and branched alkanes. In *Petroleum Microbiology*, ed. R. M. Atlas. New York, NY: MacMillan Publishing Company.

Smith, J. S., Eng, L. & Shepperd, D. A. 2001. Age-dating oil: Is Christensen and Larsen applicable? *The Chemist* March/April Issue, 9– 12.

Smith, M. R. 1990. The biodegradation of aromatic hydrocarbons by bacteria. *Biodegradation* 1:191–206.

Solevic, T., Jovancicevic, B., Vrvic, M. & Wehner, H. 2003. Oil pollutants in alluvial sediments—influence of the intensity of contact with ground waters on the effect of microorganisms. *J. Serbian Chem. Soc.* 68:227–234.

Song, C. 2000. Introduction to chemistry of diesel fuels. In *Chemistry of Diesel Fuels*, eds. Song, C., Hsu, C. S. & Mochida, I. New York, NY: Taylor & Francis.

Song, H-G., Wang, X. & Bartha, R. 1990. Bioremediation potential of terrestrial fuel spills. *Appl. Environ. Microbiol.* 56:652– 656.

de Souza, E. S. & Triguis, J. A. 2006. Degradação do petróleo em derrames no mat—intemperismo e biorremediacão [in Portuguese]. In *3º Congresso Brasileiro de P&D em Petroleo e Gas.*

Srinivasan, G., Tartakovsky, D. M., Robinson, B. A. & Aceves, A. B. 2007. Quantification of uncertainty in geochemical reactions. *Water Resour. Res.* 43:W12415.

State of New Jersey. 1986. Underground storage tanks containing hazardous substances; legislative findings and declarations. N.J.A.C 58:10A- 21, effective September 3, 1986.

Stout, S. A., Uhler, A. D., McCarthy, K. J. & Emsbo-Mattingly, S. D. 2002a. Invited commentary of the Christensen and Larsen technique. *Environ. Forensics* 3:9–11.

Stout, S. A., Uhler, A. D., McCarthy, K. J. & Emsbo-Mattingly, S. 2002b. Chemical fingerprinting of hydrocarbons. In *Introduction to Environmental Forensics*, eds. Morrison, R. D. & Murphy, B. M. San Diego, CA: Academic Press.

Stout, S. A., Uhler, A. D., and McCarthy, K. J. 2006. Chemical characterization and sources of distillate fuels in the subsurface of Mandan, North Dakota. *Environ. Forensics* 7:267–282.

Stout, S. A. & Douglas, G. S. 2007. Age-dating diesel fuel: A case study averse to the Christensen and Larsen method. In *The Annual International Conference on Soil, Sediment & Water*. Amherst, MA: Association of Environmental Health and Soils.

Stout, S. A. & Uhler, A. D. 2006. Causation of variable n-alkylcyclohexane distributions in distillate nonaqueous phase liquids from Mandan, North Dakota. *Environ. Forensics* 7:283–287.

Stout, S. A. & Wang, Z. 2007. Chemical fingerprinting of spilled or discharge petroleum — Methods and factors affecting petroleum fingerprints in the environment. In *Oil Spill Environmental Forensics,* eds. Z. Wang & S. A. Stout. San Diego, CA: Academic Press, 1–72.

Sun, P., Gao, Z., Zhou, Q., Zhao, Y., Wang, X., Cao, X. & Li, G. 2009. Evaluation of the oil spill accident in Bohai Sea, China. *Environ. Forensics* 10 (4): 308-316.

Sukol, R., Woolson, E. & Thompson, W. 1988. Fate and Persistence in Soil of Selected Toxic Organic Chemicals. U. S. Environmental Protection Agency Project Summary EPA/600/S6-87/003.

Swannell, R. P. J., Lee, K. & McDonaugh, M. 1996. Field evaluations of marine oil spill bioremediation. Microbiology Reviews 60:342–365.

Swindell, A. L. & Reid, B. J. 2006. Influence of diesel concentration on the fate of phenanthrene in soil. *Environ. Poll.* 140:79–86.

Teh, J. S. & Lee, K. H. 1973. Utilization of n-alkanes by Cladosporium resinae. *Appl. Microbiol.* 25:454–457.

Wade, M. J. 2001. Age-dating diesel spills: Using the European empirical time-based model in the USA. *Environ. Forensics* 2:347– 358.

Wade, M. J. 2002. Invited commentary of the Christensen and Larsen technique *Environ. Forensics* 3:13.

Wade, M. J. 2005. The use of isoprenoid ratios to calculate percentage mixing of different distillate fuels released to the environment. *Environ. Forensics* 6:187–196.

Walker, J. D., Colwell, R. R. & Petrakis, L. 1976. Biodegradation rates of components of petroleum. *Can. J. Microbiol.* 22:1209– 1213.

Wang, Z. & Fingas, M. 1995a. Use of methyldibenzothiophenes as markers for differentiation and source identification of crude and weathered oils. *Environ. Forensics* 29:2842–2849.

Wang, Z. & Fingas, M. 1995b. Study of the effects of weathering on the chemical composition of a light crude oil using GC/MS GC/FID. *J. Microcolumn Sep.* 7 (6): 617-639.

Wang, Z., Hollebone, B. P., Fingas, M., Fieldhouse, B., Sigouin, L., Landriault, M., Smith, P., Noonan, J. & Thouin, G. 2003. Characteristics of spilled oils, fuels, and petroleum products: 1. Composition and properties of selected oils. EPA/600/R-03/072.

Wang, Z., Yang, C. & Fingas, M. 2005. Characterization, weathering, and application of sesquiterpanes to source identification of spilled lighter petroleum products. *Environ. Sci. Technol.* 39:8700–8707.

Whyte, L. G., Hawari, J., Zhou, E., Bourbonniere, L., Inniss, W. E. & Greer, C. W. 1998. Biodegradation of variable-chain-length alkanes at low temperatures by a psychrotrophic Rhodococcus sp. *Appl. Environ. Microbiol.* 64:2578–2584.

Zaidi, B. R. & Imam, S. H. 1999. Factors affecting microbial degradation of polycyclic aromatic hydrocarbon phenanthrene in the Caribbean coastal water. *Marine Poll. Bull.* 38 (8): 737-742.

Zajic, J. E., Supplisson, B. & Volesky, B. 1974. Bacterial degradation and emulsification of no. 6 fuel oil. *Environ. Sci. Technol. 8:664–668.*

Zemo, D. 2007. Forensic tools for petroleum hydrocarbon releases. *SW Hydrol.* July/August:26–35.

Zibiske, L. M., and Risser, J. A. 1986. Effects of soil texture on respiration and metal solubility in heating oil-amended soils. *Bull. Environ. Contam. Toxicol.* 36:540–547.

Zobell, C. E. 1946. Action of microorganisms on hydrocarbons. *Bacteriol. Rev.* 10:1–49.

Zytner, R G, Salb, A C. & Stiver, W H. 2006. Bioremediation of diesel fuel contaminated soil: Comparison of individual compounds to complex mixtures. *Soil Sed. Contam.* 15:277–297.

7

Monitoring of Heavy Metal Concentration in Groundwater of Mamundiyar Basin, India

Imran Ahmad Dar[1], K. Sankar[1], Dimitris Alexakis[2] and Mithas Ahmad Dar[1]
[1]Department of Industries and Earth Sciences, Tamil University- Thanjavur
[2]Centre for the Assessment of Natural Hazards and Proactive Planning, Laboratory of
Reclamation Works and Water Resources Management
National Technical University of Athens, Athens
[2]Greece
[1]India

1. Introduction

Heavy metals designate a group of elements that occur in natural system in minute concentration and when present in sufficient quantities and are toxic to living organisms. The behavior of trace metals in groundwater is complicated and is related to source of group water and the bio-geochemical process in elemental conditions.

It is often assumed that natural, uncontaminated waters from deep (bedrock) wells are clean and healthy (Banks et al., 1998b). This is usually true with regards to bacteriological composition. The inorganic chemical quality of these waters is, however, rarely adequately tested before the wells are put into production. Due to variations in the regional geology and water rock interactions, high concentrations of many chemical elements can occur in such waters. During the last 5–10 years several studies have shown that wells in areas with particular geological features yield water that does not meet established drinking water norms (e.g. Varsanyi et al., 1991; Bjorvatn et al., 1992, 1994; Edmunds and Trafford, 1993; Banks et al., 1995a,b, 1998a; Sæther et al., 1995; Reimann et al., 1996; Edmunds and Smedley, 1996; Smedley et al., 1996; Williams et al., 1996; Morland et al., 1997, 1998; Midtga°rd et al., 1998; Misund et al., 1999; Frengstad et al., 2000) without any influence from anthropogenic contamination. These studies also document that quite a number of elements for which no drinking water guideline values (GL) or maximum acceptable concentration limits (MAC) have been established can occur at unpleasantly high levels in natural well waters (e.g. Be, Th, Tl). In Norway, F and radon (Rn) are the most problematic elements (see Frengstad et al., 2000) in terms of possible health effects. In Hungary, Bangladesh and India, arsenic represents one of the most drastic examples of unwanted natural chemical 'contamination' of groundwater. Several 100 000 people in these regions suffer skin cancer due to high As concentrations in drinking water from drilled wells (Chatterjee et al., 1995; Das et al., 1995; Smith et al., 2000; Smedley and Kinniburgh, 2002).

It has been established that various trace elements have certain health on living organisms (WHO, 1984). But the extent to which these elements affect health of living organisms depends on the chemical characteristics and the concentration of the element in the water consumed. Furthermore, the time of exposure will also determine the level of the element on the organism. Some elements are biocumulative and therefore get increased with time in the body. The present paper reports analytical results for 6 chemical elements (trace elements) from 50 sampling stations of Mamundiyar basin, India.

2. Geography and geology of the study area

Mamundiyar basin, India lies in hard rock terrain. Groundwater is available only in weathered and fractured zones. In this area assured surface water supplies are nominal and most of the farmers depend on groundwater for drinking and irrigation purposes. Average annual rainfall is around 464 mm which is mostly lost as surface runoff and evaporation. Only one-fifth of it is recharging to groundwater. Therefore, groundwater development assumes great significance in improving the quality of life of the most deprived and vulnerable people of this basin by improving their access to safe drinking water.

The Mamundiyar basin extends over approximately 720 km^2 and lies between 10^0 25` and $10^0$ 40`N latitudes and 78^0 10` and $78^0$ 30` E longitudes in the southern part of Tamilnadu, India (Fig. 1). Mamundiyar River originates at an altitude of 315 m above Irungadu group of hills and joins Ariyavur River near Maravanur about 25 Km south-west of Tiruchirapalli. The western, north-western and south-western parts are characterized by the presence of residual hills. The basin is generally hot and dry except during winter season. The mean maximum monthly temperature varies from 37^0C in May to 29^0C in December. While as mean minimum monthly temperature ranges from 27^0C in June and 20^0C in January. The area receives an average annual rainfall of about 464 mm. The surface runoff goes to stream as instant flow. Rainfall is the direct recharge source and the irrigation return flow is the indirect source of groundwater in the Mamundiyar hydrographic basin. The study area depends mainly on the North-east monsoon rains which are brought by the troughs of low pressure established in the South Bay of Bengal.

Several digital image processing techniques, including standard color composites, intensity-hue-saturation (IHS) transformation and decorrelation stretch (DS) were applied to map rock types. The statistical technique adopted by Sheffield (1985) was employed to select the most effective Three-band color composite image. The band combination 1, 4 and 5 is the best triplet and was used to create color composites with Landsat TM bands 5, 4 and 1 in red, green and blue, respectively. IHS transformation and DS were also applied to the selected band combination in order to enhance the difference between rock types. Better contrast was obtained due to color enhancement and this facilitated visual discrimination of various rock types. Eleven lithologic units were mapped and could be distinguished by distinct colors in the processed images. These are: Ultramafics, Hornblende biotite gneiss, Basic rocks, Charnockite, Pyroxene granulite, Pink magmatite, Quartzite, Pegmatite vein, Quartz vein, Granite, and Calc granulite and limestone. Fig. 2 is a map of the interpreted distribution of rock types Mamundiyar basin (Dar et. el, 2010).

3. Sampling

Most samples reported here were taken from drinking water wells in small villages and settlements scattered throughout the Mamundiyar basin, India. Factory new, unwashed 100-ml high-density polyethylene (HDPE) bottles were used for sampling. Different brands of plastic bottles had previously been thoroughly checked for possible contamination (Reimann et al., 1999a). No risk of contamination from such bottles was found for the

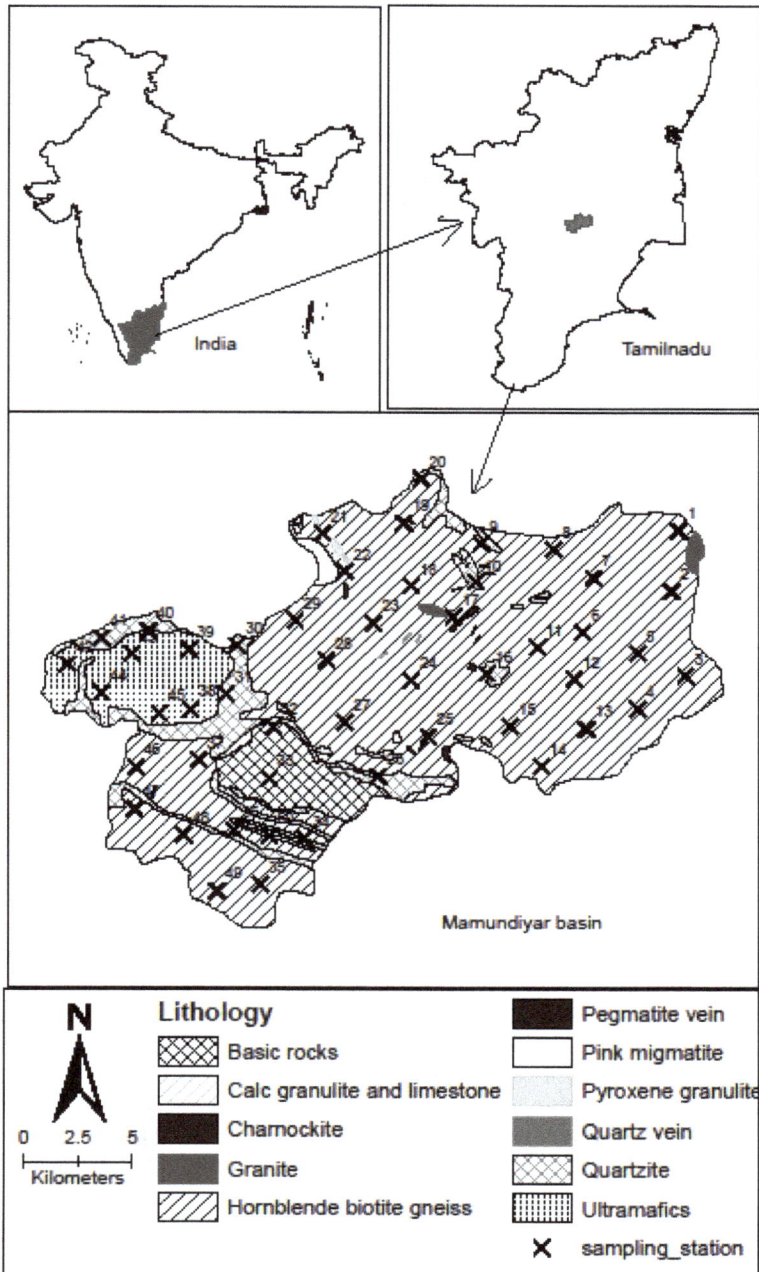

Fig. 1.

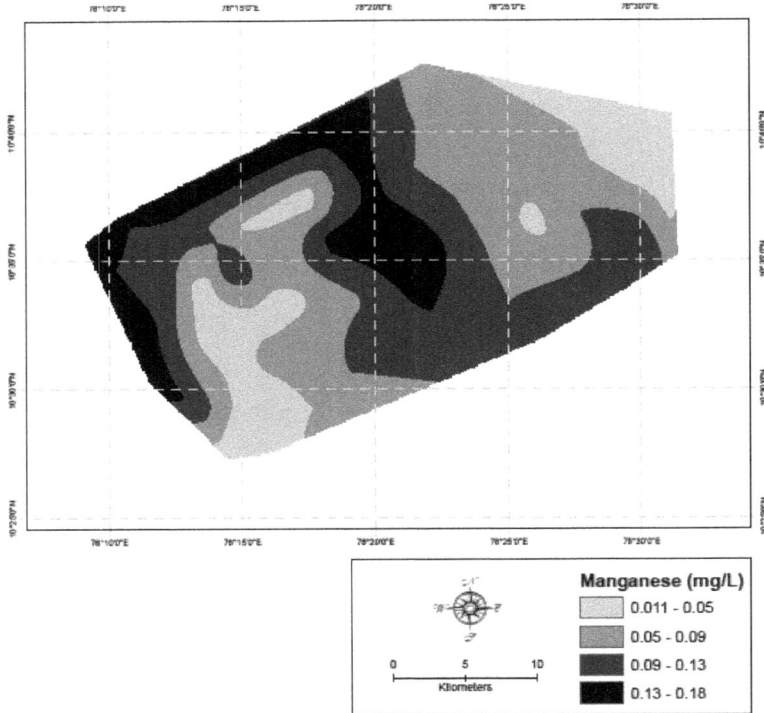

Fig. 2.

parameters reported here, as long as the bottles are thoroughly rinsed with water prior to sampling. In the field the bottles were rinsed three times with running water and then filled to the top.

Sampling took place directly at the tap or the wellhead. In order to collect fresh well water, the water was left running for at least 5 min or until temperature and conductivity remained stable. In most cases, each of these wells supplies more than 100 people with their daily drinking water. The water, therefore, never accumulates over longer periods in the well.

Two 100-ml bottles were collected at each site. The first sample, which was intended for anion analyses, was left unfiltered and unacidified. The unfiltered water of the second sample was acidified with 2 ml of concentrated nitric acid (Merck, Ultrapure). This second sample was used later for cation analysis. The acid was tested for its trace element content using the same analytical procedure as for the water samples. In the field, the samples were stored in a cool box and in the evening transferred to a refrigerator, where they were stored until shipment to the laboratory.

4. Analysis

The trace elements analyzed included manganese (Mn), iron (Fe), chromium (Cr), copper (Cu), zinc (Zn) and boron (B).

5. Results and discussion

The result of the analysis of 50 groundwater sampling stations is shown in tables 1 and 2.

Sample station	Zinc	Copper	Iron	Manganese	Chromium	Boron
1	0.01	0.02	0.03	0.011	0.001	0.38
2	0.01	0.2	0.02	0.012	0.001	0.35
3	0.04	0.03	0.03	0.07	0.002	0.31
4	0.03	0.03	0.04	0.13	0.001	0.31
5	0.04	0.03	0.03	0.12	0.001	0.4
6	0.01	0.02	0.01	0.09	0.001	0.55
7	0.07	0.03	0.02	0.05	0.002	0.56
8	0.07	0.02	0.02	0.06	0.001	0.25
9	0.006	0.03	0.04	0.08	0.001	0.38
10	0.07	0.02	0.06	0.05	0.001	0.55
11	0.03	0.03	0.06	0.04	0.001	0.31
12	0.04	0.04	0.02	0.07	0.001	0.31
13	0.02	0.03	0.2	0.13	0.002	0.4
14	0.05	0.01	0.03	0.12	0.002	0.41
15	0.002	0.02	0.03	0.09	0.002	0.32
16	0.1	0.02	0.04	0.09	0.001	0.48
17	0.1	0.02	0.04	0.09	0.001	0.55
18	0.1	0.02	0.04	0.09	0.001	0.41
19	0.1	0.02	0.04	0.09	0.001	0.31
20	0.01	0.01	0.03	0.06	0.001	0.4
21	0.07	0.02	0.04	0.18	0.001	0.12
22	0.08	0.03	0.03	0.15	0.002	0.18
23	0.002	0.03	0.03	0.14	0.001	0.38
24	0.07	0.01	0.02	0.17	0.001	0.55
25	0.08	0.01	0.03	0.13	0.001	0.31
26	0.05	0.02	0.01	0.12	0.002	0.31
27	0.002	0.03	0.03	0.08	0.001	0.4
28	0.07	0.01	0.02	0.15	0.001	0.23
29	0.05	0.03	0.03	0.02	0.002	0.31
30	0.08	0.01	0.04	0.04	0.001	0.38
31	0.07	0.02	0.06	0.13	0.001	0.55
32	0.08	0.01	0.01	0.02	0.001	0.41
33	0.07	0.03	0.01	0.07	0.002	0.31
34	0.01	0.02	0.02	0.05	0.001	0.4
35	0.07	0.02	0.02	0.04	0.002	0.32
36	0.02	0.03	0.03	0.03	0.001	0.38
37	0.08	0.01	0.12	0.02	0.002	0.42
38	0.1	0.02	0.02	0.02	0.001	0.44
39	0.1	0.01	0.03	0.1	0.001	0.47
40	0.09	0.03	0.21	0.15	0.001	0.38
41	0.07	0.01	0.02	0.14	0.002	0.55
42	0.09	0.02	0.21	0.17	0.001	0.31
43	0.1	0.01	0.03	0.12	0.001	0.31
44	0.09	0.03	0.03	0.13	0.001	0.4
45	0.07	0.03	0.04	0.1	0.002	0.38
46	0.05	0.01	0.02	0.15	0.001	0.55
47	0.02	0.02	0.02	0.17	0.001	0.31
48	0.07	0.01	0.03	0.12	0.001	0.31
49	0.08	0.03	0.03	0.02	0.001	0.4
50	0.05	0.03	0.02	0.02	0.001	0.42

Table 1.

Parameters	z	Cu	Fe	Mn	Cr	B
Minimum	0	0.01	0.01	0.01	0	0.12
Maximum	0.1	0.2	0.21	0.18	0	0.56
Range	0.1	0.19	0.2	0.17	0	0.44
Mean	0.06	0.03	0.04	0.09	0	0.38
Median	0.07	0.02	0.03	0.09	0	0.38
First quartile	0.03	0.01	0.02	0.05	0	0.31
Third quartile	0.08	0.03	0.04	0.13	0	0.42
Standard error	0	0	0.01	0.01	0	0.01
95% confidence interval	0.01	0.01	0.01	0.01	0	0.03
99% confidence interval	0.01	0.01	0.02	0.02	0	0.04
Variance	0	0	0	0	0	0.01
Average deviation	0.03	0.01	0.03	0.04	0	0.07
Standard deviation	0.03	0.03	0.05	0.05	0	0.1
Coefficient of variation	0.57	1.06	1.09	0.55	0.35	0.26
Skew	-0.4	6.01	3.09	0	1.13	0.06
Kurtosis	-1.2	40.1	9.06	-1.2	-0.8	0.2
Kolmogorov-Smirnov stat	0.22	0.39	0.38	0.13	0.46	0.15
Critical K-S stat, alpha=.10	0.17	0.17	0.17	0.17	0.17	0.17
Critical K-S stat, alpha=.05	0.19	0.19	0.19	0.19	0.19	0.19
Critical K-S stat, alpha=.01	0.23	0.23	0.23	0.23	0.23	0.23

Table 2.

6. Manganese

The U.S. Public Health Service Drinking Water Standards of 1925, 1942, and 1946 included manganese with iron for a combined maximum level of 0.30 mg/L, but in 1962 the regulations included, in addition, a maximum concentration of 0.05 mg/L for manganese. The USEPA adopted the 0.05 mg/L of the USPHS as recommendation, and issued a secondary standard in 1989. The WHO recommended 0.05 mg/L (maximum acceptable) and 0.50 mg/L as maximum allowable. The European Community (1980) used a guide value of 0.02 mg/L and a maximum of 0.05 mg/L.

Figure 2 shows the spatial distribution of manganese through Natural Nearst Neighbor interpolation technique. It's quite obvious from the map that manganese tends to dominate the central part of the Mamundiyar basin, and its concentration diminishes radially outwards from the centre. The highest concentration of manganese found was 0.18; which means all the samples fall within the permissible limit set by WHO.

7. Iron

Since the standards for iron have been set for less than 0.3 mg/L, acceptability of water sources was a condition for meeting this concentration. Groundwater exceeding this limit may need a treatment to meet the standard at the distribution system. Groundwater containing soluble iron may remain clear when pumped out, but exposure to air will cause precipitation of iron due to oxidation, with a consequence of rusty color. The presence of iron bacteria may clog well screens particularly when sulphate compounds in addition to iron may be subjected to chemical reduction. Solubility of iron is increased by a low pH (<5).

High turbidity may help to keep acid- soluble iron in suspension. Iron in raw or potable water may be either ferrous or ferric or both and categorized as in solution, in colloidal state, in organic or inorganic compounds or in the form of coarse suspended or settled particles. The 1925 , 1942, 1946 and 1962 regulations of the U.S Public Health Service always reported the maximum concentration for iron as 0.30 mg/L. the USEPA did not include iron in the National Drinking Water Quality Regulations, but maintained in the secondary Drinking Water Regulations of 1989 the limit of 0.3 mg/L based on aesthetic and taste consideration. WHO (1963) also adopted a 0.3 mg/L as a maximum acceptable level and 1.0 as maximum allowable. The European Community adopted in 1980 a guide of 0.05 mg/L and a maximum of 0.20 mg/L; WHO (1984 and 1993) recorded a guideline of 0.30 mg/L. USEPA (1979 and 1991) confirmed the original ruling for iron as a contaminant to be included in the Secondary Drinking Water Standards with a level of 0.3 mg/L as the final rule.

Figure 3 shows the spatial distribution of iron using interpolation method in GIS environment. It's quite evident from the map that iron contaminations occur at few locations, mainly around the Kadavur (western) region of the Mamundiyar basin. In each patch, the concentration of iron was found to decrease/diminish radially outwards from centre. That means, at the point of rock (iron bearing) - water interaction, the concentration of Iron is maximum; and as the distance increases from the interface of rock (bearing iron element) and water, the concentration also gets decreased because of dilution factor. The values of iron were within the permissible limit of drinking water standards.

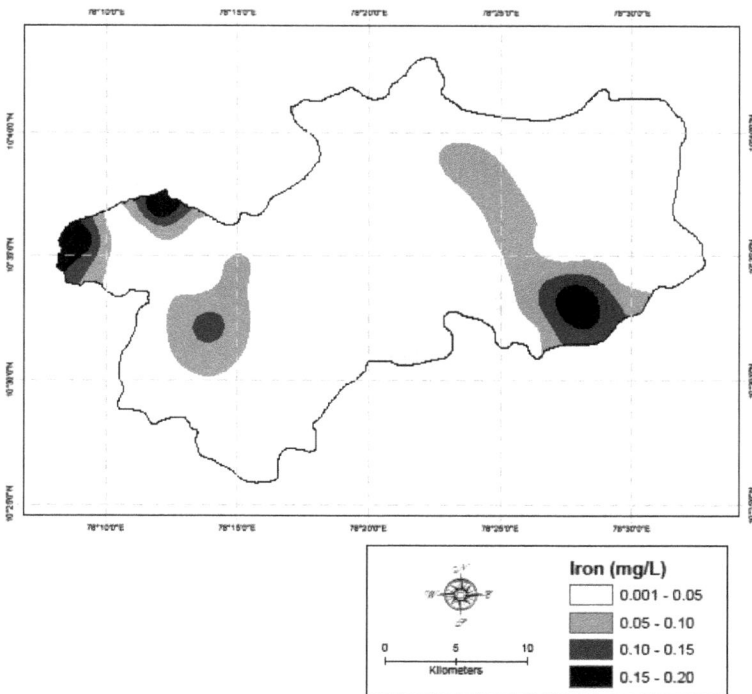

Fig. 3.

8. Chromium

It's a naturally occurring metal in drinking water.

USPHS 1925   = not stated
USPHS 1942   = 0 mg/L as hexavalent
USPHS 1946   = 0.05 mg/L
USPHS 1962   = 0.05 as hexavalent
WHO guidelines   = 0.05 mg/L (as Cr^{6+} and total chromium)
European Community   =0.05 mg/L (as Cr^{6+} and total chromium)
MCLG and MCL
(USEPA, 1989)   = 0.1 mg/L (proposed)
MCLG and MCL
(USEPA, 1991)   = 0.1 mg/L (final; effective 7/30/1992)

The spatial distribution map of chromium is shown in figure 4; which is created using Nearest Neighbors interpolation technique. It's clear from the figure that the maximum concentration (0.002 mg/L) of chromium occurs in the southern region of Mamundiyar basin; and its concentration decreases towards the north-west, due to dilution. The concentration of chromium wherever recorded is well within the limits of drinking water standards prescribed by WHO.

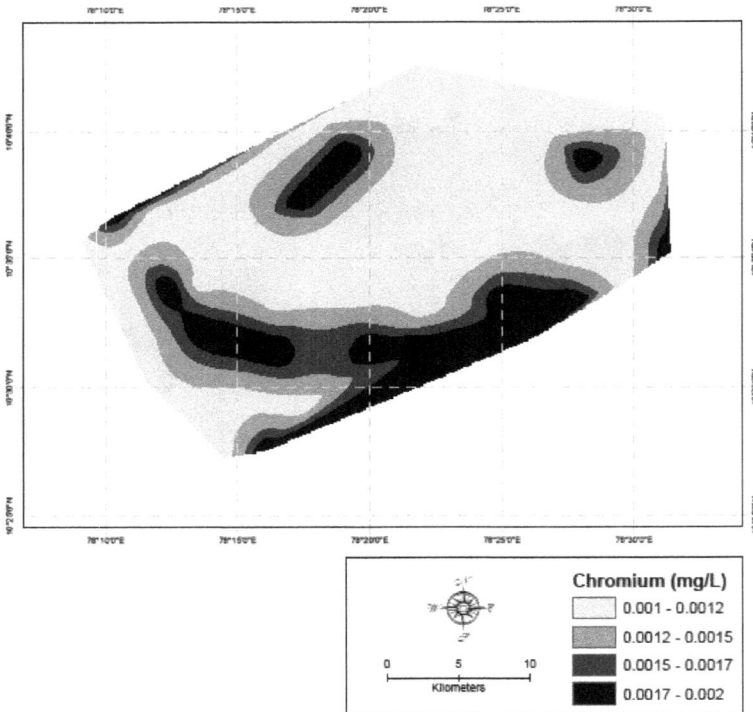

Fig. 4.

9. Copper

USPHS 1925	= 0.2 mg/L
USPHS 1942	= 3 mg/L
USPHS 1962	= 1 mg/L
WHO guidelines	= 1 mg/L (1.5 mg/L excessive)
European Community	=0.1 mg/L
MCLG and MCL	
(USEPA, 1988)	= 1.3 mg/L (proposed)
MCLG and MCL	
(USEPA, 1991)	= 1.3 mg/L at the consumer`s tap (final revised regulations for

lead and copper according to the New Lead and Copper Rule as requested by the Safe Drinking Water Act- Revision of 1986).

Copper can exist in aquatic environment in three forms namely soluble, colloidal and particulate. It is found in less quantity as an essential element for organisms. Excess of copper in human body is toxic and causes hypertension and produces pathological changes in brain tissues. Excessive ingestion of copper is responsible for specific disease of the bone (Krishnamurthy, C.R. and V. Pushpa. 1995). The spatial distribution map of copper (figure 5) is prepared using interpolation technique in GIS environment. It's quite obvious from the map that the maximum concentration of copper is present at eastern edge of Mamundiyar basin. In the present study, the values of copper are showed within the limit of drinking water standards.

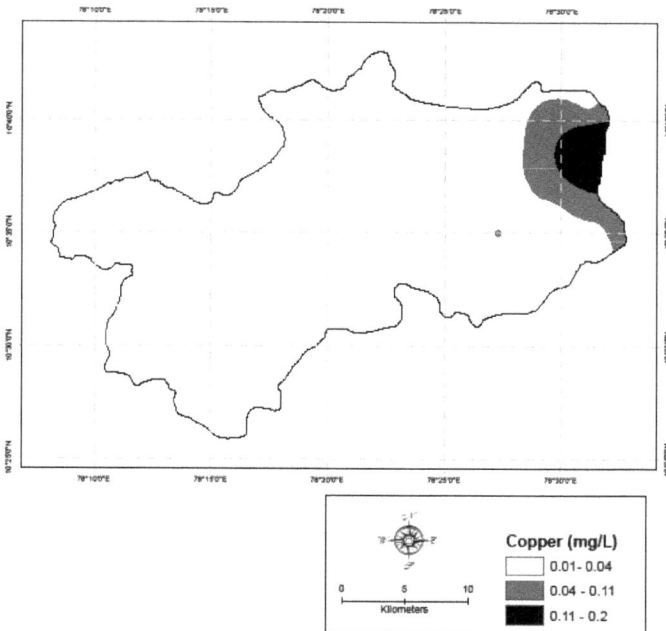

Copper (mg/L)
☐ 0.01- 0.04
▨ 0.04 - 0.11
■ 0.11 - 0.2

Fig. 5.

10. Zinc

The USPHS recommended a maximum zinc concentration of 15.0 mg/L in 1942 and 1946 standards, and 5.0 mg/L in the 1962 standards.USEPA recommended 5 mg/L in 1980 and a SMCL of 5 mg/L in 1989. WHO (1971) recommended 5 mg/L with a maximum of 15mg/L. the European Community advised 0.1 mg/L, with a maximum of 1.5 mg/L. the WHO (1984) adopted a guideline of 5 mg/L based on the taste consideration. USEPA (1991) issued a final status for Zinc as a Secondary Drinking Water Standard (SDWS) of 5 mg/L, confirming the final rule of 5 mg/L issued in 1980.

The spatial distribution map of zinc (figure 6) is prepared using interpolation technique in GIS environment. It's quite obvious from the map that the maximum concentration of zinc (0.5 MG/L) is present at the Kadavur (western) and central part of Mamundiyar basin. In the present study the values of zinc are showed within the limit of drinking water standard.

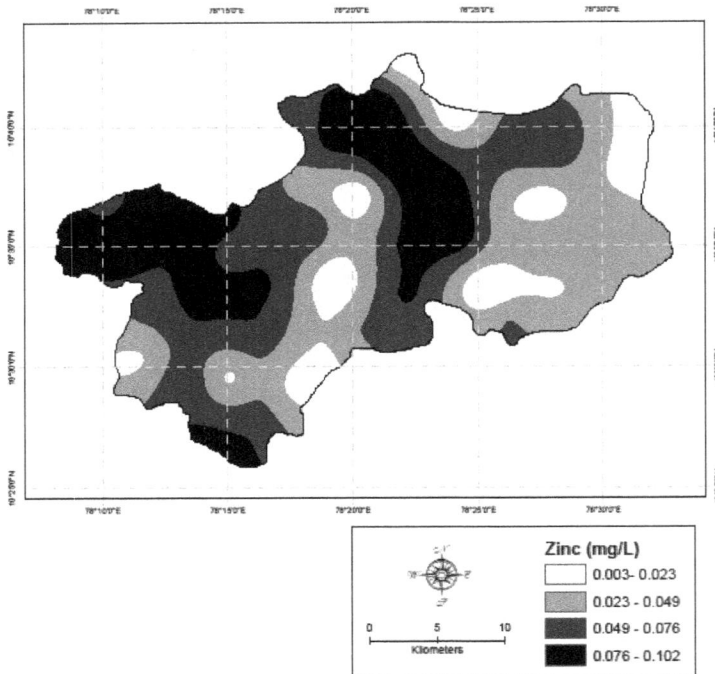

Fig. 6.

11. Boron

Spatial distribution map of boron (figure 7) depicts that maximum concentration of boron in patches at Central, Eastern and western part of Mamundiyar basin. Boron concentration varied between 0.11 to 0.56, indicating that the samples fall within the permissible limit set by WHO. Overall, boron dominates the trace metal pool of the Mamundiyar basin as shown graphically in figure 8.

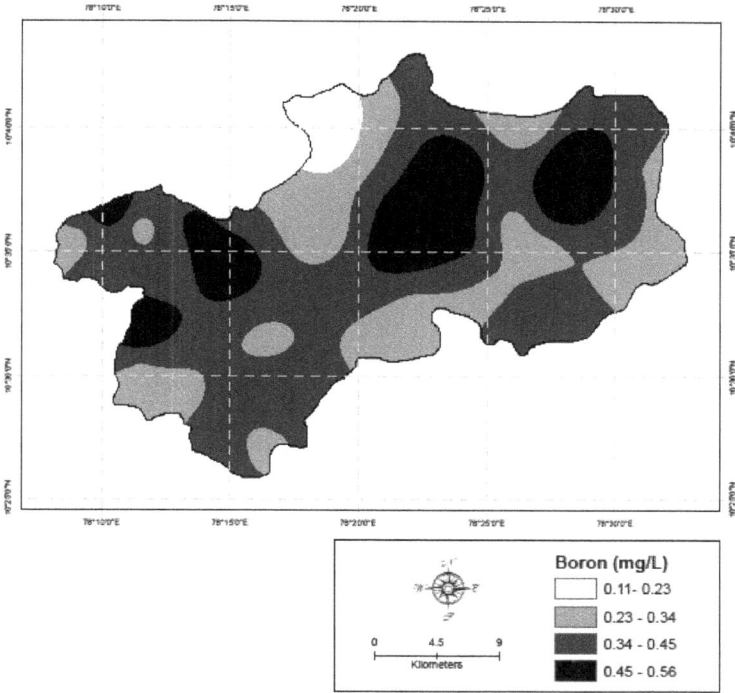

Fig. 7.

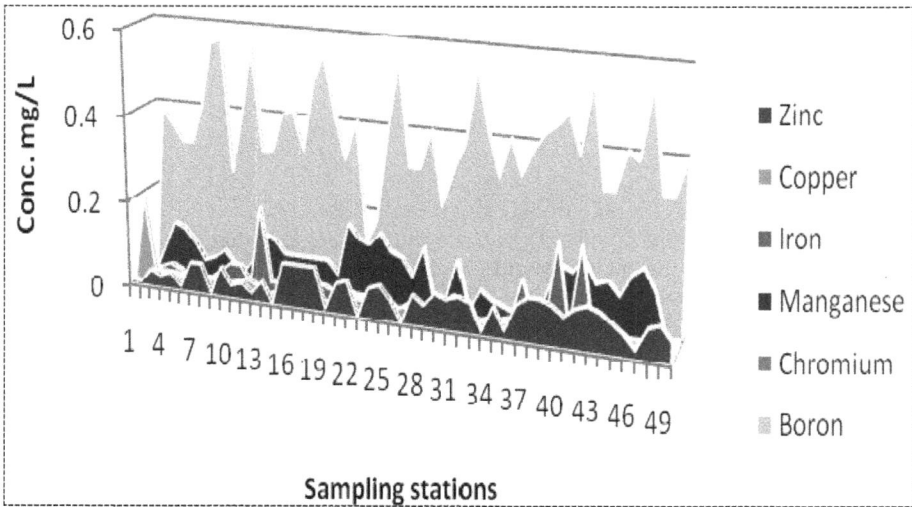

Fig. 8.

The correlation between these parameters is shown in table 3. The highest positive correlation (0.244) was found between iron and manganese, followed by iron and zinc (0.138). The lowest positive correlation (0.047) was found between iron and chromium. While as the highest negative correlation (- 0.293) was found between zinc and copper. Moreover, boron was found to show the negative correlation with all the parameters except zinc.

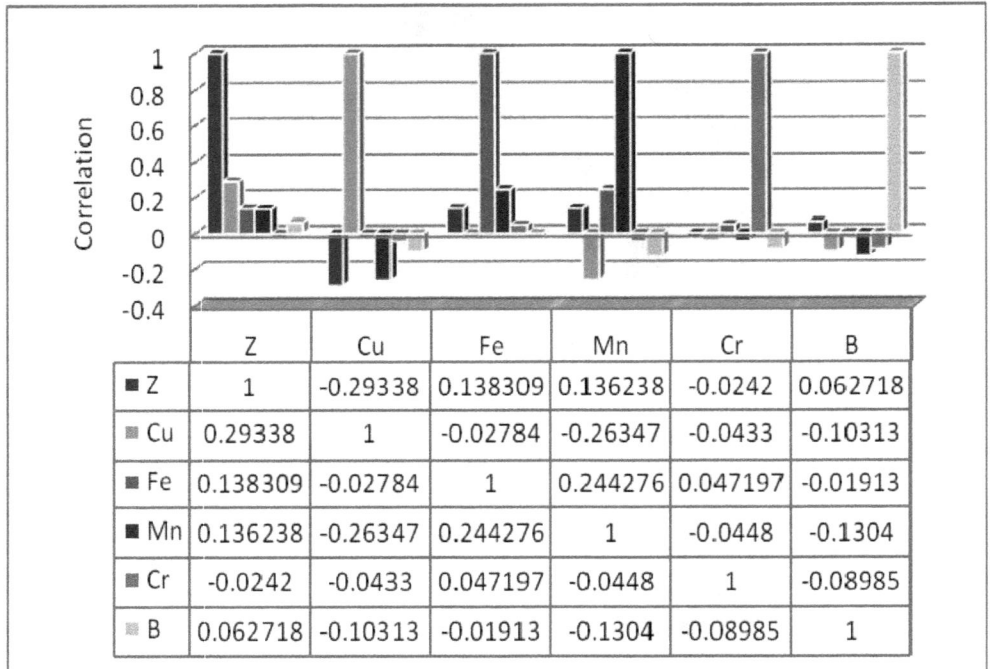

	Z	Cu	Fe	Mn	Cr	B
■ Z	1	-0.29338	0.138309	0.136238	-0.0242	0.062718
▧ Cu	0.29338	1	-0.02784	-0.26347	-0.0433	-0.10313
■ Fe	0.138309	-0.02784	1	0.244276	0.047197	-0.01913
■ Mn	0.136238	-0.26347	0.244276	1	-0.0448	-0.1304
▧ Cr	-0.0242	-0.0433	0.047197	-0.0448	1	-0.08985
▧ B	0.062718	-0.10313	-0.01913	-0.1304	-0.08985	1

Table 3.

12. Conclusion

The concentrations of the investigated heavy metals (Mn, Fe, Cr, Cu, Zn and B) in the drinking water samples from Mamundiyar basin, India were found below the guidelines for drinking waters given by the WHO (World Health Organization), EC (Europe Community), EPA (Environment Protection Agency). It was concluded that drinking waters in Mamundiyar contain low heavy metal levels.

13. References

Banks D, Frengstad B, Midtga°rd AK, Krog JR, Strand T. The chemistry of Norwegian groundwaters: I. The distribution of radon, major and minor elements in 1604 crystalline bedrock groundwaters. Sci Total Environ 1998;222:71 –91.

Banks D, Midtga°rd AK, Morland G, Reimann C, Strand T, Bjorvatn K, Siewers U. Is pure groundwater safe to drink? Natural 'contamination' of groundwater in Norway. Geol Today 1998;14(3):104 –113.

Banks D, Reimann C, Røyset O, Skarphagen H. Natural concentrations of major and trace elements in some Norwegian bedrock groundwaters. Appl Geochem 1995;10:1 – 16.

Banks D, Røyset O, Strand T, Skarphagen H. Radioelement (U,Th, Rn) concentrations in Norwegian bedrock groundwaters. Environ Geol 1995;25:165 –180.

Bjorvatn K, Ba°rdsen A° , Thorkildsen AH, Sand K. Fluorid I norsk grunnvann—en ukjent helsefaktor wFluoride in Norwegian drinking water—an unknown health factorx. Vann 1994;2:120 –128. in Norwegian.

Bjorvatn K, Thorkildsen AH, Holteberg S. Sesongmessige variasjoner i fluoridinholdet i sør og vestnorsk grunnvann [Seasonal variations of the fluoride content in south and west Norwegian groundwaters]. Den norske tannlegeforenings tidende 1992;102:128 –133. in Norwegian

Chatterjee A, Das D, Mandal BK, Chowdhurry TR, Samanta G, Chakraborti D. Arsenic in groundwater in six districts of West Bengal, India: the biggest arsenic calamity in the world. Part 1: arsenic species in drinking water and urine of affected people. Analyst 1995;120:643 –650.

Dar IA, Sankar K, Dar MA, Remote sensing technology and geographic information system modeling: An integrated approach towards the mapping of groundwater potential zones in Hardrock terrain, Mamundiyar basin. Journal of Hydrology (2010) 394: 85-295

Das D, Chatterjee A, Mandal BK, Samanta G, Chakraborti D. Arsenic in groundwater in six districts of West Bengal, India: the biggest arsenic calamity in the world. Part 2: arsenic concentration in drinking water, hair, nails, urine, skin scale and liver tissue (biopsy) of the affected people. Analyst 1995;120:917 –924.

Edmunds WM, Smedley PL. Groundwater geochemistry and health: an overview. In: Appleton JD, Fuge R, mccall GJH, editors. Environmental geochemistry and health. Geological Society Special Publication 113 1996. P. 91 –105.

Edmunds WM, Trafford JM. Beryllium in river baseflow, shallow groundwaters and major aquifers of the UK. Appl Geochem 1993;2(Suppl):223 –233.

European Union. 80y778veec Council Directive of 15 July 1980 relating to the quality of water intended for human consumption. Official Journal of the European Community 1980. P. L229y11 –L229y29.

European Union. Council Directive 98y83yec of 3 November 1998 on the quality of water intended for human consumption. Official Journal of the European Community 1998. P. L330y32 –L330y54.

Frengstad B, Midtga°rd AK, Banks D, Krog JR, Siewers U. The chemistry of Norwegian groundwaters. III. The distribution of trace elements in 476 crystalline bedrock groundwaters, as analysed by ICP-MS techniques. Sci Total Environ 2000;246: 21 –40.

Krishnamurthy, C.R. and V. Pushpa. 1995. Toxic metals in the Indian Environment. Tata McGraw Hill Publishing Co. Ltd., New Delhi. pp 280.

Midtga°rd AK, Frengstad B, Banks D, Krog JR, Strand T, Siewers U. Drinking water from crystalline bedrock aquifers— not just H2O. Min Soc Bull 1998;121:9 –16.

Misund A, Frengstad B, Siewers U, Reimann C. Natural variation of 66 elements in European mineral waters. Sci Total Environ 1999;243y244:21 –41.

Morland G, Reimann C, Strand T, Skarphagen H, Banks D, Bjorvatn K, Hall GEM, Siewers U. The hydrogeochemistry of Norwegian bedrock groundwater-selected parameters (ph, Fy, Rn, U, Th, B, Na, Ca) in samples from Vestfold and Hordaland, Norway. NGU Bull 1997;432:103 –117.

Morland G, Strand T, Furuhaug L, Skarphagen H, Banks D. Radon concentrations in groundwater from Quaternary sedimentary aquifers in relation to underlying bedrock geology. Ground Water 1998;36:143 –146.

Reimann C, Hall GEM, Siewers U, Bjorvatn K, Morland G, Skarphagen H, Strand T. Radon, fluoride and 62 elements as determined by ICP-MS in 145 Norwegian hardrock groundwaters. Sci Total Environ 1996;192:1 –19

Reimann C, Siewers U, Skarphagen H, Banks D. Does bottle type and acid washing influence trace element analyses by ICP-MS on water samples? A test covering 62 elements and four bottle types: high-density polyethene (HDPE), polypropene (PP), fluorinated ethene propene copolymer (FEP) and perfluoroalkoxy polymer (PFA). Sci Total Environ 1999;239:111 –130.

Sæther O, Reimann C, Hilmo BO, Taushani E. Chemical composition of hard- and softrock groundwaters from central Norway with special consideration of fluoride and Norwegian drinking water limits. Environ Geol 1995;26(3):147 – 156.

Smedley PL, Edmunds WM, Pelig-Ba KB. Mobility of arsenic in groundwater in the Obuasi gold-mining area of Ghana: some implications for human health. In: Appleton JD, fuger, mccall GJH, editors. Environmental geochemistry and health. Geological Society Special Publication 113 1996. P. 163 –181.

Smedley PL, Kinniburgh DG. A review of the source, behavior and distribution of arsenic in natural waters. Appl Geochem 2002;17(5):517 –568.

Smith AH, Lingas EO, Rahman M. Contamination of drinking water by arsenic in Bangladesh: a public health emergency. Bull WHO 2000;78(9):1093 –1101.

USEPA, 1989. National Primary and Secondary Drinking Water Regulations, Proposed Rule, Fed. Reg. (Vol. 54, No. 97).

USEPA, 1991. Safe Drinking Water Act, 1991 Amendments, EPA 570/9-86-002. Washington, D.C.

USPHS, Drinking water standards, Unitede States Public Health Services, 1987, Washington DC.

Varsanyi I, Fodre Z, Bartha A. Arsenic in drinking water and mortality in the Southern Great Plain, Hungary. Environ Geochem Health 1991;13:14 –22

WHO. Guidelines for drinking water quality. Geneva: World Health Organisation, 1993.

WHO. Guidelines for drinking-water quality. Addendum to vol. 1. Recommendations, 2nd ed. Geneva: World Health Organisation, 1998. P. 10 –11.

WHO: 1984, *Guidelines for Drinking Water Quality*, Geneva.

Williams M, Fordyce F, Paijiprapapon A, Charoenchaisri P. Arsenic contamination in surface drainage and groundwater in part of the Southeast Asian tin belt, Nakhon Si Thamarat Province, southern Thailand. Environ Geol 1996;27:16 –33.

8

Radiolarian Age Constraints of Mid-Cretaceous Black Shales in Northern Tunisia

Ben Fadhel Moez[1], Soua Mohamed[2], Zouaghi Taher[1],
Layeb, Mohsen[3], Amri Ahlem[1] and Ben Youssef Mohamed[1]
[1]CERTE, Technopole de Borj Cédria,
[2]Entreprise Tunisienne d'Activités Pétrolières,
ETAP-CRDP 4 Rue des Entrepreneurs, 2035 la Charguia II,
[3]ISMP, Tunis,
Tunisia

1. Introduction

Mid-Cretaceous pelagic deposits outcropping in Northern Tunisia include organic-rich beds locally associated with high abundance of radiolarian microfauna, which are interpreted as the signature of the two global oceanic anoxic events OAE1 and OAE2 (Talbi, 1991; Saïdi & Belayouni; 1994; Caron et al., 1999; Amédro et al., 2005, Heldt et al., 2008; Khazri et al., 2009; Soua et al., 2009; Robascynski et al., 2010; Ben Fadhel et al., 2011). Several studies have stated the close association between organic-rich sediments and radiolarian in the Atlantic and Tethyan realms (Marcucci-Passerini et al, 1991; O'Dogherty, 1994; Erbacher & Thurow, 1998; Danelian et al., 2004, 2007).

In North African margins, the radiolarian biostratigraphy have focused upon radiolarian-bearing Jedidi Formation which has been thoroughly discussed by Cordey et al, (2005) and Boughdiri et al, (2007). The first attempts at dating radiolarian series in Northern Tunisia show that radiolarian associated with carbonate-siliceous beds, have yielded useful diagnostic radiolarian assemblages (Cordey et al., 2005; Soua et al., 2006; Ben Fadhel et al., 2010).

Albian and Cenomanian-Turonian black shales of Northern Tunisia were considered to have good generative oil source rock (Layeb, 1990; Saidi & Belayouni, 1994; Bechtel et al., 1998; Ben Fadhel et al., 2011). In this overall context, the restudy and high-resolution biostratigraphy of Albian black shale beds of Lower Fahdene Formation and C/T cherty beds of organic-rich Bahloul Formation outcropping in Northern Tunisia domain have yielded well-preserved and age-diagnostic radiolarians species.

The aim of this paper is to: 1) give new illustrations of radiolarian taxa recovered from albian pelagic deposits of north african margins 2) establish a direct age of black shales using radiolarian assemblages 3) compare the radiolarian assemblages with time equivalent investigated in tethyan and east Pacific domains.

2. Geological setting

The area of investigation is located in Northern Tunisia (Fig. 1). Three sections are selected in this study on the basis of occurrence of organic and radiolarian-rich layers:

1. The Jebel Srassif area (Fig. 1a) is located in the northwestern extremity of the 'Dome Belt', a complex structure linked to Triassic extrusions and strike-slip faults. According to Chikhaoui et al., (1991) and Chikhaoui & Turki (1996), the observed structural complexity was the result of the extensive tectonic movement, which led to the extrusion of Triassic evaporites during the Albian–Aptian period. Consequently, halokinetic and tilted blocks movements are responsible for the horst and- graben architecture. The so-called 'tectonic corners', described by the previous authors, are induced by the reactivation of strike slip faults during the Tertiary compressive phase.
    The Jebel Srassif section belongs to the subsiding basin of the Mellegue 'paleograben' (Chikhaoui et al., 1991) bordered by two structural highs: Koumine to the west and Nebeur to the east.
    Cretaceous successions are characterized by a thick pile of Aptian to Campanian pelagic sequences, which are affected by multiple non-depositional unconformities and condensed layers (Chikhaoui, 1988)

2. The Fadeloun-Garci-Mdeker structure (fig.1b) in which belongs the Jebel Garci section, is composed of three anticlines, trending North South and considered as the northern prolongation of the N-S axis (Saadi, 1990). The anticlines are separated from the Atlasic domain by the Zaghouan thrust, which its North-eastern part becomes south-verging, commonly defined as the Chérichira-Kondar thrust (Khomsi et al., 2004).
    The Cretaceous sedimentation was under the control of syn-sedimentary faults trending N140-160 reflected by chaotic and gravitational deposits (Saadi, 1991). Early Cretaceous successions show northward, reduced thickness and affected by hiatus and extreme condensations in Hammam Zriba (Saadi, 1990). The motion of a corridor trending north-south by N140-160 faults has led to the compartmentalization of the seafloor in losangic basins (Saadi, 1991).
    During the Valanginian – Barremian time span, theses basins were supplied by siliciclastic deposits while condensed sedimentation occupied uplifted horsts (Biely et al., 1973; Saadi et al., 1994)

3. The Oued Kharroub section (Fig. 1c) is located in the Atlas domain (Northern and Central Tunisia), characterized by various facies of Cenomanian-Turonian transition (C-T) deposits, including benthic fauna-rich carbonates e.g. Zebbag Formation by Burollet (1956) and Gattar Formation by Boltenhagen and Mahjoub (1974) and organic-rich black shales with pelagic fauna (e.g. Bahloul Formation by Burollet, 1956) respectively of shelf and slope in the southern margin of the Tethyan realm. During this period and since the Jurassic, this domain has been influenced by the opening of the Tethyan palaeosea, its deepening as well as its southern margin migration. Generally speaking, the Bargou area, connected palaeogeographically to central Tunisia, is characterized by (1) emerged palaeohighs displaying gaps and discontinuities (Turki, 1985) and (2) subsiding zones affected by deep-water sedimentation. This area is dominated by N140° and N70° trend faults limiting several blocks. Cretaceous sedimentation varies on both sides. Its structural evolution may be summarized as follow : (1) during the late Jurassic to early Cretaceous, the area was subjected to a major extensional phase that delimited horst and graben systems (Martinez & Truillet, 1987) (2) In the uppermost Aptian, a regional compressional pulsation affecting the north-African platform had resulted from a transpressional scheme (Ben Ayed & Viguier, 1981) (3) New NNE-SSW trend anticline structures appeared attested by the Albian Fahdene Formation onlap features on the reefal aptian Serj deposits in subsurface (Messaoudi & Hammouda, 1994) or upper

Aptian - Albian unconformity in outcrops (Ouahchi et al., 1998). (4) During the Albian, the geodynamic evolution is marked by the sealing of lower Cretaceous structures during an extensional phase that persisted to form graben systems promoting organic-rich and siliceous strata deposition throughout upper Cenomanian to Lower Turonian times (Soua et al., 2009). The major faults in this area are represented by N140° and N70° trend features. The Bahloul thickness is significantly variable in this area. It may varies from 10m to 40m in thickness (Layeb & Belayouni, 1989; Soua & Tribovillard, 2007). Uniquely, in this area, the top of the Bahloul represents many cenomanian olistolith levels (Soua et al., 2006) marking syndepositional tectonic activities (Turki, 1985).

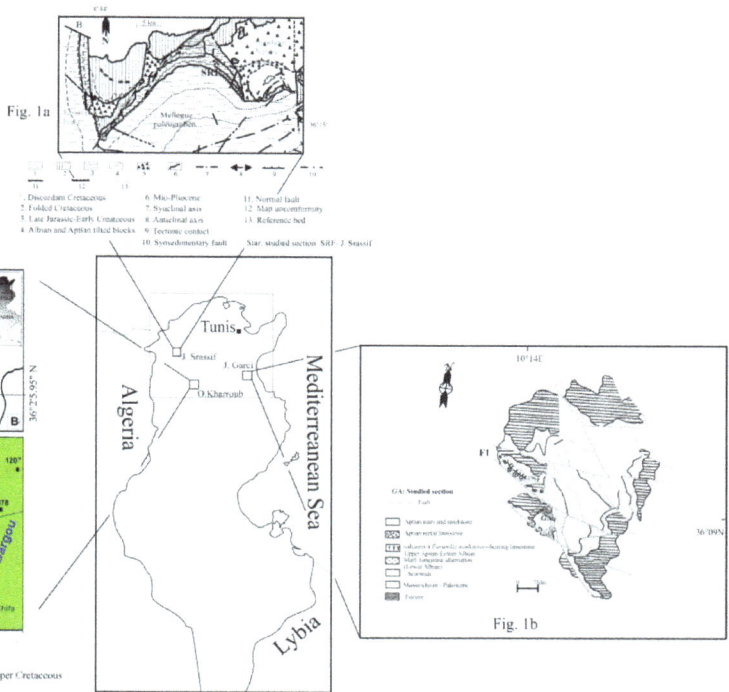

Fig. 1. Geological map of the studied sections (After Chikhaoui et al., 1991; Meddeb, 1986)

3. Result

3.1 Jebel Garci section

The condensed section of Jebel Garci (Fig. 2) begins with orbitolinids-rich green to gray clay alternating with discontinuous sandy limestone beds which are attributed to the Hameima Formation. The clay intervals have also provided fragments of rudist and bryozoans (GA1). The upper part contains olistolites deposits that gradually pass to a reefal limestone which is outlined at the top by burrowed hardground.

The next successions (GA9 - GA23) which correspond to the "Allam" Member consist of centimeter-thick grey to dark laminated limestone bed and organic-rich black marl intervals.

Upwards, the succession becomes rhythmic and the marly intervals increase in thickness in opposition to limestone beds. The microfauna content yields depauperate planktic foraminiferal assemblages and radiolarian rich microfauna.

Fractures related to a strike slip fault outlining the black hales unit are onlapped by a marly intervals and gray limestone beds alternation (GA24-GA27).

The organic-rich beds (GA17 – GA 23) have released a moderately to well-preserved and age-diagnostic radiolarian species. Twenty nine species were recorded in the studied section.

The radiolarians appear with few discrete taxa within GA6 level. It provides an assemblage composed of *Holocryptocanium barbui* Dumitrica, *Spongostichomitra elatica* (Aliev), *Pseudoeucyrtis hanni* (Tan), *Archeodictyomitra vulgaris* Pessagno, *Thanarla brouweri* (Tan), *Stichomitra simplex* (Smirnova and Aliev), *Angulobracchia portmanni* Baumgartner, *Thanarla pacifica* Nakaseko and Nishimura. They become diversified and abundant within GA7. It yields an association of *Dictyomitra* aff. *gracilis* (Squinabol), *Dictyomitra communis* Squinabol, *Dictyomitra montisserei* (Squinabol), *Pseudodictyomitra lodogaensis* Pessagno, *Thanarla praeveneta* Pessagno, *Archaeodictyomitra* aff. *A. vulgaris* Pessagno, *Hiscocapsa sp.*, *Thanarla aff.pulchra* (Squinabol), *Spongostichomitra elatica* (Aliev), *Thanarla brouweri* (Tan), *Angulobracchia portmanni* Baumgartner, *Stichomitra simplex* (Smirnova and Aliev), *Stichomitra communis* Squinabol

GA15 sample provided very diversified and abundant radiolarian population. It is composed by *Dictyomitra montisserei* (Squinabol), *Holocryptocanium barbui* Dumitrica *Pseudodictyomitra lodogaensis* Pessagno, *Stichomitra simplex* (Smirnova and Aliev), *Pseudoeucyrtis hanni* (Tan), *Diacanhocapsa sp.*, *Hiscocapsa grutterinki* (Tan) *Angulobracchia portmanni* Baumgartner, *Stichomitra communis* Squinabol, *Pseudodictyomitra paronai* (Aliev), *Cryptamphorella conara* (Foreman)

GA18 provided *Dictyomitra gracilis* (Squinabol), *Thanarla conica* (Squinabol)

The upper part of black shales (G17-23), composed by rhythmic bundles of limestone and marl beds, is characterized by a decrease of radiolarian abundance. The sample GA20 has released a radiolarian assemblages composed of *Thanarla brouweri* (Tan), *Spongostichomitra phalanga* O' Dogherty, *Pseudodictyomitra paronai* (Aliev), *Dictyomitra communis* Squinabol, *Holocryptocanium barbui* Dumitrica, *Pseudodictyomitra lodogaensis* Pessagno, *Dictyomitra gracilis* (Squinabol), *Dictyomitra montisserei* (Squinabol), *Spongostichomitra elatica* (Aliev), *Pseudodictyomitra paronai* (Aliev)

Although the uppermost beds have yielded (GA24-27) benthic foraminiferal-rich assemblages, we identified well-preserved radiolarian population (GA24) composed of *Pessagnobrachia rara* (Squinabol), *Stichomitra communis* Squinabol, *Dictyomitra montisserei* (Squinabol), *Cryptamphorella conara* (Foreman), *Holocryptocanium barbui* Dumitrica, *Pseudodictyomitra lodogaensis* Pessagno, *Torculum coronatum* (Squinabol), *Xitus spicularius* (Aliev), *Obeliscoites vinassai* (Sqinabol), *Hiscocapsa asseni* (Tan), *Thanarla pulchra* (Squinabol), *Dictyomitra communis* Squinabol, *Hiscocapsa grutterinki* (Tan)

Marly interval of the top GA27 have released an assemblage of *Holocryptocanium barbui* Dumitrica, *Stichomitra simplex* (Smirnova and Aliev), *Pseudodictyomitra paronai* (Aliev), *Dactyliosphaera maxima* (Pessagno)

3.2 Jebel Srassif section

The base of Jebel Srassif section (Fig. 3) which constitutes the "Marnes Moyennes" Member, consists of 130 meter-thick alternations of grey marl and limestone, which become dark and laminated at the top. A cyclic marl/limestone bundles (10m) can be distinguished having an

Legend

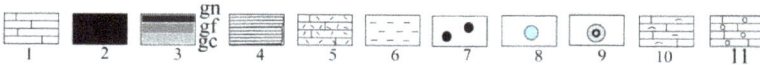

1. Limestone 2. Black shales 3. Marl (gc: light gray, gf:dark to gray, gn:gray to black) 4. Thin laminated limestone
5. Siliceous limestone 6. Shales 7. Iron oxide level 8. Olistolith layers 9. Ammonites horizon 10. Reefal limestone 11. Sandy limestone

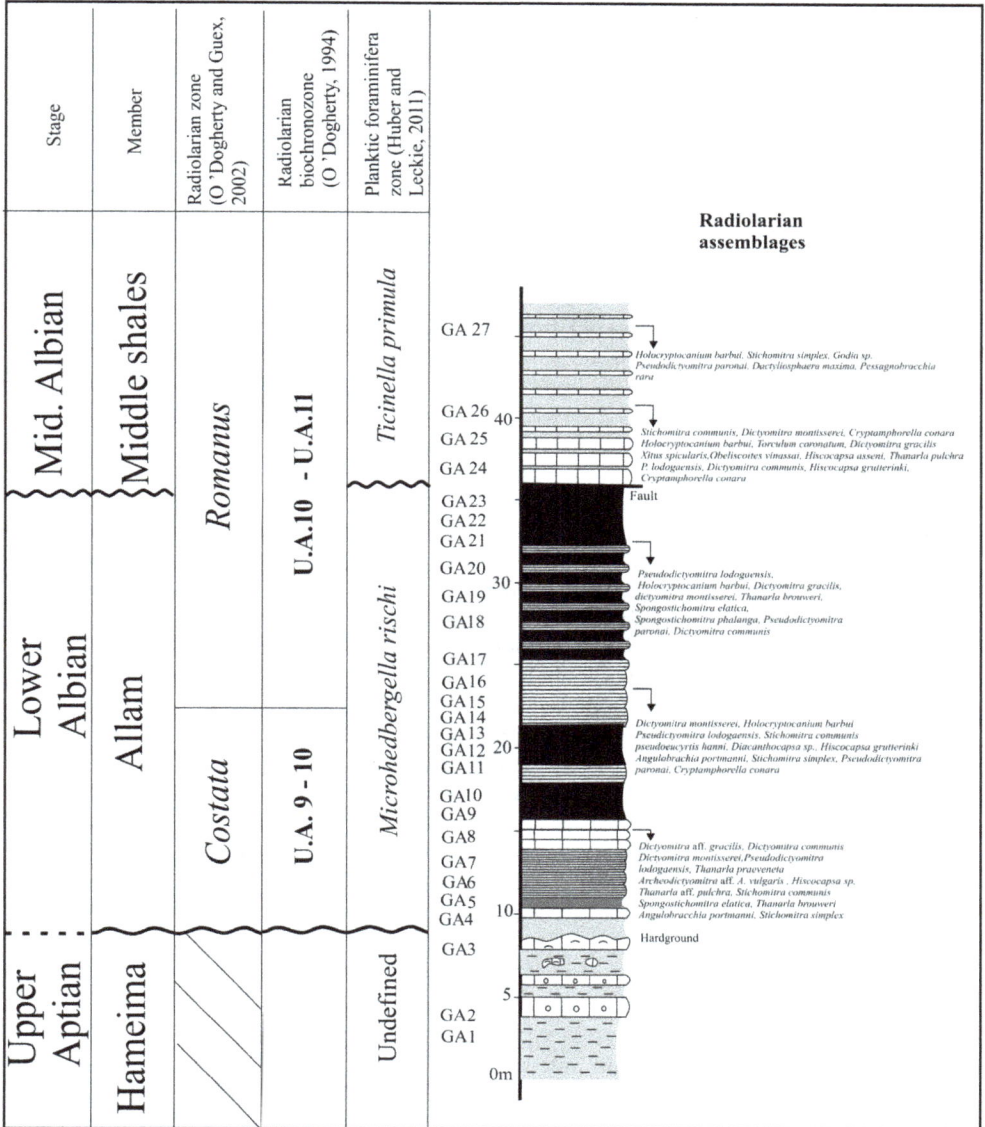

Fig. 2. Jebel Garci section

organic and radiolarian-rich mudstone texture. It is capped by a thick organic-rich limestone bed (20 m) characterized by bituminous odor and yellowish color in patina. This level corresponds to the Mouelha member (Burollet, 1956). The 40 meters of the top consist of an alternation of grey limestones and dark grey-ochre marls yielding septarian nodules characterizing the Defla member. They are overlain by a succession of lenticular limestone beds and grey marl of Azreg member (50 m).

Three samples were selected, based on the good preservation of the faunal assemblages. Among 35 radiolarian morphotypes, only 23 species were figured. Biostratigraphic analysis of the fossil record and planktic foraminifer's zones (Fig. 3, and Plates 1 and 2) correlation allow us to distinguish the following three radiolarian assemblages:

1. The sample 37 has provided a diversified radiolarian fauna with the co-occurrence of *Dictyomitra montisserei* (Squinabol), *Obeliscoites perspicuus* (Squinabol), *Tubilustrium transmontanum* O'Dogherty, *Dictyomitra gracilis* (Squinabol), *Holocryptocanium barbui* Dumitrica, *Stichomitra* aff. *navalis* O'Dogherty, *Cryptamphorella conara* (Foreman), *Torculum dengoi* (Schmidt-Effing), *Stichomitra communis* Squinabol, *Torculum coronatum* (Squinabol), *Distylocapsa micropora* (Squinabol), *Patellula verteroensis* (Pessagno), *Godia concava* (Li &Wu).

2. Radiolarian assemblage recovered from sample 62 is highly diversified at the top of Mouelha blackshales. Likewise, it records an acme of species belonging to Hagiastridae and Cavaspongiidae taxa. This interval shows the co-occurrence of *Dispongotripus acutispinus* Squinabol, *Dactyliosphaera maxima* (Pessagno), *Pessagnobrachia* sp., *Cavaspongia euganea* (Squinabol), *Cryptamphorella conara* (Foreman), *Pessagnobrachia rara* (Squinabol), *Dorypyle communis* (Squinabol), *Pseudodictyomitra paronai* (Aliev), *Pseudodictymitra* sp., *Torculum coronatum* (Squinabol), *Holocryptocanium tuberculatum* Dumitrica, *Distylocapsa micropora* (Squinabol), *Obeliscoites perspicuus* (Squinabol), *Dactyliosphaera acutispina* Squinabol, *Dictyomitra gracilis* (Squinabol), *Thanarla spoletoensis* O'Dogherty, *Dactyliosphaera lepta* (Foreman), *Patellula verteroensis* (Pessagno), *Savaryella novalensis* (Squinabol), *Savaryella quadra* (Foreman), *Pessagnobrachia fabianii* (Squinabol), *Stichomitra communis* Squinabol, *Holocryptocanium barbui* Dumitrica, *Xitus* aff. *spicularius* (Aliev), *Torculum coronatum* (Squinabol), *Crolanium* aff. *spineum* (Pessagno),

3. 3. Sample 68 is characterized by the abundance of cryptocephalic nassellaria (Holocryptocanium). Moreover, we notice the first occurrence and bloom of *Mallanites triquetrus*. This interval shows the co-occurrence of *Xitus mclaughlini* (Pessagno), *Hexapyramis pantanelli* Squinabol, *Mallanites triquetrus* (Squinabol), *Thanarla spoletoensis* O'Dogherty, *Dictyomitra montisserei* (Squinabol), *Godia concava* (Li & Wu), *Cryptamphorella conara* (Foreman), *Torculum coronatum* (Squinabol), *Cavaspongia euganea* (Squinabol), *Distylocapsa micropora* (Squinabol), *Dactyliosphaera maxima* (Pessagno), *Holocryptocanium barbui* Dumitrica, *Dactyliodiscus longispinus* (Squinabol), *Dispongotripus acutispinus* Squinabol.

3.3 Oued Kharroub section

The outcrop (Fig. 4) is composed mainly by dark clayey limestone and organic-rich black shales with abundant planktic foraminifera. These organic-rich deposits include siliceous beds with abundant radiolarians, an equivalent to "Livello Bonarelli" bed marker (Marcucci Passerini et al, 1991; Salvini and Marcucci Passerini, 1998; Premoli-Silva et al, 1999; Scopelliti et al, 2004; Musavu-Moussavou et al, 2007)

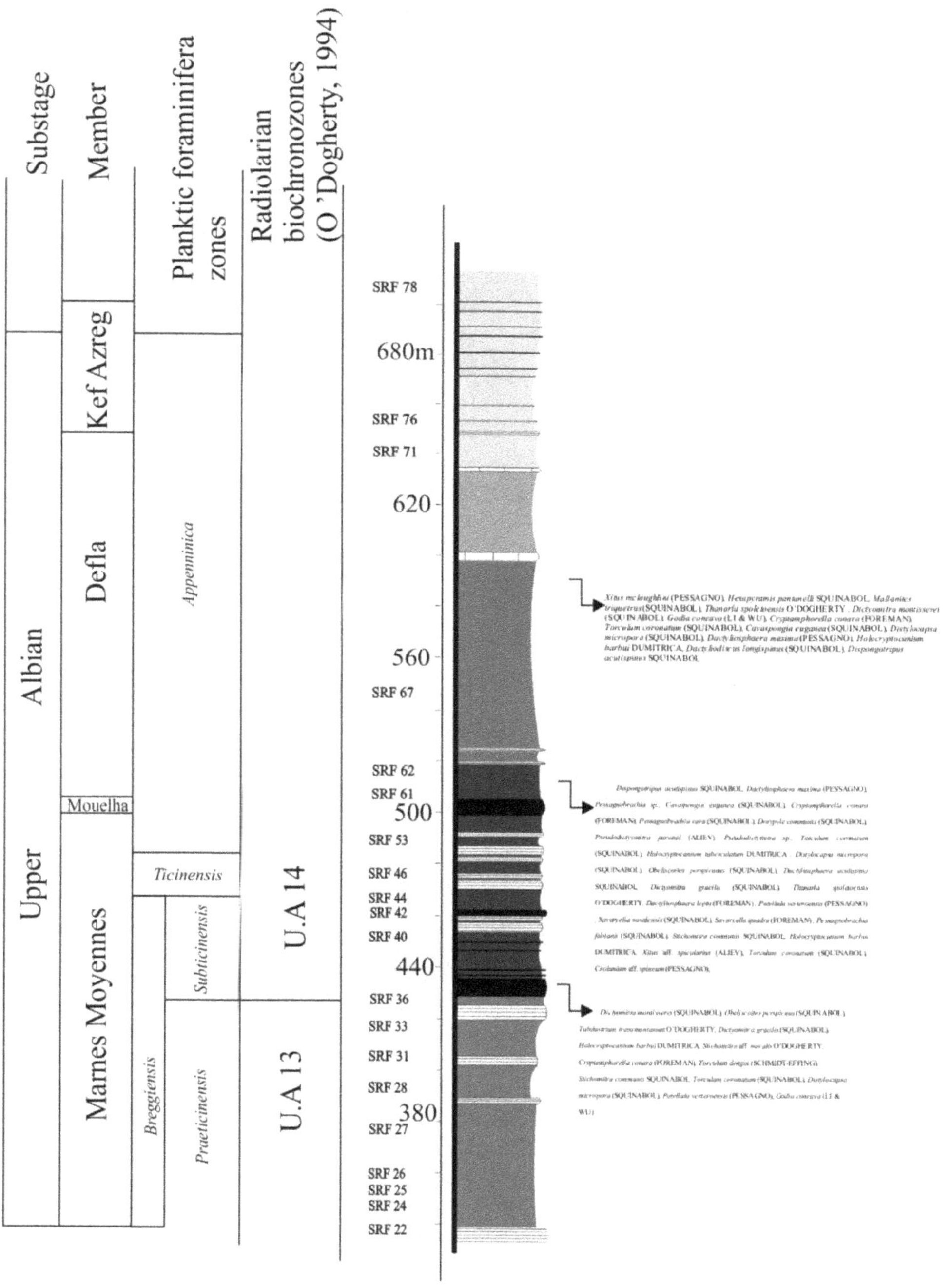

Fig. 3. Jebel Srassif section

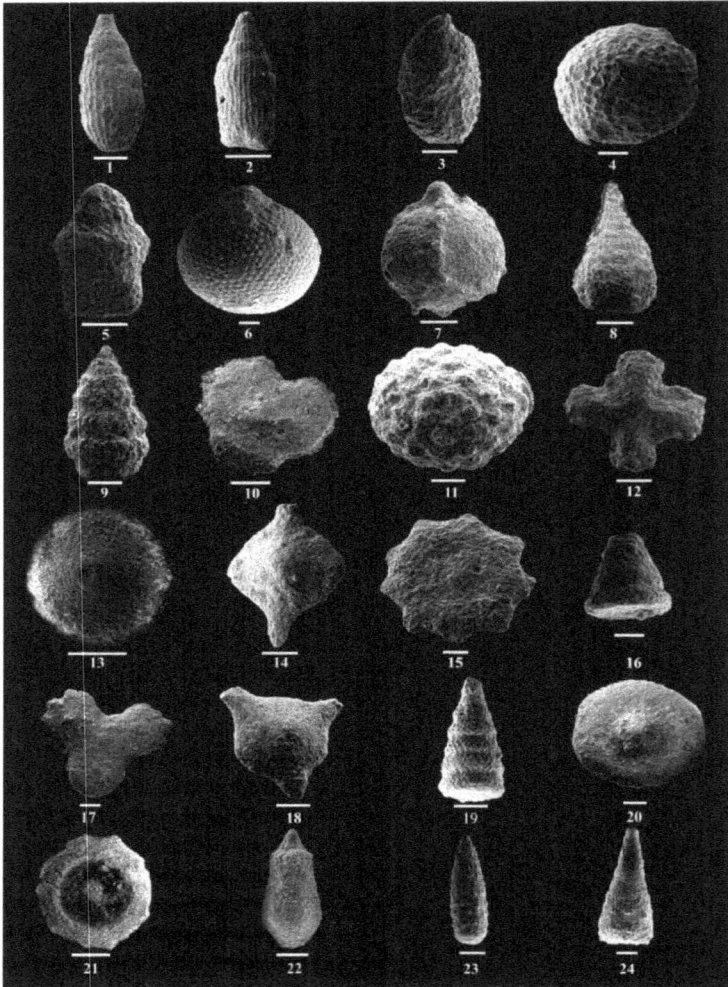

Plate 1. 1 – *Dictyomitra gracilis* (SQUINABOL), scale bar: 50m, sample 62. 2 – *Dictyomitra montisserei* (SQUINABOL), scale bar: 50m, sample 68. 3 – *Tubilustrium transmontanum* O'DOGHERTY, scale bar: 50m, sample 37. 4 – *Holocryptocanium barbui* DUMITRICA, scale bar: 50m, sample 37. 5 – *Stichomitra* aff. *navalis* O'DOGHERTY, scale bar: 50m, sample 37. 6 – *Cryptamphorella conara* (FOREMAN), scale bar: 50m, sample 37. 7 – *Mallanites triquetrus* (SQUINABOL), scale bar: 100m, sample 68. 8 – *Dictyomitra gracilis* (SQUINABOL), scale bar: 100m, sample 62. 9 – *Xitus* aff. *spicularius* (ALIEV), scale bar: 100m, sample 62. 10 – *Dispongotripus acutispinus* SQUINABOL, scale bar: 100m, sample 62. 11 – *Holocryptocanium tuberculatum* DUMITRICA, scale bar: 50m, sample 62. 12 – *Savaryella quadra* DUMITRICA, scale bar: 100m, sample 62. 13 – *Patellula verteroensis* (PESSAGNO), scale bar: 150m, sample 37. 14 – *Dactyliosphaera lepta* (FOREMAN), scale bar: 50m, sample 62. 15 – *Dactyliodiscus longispinus* (SQUINABOL), scale bar: 50m, sample 68. 16 – *Torculum dengoi* (SCHMIDT-EFFING) Scale bar:

50m, sample: 37. 17 – *Pessagnobrachia* sp., scale bar: 100m, sample 62. 18 – *Cavaspongia euganea* (SQUINABOL), scale bar: 100m, sample 62. 19 – *Stichomitra communis* SQUINABOL, scale bar: 50m, sample 62. 20 – *Dactyliosphaera maxima* (PESSAGNO), scale bar: 100m, sample 62. 21 – *Godia concava* (LI &WU), scale bar: 100m, sample 37. 22 – *Torculum coronatum* (SQUINABOL), scale bar: 150m, sample 62. 23 – *Crolanium* aff. *spineum* (PESSAGNO), scale bar: 100m, sample 62. 24 – *Obeliscoites perspicuus* (SQUINABOL), scale bar: 100m, sample 62.

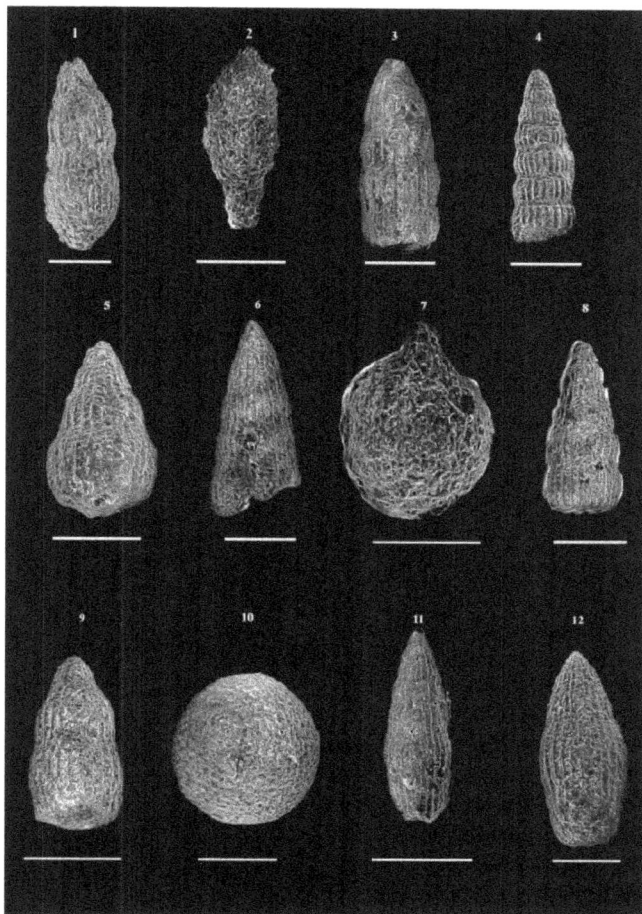

Plate 2. 1 - *Dictyomitra gracilis* (SQUINABOL), 100μm, GA18. 2 - *Pseudoeucyrtis hanni* (TAN), 100μm, GA15. 3 - *Dictyomitra montisserei* (SQUINABOL), 100μm, GA21. 4 - *Pseudodictyomitra lodogaensis* PESSAGNO, 100μm, GA21. 5 - *Thanarla praeveneta* PESSAGNO, 100μm, GA8. 6 - *Archeodictyomitra* aff. *vulgaris* PESSAGNO, 100μm, GA8. 7 - *Hiscocapsa sp.* 100μm, GA8. 8 - *Dictyomitra communis* (SQUINABOL), 100μm. GA8. 9 – *Thanarla* aff.*pulchra,*(SQUINABOL), 100μm, GA8. 10 - *Holocryptocanium barbui* DUMITRICA, 100μm, GA20. 11 - *Dictyomitra gracilis* (SQUINABOL), 200μm, GA9. 12 - *Thanarla brouweri* (TAN), 100μm, GA20.

A total of twenty five of radiolarian species are recognized belonging to nassellarians and spumellarians with maximum of eighteen (18) species in sample OKS 11. Their differential stratigraphical range and relative abundance allow to distinct two successive assemblages (R$_I$ and R$_{II}$) through the C-T transition.

Although, the studied radiolarian species do not exhibit a good potential for biostratigraphic dating, the section is calibrated either by foraminifers and ammonites.

The R$_{II}$ assemblage spans the upper part of the OAE-2 interval and the organic-poor deposits overlying this interval. It is characterized by a decrease trend of the nassellarian relative abundances (from 87% to 42%). Therefore, maybe dissolution or bad preservation conducted to the absence of this group close to the base of the upper half of the section, across the OKS40-OKS45 samples interval. Many species show rapid and gradual disappearing following a stepwise-like pattern (e.g. *Guttacapsa* sp., *Spongostichomitra elatica*, *Novixitus* sp., *Stichomitra stocki*, *Mita gracilis*, *Pseudodictyomitra pseudomacrocephala*, *Thanarla pacifica D. montisserei*).

About the associated spumellarians, several species from the R$_I$ assemblage persisted more or less long time (e.g. *Archaeocenosphera* aff. *vitalis*, *Crucella messinae*, *Praeconocaryomma lipmanae*, *Rhopalosyringium hispidum Pyramispongia glascockensis* Pessagno., *Cavaspongia euganea* (Squinabol),, *C. Californiaensis* Campbell and Nishimura, *Pseudoeucyrtis spinosa* (Squinabol), *Archaeocenosphaera ? mellifera* O'Dogherty,). Nevertheless, very few species of nassellarians first occurred across the upper half part of the studied section. All these species are represented by dwarf and poorly preserved specimens.

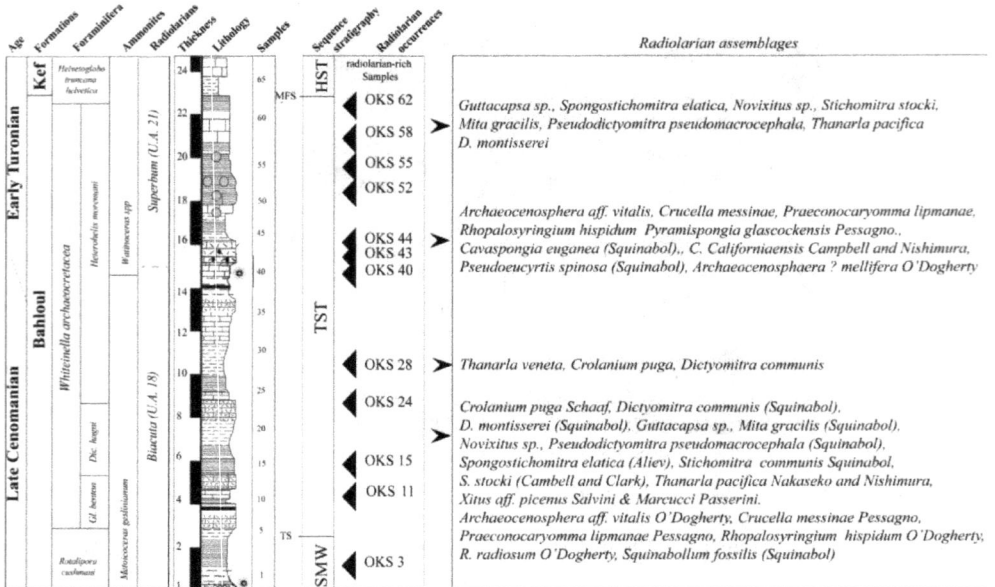

Fig. 4. Oued Kharroub section

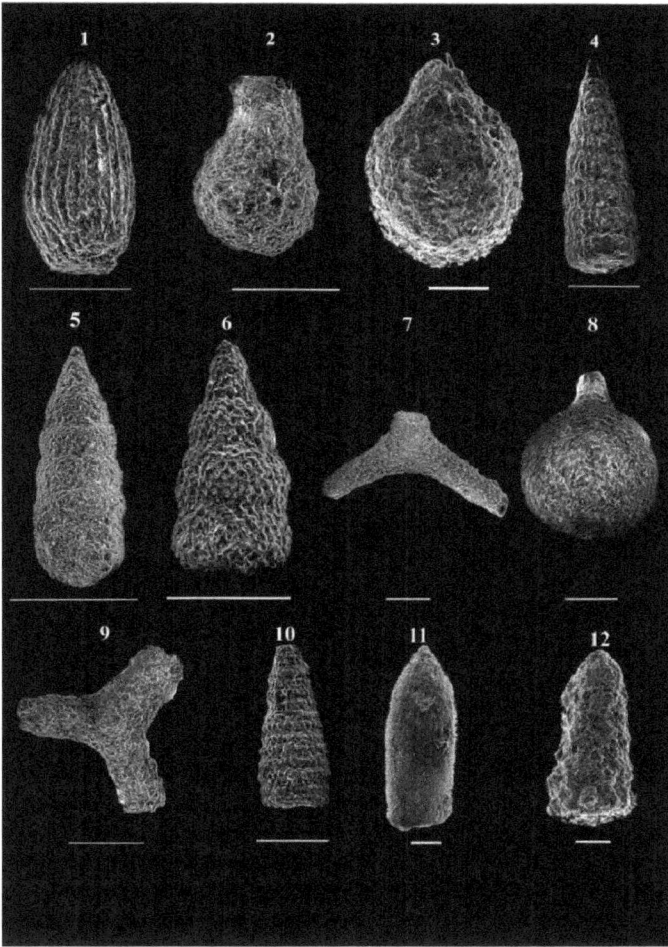

Plate 3. 1 - *Thanarla conica* (SQUINABOL), 100μm, GA18. 2 - *Diacanthocapsa sp.* 200μm, GA15. 3 - *Cryptamphorella conara* (FOREMAN), 50μm, GA22. 4 - 8 – *Pseudodictyomitra paronai* (ALIEV), 100μm, GA20. 5 – *Torculum coronatum* (SQUINABOL), 200μm, GA21. 6 - *Stichomitra simplex* (SMIRNOVA et ALIEV), 100μm, GA15. 7 - *Angulobracchia portmanni* BAUMGARTNER, 100μm, GA15. 8 - *Hiscocapsa* aff. *grutterinki* (TAN) 50μm, GA23. 9 – *Pessagnobrachia rara* (SQUINABOL), 100μm, GA27. 10 - *Xitus spicularius* (ALIEV), 100μm, GA25. 11 - *Spongostichomitra elatica* (ALIEV), 50μm, GA20. 12 - *Stichomitra communis* SQUINABOL, 50μm, GA8.

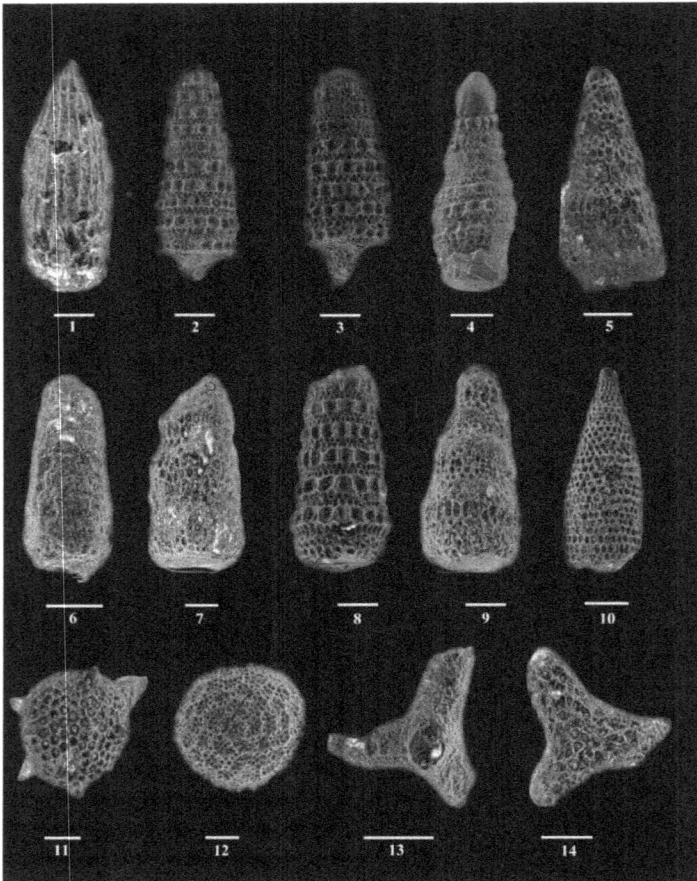

Plate 4. 1- *Mita gracilis* (?) (Squinabol), scale bar: 100, OKS-40. 2- *Novixitus* (?) sp., scale bar: 100, OKS-43;3-*Novixitus* sp. Scale bar: 100, OKS-42. 4- *Pseudodictyomitra pseudomacrocephala* (Squinabol), scale bar: 100, sample OKS-11. 5- *Guttacapsa* sp. Scale bar: 50, OKS-24. 6- indetermined species, scale bar: 100, OKS-24. 7- indetermined species, scale bar: 50, sample OKS-58. 8- *Pseudodictyomitra* sp. Scale bar: 100, OKS-15. 9- *Xitus* aff. *picenus* Salvini and Marcucci Passerini, scale bar: 50, OKS-40. 10- *Phalangites* (?) sp. Scale bar: 50, OKS-58; 11- *Squinabollum fossile* (Squinabol), scale bar: 50, OKS-28. 12- *Archaeocenosphaera? mellifera* O'Dogherty, scale bar: 50, OKS-62. 13- *Cavaspongia* sp., scale bar : 50, OKS-3. 14- *Cavaspongia californiaensis* Pessagno, scale bar: 50, OKS-58.

4. Discussion

Detailed analysis of radiolarian assemblages allows us to attribute a biostratigraphic framework for the organic-rich beds. In the following section, we used zonal scheme proposed by O'Dogherty (1994) for tethyan realms. The age-diagnostic assemblages are discussed and compared with time equivalent investigated in adjacent tethyan domains.

O'Dogherty (1994) proposed a radiolarian zonation for the Albian based on Unitary Associations. He described for the Upper Albian to the base of the Cenomanian the Spoletoensis zone divided into three radiolarian subzones: the Romanus, Missilis and Anisa subzones. Bak (1995) established a radiolarian zonation (H. barbui – H. geysersensis) for the Albian–Cenomanian of northern Tethyan domains, based on the co-occurrence of *Holocryptocanium barbui* DUMITRICA, *Holocryptocanium geysersensis* PESSAGNO, *Novixitus weyli* SCHMIDT-EFFING, *Squinabollum fossile* (SQUINABOL), *Crymptamphorella macropora* DUMITRICA, *Hemicryptocapsa tuberosa* DUMITRICA

The first appearance of *B. breggiensis*, recorded within Upper Albian basal intervals, coincides with first appearance of radiolarian species *Tubilustrium transmontanum* O'DOGHERTY, which is confined with the upper part of the Romanus subzone (O'Dogherty, 1994). An assemblage containing *Stichomitra navalis* and *Torculum coronatum* was recorded also within this subzone.

Babazadeh & de Wever (2004) described a radiolarian assemblage yielding the co-occurrence of *Dictyomitra gracilis*, *Holocryptocanium barbui* and *Dictyomitra montisserei* and assigned it to Middle–Late Albian age. Nevertheless, the presence of *T. dengoi*, whose first appearance coincides with the Missilis – Anisa subzones boundary (O'Dogherty, 1994), allows rejuvenating the assemblage age.

Samples recovered from the succession overlying the Mouelha Member blackshales show an assemblage composed of *Cryptamphorellla conara* Dumitrica, *Pessagnobrachia* sp., and *Thanarla spoletoensis* O'DOGHERTY, which correspond to the lower part of the *Appenninica* zone and the middle part of the Anisa subzone of O'Dogherty (1994). Although the coexistence of *D. lepta*, *Stichomitra communis* and *Patellula verteroensis* is assigned to early Late Cenomanian age (Erbacher, 1998), this assemblage possibly characterize the Late Albian taking into account the presence of *D. maxima* whose last occurrence is coeval with the base of Anisa subzone (O'Dogherty, 1994).

Samples recovered from basal beds (GA2-GA6) show high abundance of *Pseudodictyomitra lodogaensis* and contain some early Cretaceous taxa from *Turbocapsula* Zone such as *A.portmanni* and *Th. pacifica* (O' Dogherty, 1994; Erbacher and Thurow, 1998; Danelian et al., 2007; Michalik et al., 2008). Thus, a late Aptian age of these beds could not hitherto be ruled out.

According to Erbacher & Thurow (1998), the first occurrence of *Pseudodictyomitra lodogaensis* coincides with the upper part of *G. algerianus* Zone. Its last occurrence coincides with the Aptian-Albian boundary and the first occurrence of *Mita gracilis* (= *Dictyomitra gracilis*). This taxon is also reported from the Albian to Cenomanian deposits of the Atlantic domain, California and Pacific realms (Thurow, 1988; Karminia, 2006; Palechek et al., 2010).

It is possible that black shale unit of Jebel Garci could underlines the Aptian-Albian boundary. In fact, Danelian (2008) have reported the presence of *Thanarla praeveneta* from the Upper Aptian – Lower Albian bed which occurs in GA7 beds underlying the black shale successions.

On the other side, Slazcka et al., (2009) described an assemblage containing *Angulobracchia portmanni* Baumgartner, *Dictyomitra communis* (Squinabol), *Hiscocapsa asseni* (Tan), *Pseudodictyomitra lodogaensis* Pessagno, *Pseudoeucyrtis hanni* (Tan), almost similar to GA7 taxa. These authors attributed the assemblage to *Costata* zone that is confined to UA6-9 biochronozones of mid to late Aptian age (O'Dogherty, 1994).

It is noteworthy to point the coexistence of Albian species in all samples such as *D.montisserei* and *D.gracilis* with Aptian taxa particularly in GA7, GA 15 and GA26.

In that score, an assemblage recovered from Mid Cretaceous outcrops of Northern Tethys margins was described by Danelian et al., (2007), shows the co-occurrence of *P. lodogaensis, Dictyomitra gracilis, Thanarla brouweri, Archaeodictyomitra* aff.*vulgaris* assigning it to the early Albian UA10-11 biochronozone. Danelian et al (2004) consider that an early Albian age of Dercourt Member cannot be ruled out despite the presence of *Angulobracchia portmanni* and *pseudoeucyrtis hanni* characteristic of U.A.9. These species are observed hitherto within assemblage from GA15, associated with *Dictyomitra montisserei.*

Kurilov & Vishnevskaya (2011) described an assemblage extracted from Early Cretaceous outcrops of Pacific domain that does not differ from GA21. It contains *Thanarla brouweri, Pseudodictyomitra paronai, Pseudodictyomitra lodogaensis, Holocryptocanium barbui, Dictyomitra cf. montisserei, Dictyomitra communis, and Dictyomitra gracilis* indicating an early Albian age.

The sample GA26 has provided an assemblage characterized by high abundance of *Hiscocapsa asseni,* co-occurring with *D.gracilis* and *D.montisserei.* It lies with the UA10 biochronozone of Romanus zone (O'Dogherty, 1994; Danelian et al., 2004).

We suggest that lower part of black shale intervals could be assigned to the upper part of *Costata* zone (GA5 – GA14) based on the presence of Aptian taxa (i,e. *Angulobracchia portmanni, Pseudoeucyrtis hanni*). The lower part of this zone coincide with the first occurrence of *Microhedbergella praeplanispira* planktic foraminifera. Whereas the top coincide with the last occurrence of *Angulobracchia portmanni* and *Pseudoeucyrtis hanni* associated with a relative increase in abudance of Archaedictyomitrae and Williriedellidae families.

The Romanus zone (GA14 – GA27) show the dominance of high diversified nassellarian species. The assemblage recovered from GA17 is composed of *Thanarla brouweri, Archaeodictyomitra montisserei, Thanarla conica* which is attributed to the middle Albian *Mallanites* romanus subzone (U.A. 10 -11 biochronozone) (O'Dogherty, 1994; Danelian et al, 2004). However, the first occurrence of *Ticinella primula* planktic foraminifera is recorded 24 m above GA17 bed. Thus, we suggest that lower part of Romanus zone may be attributed to the Early Albian.

Studies on Cenomanian - Turonian boundary interval show that deposition of radiolarian, organic-rich sediment and large positive carbon isotopic excursion are coeval with extreme fertility conditions and correspond to a large-scale proxy that indicate a hypersiliceous period (Premoli Silva et al, 1999; Racki & Cordey, 2000)

The Bonarelli equivalent in Tunisia is commonly known by the Bahloul Formation (Burollet, 1956). In the Bargou area, the Bahloul Formation shows organic-rich intervals interbedding cherty and radiolarian limestone layers (Layeb and Belayouni 1999, Soua and Tribovillard, 2007)

Although the C/T boundary interval outcropping in the Tunisian realm was extensively studied by planktic foraminifera and ammonite biostratigraphy (Maamouri et al, 1994; Nederbragt and Fiorentino, 1999; Abdallah et al., 2000; Amédro et al, 2005), radiolaria assemblages have provided a useful tool for age calibration and subdivision of C/T organic-rich beds in this study.

Two black shale levels were identified in Oued Kharroub section:
1.   The first lies with the lower part of *Withinella archaeocretacea* planktic foraminifera zone, above the highest occurrence of *Rotalipora cushmani*
2.   The second coincides with the middle part of *Heterohelix moremani* zone

The calibration of these levels is based on age-diagnostic radiolarian recovered from biosiliceous limestone beds (Fig. 4).

The OSK 24 yields an assemblage composed of *Rhopalosyringium radiosum* O'Dogherty, *Praeconocaryomma lipmanae* Pessagno, *Acaeniotyle vitalis* O'Dogherty *Rhopalosyringium hispidum* O'Dogherty. The three first taxa have been described by Bak et al (2005) and attributed them to the late Cenomanian – early Turonian. Erbacher (1998) attribute *Rhopalosyringium radiosum* to the early Turonian, but later Musavu-Moussavou and Danelian (2006) expand its range to late Cenomanian. The assemblage contains *Xitus picenus* Salvini and Marcucci - Passerini which its range do not exceed the Silviae Zone of Bonarelli (O'Dogherty, 1994; Salvini & Marcucci-Passerini, 1998). Consequently, we assign the lower black shale beds (OKS 11) to the late Cenomanian and to upper part of *Silviae* Zone [U.A 18 biochronozone of O'Dogherty (1994)]

Many authors have stated the occurrence of *Archaeocenosphaera mellifera* O'Dogherty within Turonian strata of Boreal and northern Tethyan domains (Bandini et al., 2006; Smreckova, 2011). In East Pacific domain, this taxon, associated with *C. californaensis* and *Pyramispongia glascockensis* PESSAGNO, is recorded within the Silviae Zone of late Cenomanian age (Bragina, 2009). Salvini & Marcucci-Passerini (1998) stated that *C. californiaensis* occurs only in the base of upper assemblage C of Bonarelli Level which lies with the base Superbum Zone defined by O'Dogherty (1994). In the Atlantic domain, the last occurrence of *C. californiaensis* is recorded in the late Cenomanian just beneath the organic-rich beds related to the OAE2 (Musavu-Moussavou and Danelian, 2006). Taking into account the paleogeographic similarities between northern and southern Mediterranean Tethys margins, the radiolarian assemblage recovered from OKS44 level could be correlated with upper assemblage (Superbum Zone) of Bonarelli level in Central Italy. Thus, the second black shale lie with the upper part of *Biacuta* subzone of late Cenomanian age, if we take into consideration the position of turonian *Watinoceras* spp. ammonite (Amédro et al., 2005)

5. Conclusion

Biostratigraphic investigations of Albian and C/T boundary intervals in Northern Tunisia show that organic-rich beds are generally associated with high abundance of radiolarian fauna.

Age constraint of organic-rich sediments is established and correlated with biochronozones of O'Dogherty (1994). In the light of these results, we deduce that:

1. Black shale interval of Jebel Garci which is embedded within the "Allam" Member is assigned to the early Albian U.A.10 biochronozone. However, the latest Aptian could not be excluded for the lower part.

2. Late Albian organic-rich beds of Jebel Srassif including cyclic limestone/marl beds of "Marnes Moyennes" and Mouelha Members lie with the boundary interval between U.A. 13 and U.A. 14 biochronozones.

3. Two black shale levels embedded within Bahloul Formation are probably of late Cenomanian age and confined with the U.A.18 biochronozone. The first occurrence of turonian *Watinoceras* spp. ammonite is recorded 70 cm above the second black shale bed (OSK40)

It seems that distribution of radiolarian assemblages of albian and cenomanian-turonian boundary intervals shows some difference from those of Atlantic and east Pacific domains. Preservation index and range discrepancies of some radiolarian species could affect the subdivision resolution. Further studies on radiolarian distribution assemblages and relationships with environmental changes during Mid-Cretaceous time are needed to establish paleogeographic reconstructions of southern tethyan margins.

6. Acknowledgments

We thank Drs Moncef Saidi and Hedia Bessaies from ETAP center research for giving all facilities needed for SEM photographs. Authors gratefully acknowledge Dr. Imran Ahmad Dar, Editor-in-chief, for accepting the publication of this work.

7. References

Abdallah, H., Sassi, S., Meister, C., & Souissi, R. (2000). Stratigraphie séquentielle et paléogéographie à la limite Cénomanien–Turonien dans la région de Gafsa–Chotts (Tunisie centrale). *Cretac. Res.,* vol.21, No. 1, pp. 35–106, ISSN 01956671

Amédro F., Accarie H., & Robaszynski F. (2005). Position de la limite Cénomanien-Turonien dans la Formation Bahloul de Tunisie centrale : apports intégrés des ammonites et des isotopes du carbone (δ13 C). *Eclogae Geologicae Helvetiae,* vol. 98, No. 2, pp. 151-167, ISSN 0012-9402

Babazadeh, S. A. & De Wever, P. (2004a). Radiolarian Cretaceous age of Soulabest radiolarites in ophiolite suite of eastern Iran. *Bulletin de la Société Géologique de France,* Vol. 175, No.2, pp. 121-129, ISSN 0037-9409

Bak, M. (1995). Mid Cretaceous radiolarians from the Pieniny Klippen Belt, Carpathians. Poland. *Cretaceous Research,* vol. 16, No. 1, pp. 1–23, ISSN 01956671

Bak, M., Bak, K. & Ciurej, A. (2005). Mid-Cretaceous spicule-rich turbidites in the Silesian Nappe of the Polish Outer Carpathians: radiolarian and foraminiferal biostratigraphy. *Geological Quarterly,* vol, 49, No. 3, pp. 275-290, ISSN 1641-7291

Bandini, A.N, Baumgartner, P.O. & Caron, M. (2006). Turonian Radiolarians from Karnezeika, Argolis Peninsula, Peloponnesus (Greece). *Eclogae geol. Helv.* 99 Supplement, Vol.2, pp.1–20, ISSN 0012-9402

Biely, A., Memmi, L., Salaj, J. & (1973). Le Crétacé inferieur de la région d'Enfidaville. Découverte d'Aptien condense, *Livr. Jub. M. Solignac, Ann. Min. Geol.,* vol. 26, pp.169-l 78, ISSN: 03300013

Bechtel, A., Savin, S.M. & Hoernes, S. (1999). Oxygen and hydrogen isotopic composition of clay minerals of the Bahloul Formation in the region of the Bou Grine zinc–lead ore deposit (Tunisia): evidence for fluid–rock interaction in the vicinity of salt dome cap rock. *Chemical Geology,* Vol. 156, Issues 1-4, pp. 191-207, ISSN 0009-2541

Ben Ayed, N. & Viguier, C. (1981). Interprétation structurale de la Tunisie atlasique. *CRAS* Paris, t 292 série II pp. 1445-1448, ISSN 16310713

Ben Fadhel, M., Layeb, M. & Ben Youssef, M. (2010). Upper Albian planktic foraminifera and radiolarian biostratigraphy (Nebeur – northern Tunisia). *Comptes Rendus Palevol,* vol. 9, No. 3, pp. 73 – 81, ISSN 16310713

Ben Fadhel, M., Layeb, M., Hedfi, A. & Ben Youssef, M. (2011). Albian Oceanic Anoxic Events in northern Tunisia: Biostratigraphic and geochemical insights. *Cretaceous Research* vol. 32, No. 6, pp. 685-699, ISSN 01956671

Boltenhagen, C. & Mahjoub, M. N. (1974-79). Divers rapports inédits sur la géologie du Cretacé moyen de Tunisie centrale. Archives SEREPT.

Boughdiri, M., Cordey, M., Sallouhi, H., Maalaoui, K., Masrouhi, M. & Soussi, M. (2007). Jurassic radiolarian-bearing series of Tunisia: biostratigraphy and significance to western Tethys correlations. *Swiss Journal of Geoscience,* vol. 100, No.3, pp. 431–441, ISSN 16618726.

Bragina, L.G. (2009). Radiolarians and Stratigraphy of Cenomanian–Coniacian Deposits in the Crimean and West Sakhalin Mountains, Pt. 1: Biostratigraphic Subdivision and Correlation. *Stratigraphy and Geological Correlation*, vol. 17, No. 3, pp. 316–330, ISSN: 08695938

Burollet, P.F. (1956). Contribution a l'étude stratigraphique de la Tunisie centrale. *Ann. Mines Geol.*, Tunis, n° 18, 350p. IVpl, ISSN 03300013

Caron, M., Robaszynski, F., Amédro, F., Baudin, F., Deconinck, J.F., Hochuli, P., Salis-perch nielsen, K. & Tribovillard, N. (1999). Estimation de la durée de l'événement anoxique global au passage Cénomanien/Turonien. Approche cyclostratigraphique dans la formation Bahloul en Tunisie centrale. *Bull. Soc. géol. France*, vol. 170, No.2, pp. 145–160, ISSN 0037-9409

Chikhaoui, M. & Turki, M. (1996). Rôle et importance de la fracturation méridienne dans les déformations crétacée de la zone des diapirs (Tunisie septentrionale). *J. Afr. Earth Sci.* vol. 21, No. 2, pp. 271–280, ISSN 1464343X

Chikhaoui, M., Turki, M. & Delteil, M. (1991). Témoignage de la structurogenèse de la marge téthysienne en Tunisie, en Jurassique terminal – Crétacé (région d'El Kef Tunisie septentrional). *Geol. Mediterraneenne*, vol.XVIII, No. 3, pp. 125–133, ISSN 0397-2844

Cordey, F., Boughdiri, M. & Sallouhi, H. (2005). First direct age determination from the Jurassic radiolarian-bearing siliceous series (Jedidi Formation) of north-western Tunisia. *Comptes Rendus Geoscience*, vol. 337, No. 8, pp. 777–785, ISSN 16310713

Danelian, T., Tsikos, H., Gardin, S., Baudin, F., Bellier, J.P.& Emmanuel, L. (2004). Global and regional palaeoceanographic changes as recorded in the mid-Cretaceous (Aptian-Albian) sequence of the Ionian zone (northwestern Greece). *Journal of the Geological Society*, vol. 161, No. 4; pp. 703-709, ISSN 00167649

Danelian, T., Baudin, F., Gardin, S., Masure, E., Ricordel, C., Fili, I., Mecaj, T. & Muska, K. (2007). The record of mid Cretaceous oceanic anoxic events from the Ionian zone of southern Albania. *Revue de micropaléontologie*, vol. 50, No. 3, pp. 225–237, ISSN 00351598

Danelian, T. (2008). Diversity and biotic changes of Archaeodictyomitrid Radiolaria from the Aptian/Albian transition (OAE1b) of southern Albania. *Micropaleontology*, 54, No. 1, pp. 3-12, ISSN 00262803

Erbacher, J. (1998). Mid-Cretaceous radiolarians from the Eastern equatorial Atlantic and their paleoceanography. In: *Proceedings of the Ocean Program, Scientific Results* Mascle, J., Lohmann, G.P., Moullade, M. (Ed.), vol. 159, pp.363–373, ISSN 1096-7451 College Station, TX (Ocean Drilling Program)

Erbacher, J. & Thurow, J. (1998). Mid-Cretaceous radiolarian zonation for the North Atlantic: an example of oceanographically controlled evolutionary processes in the marine biosphere? In: *Geological Evolution of Ocean Basins: Results from Ocean Drilling Program,* Cramp, A., Macleod, C.J., Lee, S.V., Jones, E.J.W. (Ed.),. Geological Society, London Special Publications 131, 71–82, ISBN   0585235759 9780585235752

Heldt, M., Bachmann, M. & Lehmann, J. (2008). Microfacies, biostratigraphy, and geochemistry of the hemipelagic Barremian – Aptian: Influence of the OAE 1a on the southern Tethys margin. *Palaeogeography, Palaeoclimatology, Palaeoecology*, vol. 261, No. 3-4, pp. 246–260, ISSN 00310182

Kariminia, S.M. (2006). Upper Jurassic and Lower Cretaceous radiolaria biostratigraphy of California coast ranges. Ph. D. Thesis, University of Texas, Dallas, 143 pp.

Kurilov, D. V. & Vishnevskaya, V. S. (2011). Early cretaceous radiolarian assemblages from the East Sakhalin Mountains. *Stratigraphy and Geological Correlation*, vol. 19, No. 1, pp. 44-62, ISSN: 08695938.

Layeb M. & Belayouni, H. (1989). La formation Bahloul au Centre et au Nord de la Tunisie un exemple de bonne Roche mère de pétrole à fort intérêt pétrolier. Mémoires de l'ETAP, n°3, *Actes des II ème journées de géologie Tunisienne appliquée à la recherche des hydrocarbures* , pp, 489-503, OCLC 32815633, Tunis, Tunisie, Nov - 1989

Layeb, M., 1990. Etude geologique, geochemique et mineralogique, regionale, des facies riches en matiere organique de la Formation Bahloul d'age Cenomano–Turonien dans le domaine de la Tunisie Nord-Centrale. PhD Thesis, Univ. de Tunis II, Tunis, 209 pp.

Layeb, M., and Belayouni, H., 1999. Paléogéographie de la Formation Bahloul (passage Cénomanien – Turonien), *Annales des mines et de la géologie,* vol. 40, pp. 21–44, ISSN 03300013

Maamouri, A.L., Zaghbib-Turki, D., Matmati, M.F., Chikhaoui, M. & Salaj, J. (1994). La Formation Bahloul en Tunisie centro-septentrionale: variations latérales, nouvelle datation et nouvelle interprétation en termes de stratigraphie séquentielle. *J. Afr. Earth Sci.*, vol. 18,No.1, pp. 37–50, ISSN 1464343X

Marcucci Passerini, M., Bettini, P., Dainelli, J. & Sirugo, A. (1991). The "Bonarelli Horizon" in the central Apennines (Italy): radiolarian biostratigraphy. *Cretaceous Research*, vol. 12, No. 3, pp. 321–331, ISSN 01956671

Martinez C. & Truillet R. (1987). Évolution structurale et paléogéographie de la Tunisie. *Memoria de la Societa Italiana de Geologia*, vol. 38, No. 35-45, ISSN 2038-1727

Messaoudi F. & Hammouda F. (1994). Evènement structuraux et types de pièges dans l'offshore Nord-Est de la Tunisie. *Proceedings of the 4th tunisian petroleum exploration conference*, pp. 55-64, OCLC 502635466, Tunis, Tunisia, may 1994).

Michalík, J.,Soták, J., Lintnerová, O., Halásová, E., Bak, M., Skupien, P. & Boorová, D. (2008). The stratigraphic and paleoenvironmental setting of Aptian OAE black shale deposits in the Pieniny Klippen Belt, Slovak Western Carpathians. *Cretaceous research* vol. 29, No. 5-6, pp. 871-892, ISSN 01956671

Musavu-Moussavou, B. & Danelian, T. (2006). The radiolarian biotic response to Oceanic Anoxic Event 2 in the southern part of the Northern proto-Atlantic (Demerara Rise, ODP Leg 207). *Revue de Micropaléontologie*, vol. 49, No. 3, pp. 141-163, ISSN 00351598

Musavu-Moussavou, B., Danelian, T., Baudin, F., Coccioni, R. & Frohlich, F. (2007). The radiolarian biotic response during OAE2. A high-resolution study across the Bonarelli level at Bottaccione (Gubbio, Italy). *Revue de Micropaléontologie*, vol. 50, No. 3, pp. 253–287, ISSN 00351598.

Nederbragt, A. J. & Fiorentino, A. (1999). Stratigraphy and paleoceanography of the Cenomanian-Turonian Boundary Event in Oued Mellegue, north-western Tunisia. *Cretaceous Research*, vol.20, No. 1, pp. 47–62, ISSN 01956671

O'Dogherty, L. (1994). Biochronology and paleoecology of Mid-Cretaceous radiolarians from Northern Apennines (Italy) and Betic Cordillera (Spain). *Mémoires de Géologie (Lausanne)*, 21, 415pp, ISSN 1015-3578

Ouahchi, A., M'Rabet, A., Lazreg, J. Mesaoudi, F. & Ouazaa, S. 1998. Early structuring, paleoemersion and porosity development: a key for exploration of the aptian serdj carbonate reservoir in Tunisia. *Proceedings of the 6th Tunisian petroleum exploration and production conference,* pp.267-284, Tunis, Tunisia , May 5th - 9th, 1998, OCLC 704163368

Palechek, T. N., Savel'ev, D. P. & Savel'eva, O. L. (2010). Albian–Cenomanian Radiolarian Assemblage from the Smaginsk Formation, the Kamchatskii Mys Peninsula of Eastern Kamchatka. *Stratigraphy and Geological Correlation,* vol.18, No. 1, pp. 63–82, ISSN: 08695938

Premoli Silva, I., Erba, E., Salvini, G., Locatelli, C. & Verga, D. (1999). Biotic changes in Cretaceous oceanic anoxic events of the Tethys. *Journal of Foraminiferal Research,* vol.29, No.4, pp. 352–370, ISSN: 00961191.

Racki, G. & Cordey, F. (2000). Radiolarian palaeoecology and radiolarites: is the present the key to the past? *Earth-Science Reviews,* vol.52, No.1-3, pp. 83–120, ISSN 00128252

Robaszynski, F., Zagrarni, M.F., Caron, M. & Amédro, F. (2010). The global bio-events at the Cenomanian-Turonian transition in the reduced Bahloul Formation of Bou Ghanem (central Tunisia). *Cretaceous Research,* vol.31, No.1, pp. 1-15, ISSN 01956671

Saadi, J., (1990). Exemple de sédimentation syntectonique au Crétacé inférieur le long d'une zone de décrochement NS. Les structures d'Enfidha (Tunisie nord-orientale). *Géodynamique,* vol.5, No. 1, pp. 17-33, ISSN 0766-5105

Saadi, J., (1991). Sedimentation en zone mobile coulissante, l'exemple du Cretace inferieur des structures submeridiennes de la region d'Enfidha. (Prolongement septentrional de l'Axe N.S-Tunisie nord-orientale). PhD Thesis, Université de Pau,244 pp.

Saïdi, M. & Belayouni, H. (1994). Etude géologique et géochimique des roche mères albovraconienne dans le domaine de la Tunisie centro-septentrional. *Proceeding of the 4th Petroleum Exploration Conference,* No. 7, pp. 91-116, OCLC 502635466, Tunis, Tunisia, may 1994

Salvini, G. & Marcucci Pesserini, M. (1998). The radiolarian assemblages of the Bonarelli Horizon in the Umbria-Marches Apennines and Southern Alps, Italy. *Cretaceous Research,* vol. 19, No. 6, pp. 777–804, ISSN01956671

Scopelliti, G., Bellanca, A., Coccioni, R., Luciani, V., Neri, R., Baudin, F., Chiari, M. & Marcucci, M. (2004). High-resolution geochemical and biotic records of the Tethyan "Bonarelli Level" (OAE2, Latest Cenomanian) from the Calabianca-Guidaloca composite section, northwestern Sicily, Italy. *Palaeogeography, Palaeoclimatology, Palaeoecology,* vol. 208, No. 3-4, pp.293–317, ISSN 00310182

Ślączka, A., Gasiński, M. A., Bąk, M. & Wessely, G. (2009). The clasts of Cretaceous marls in the conglomerates of the Konradsheim Formation (Pöchlau quarry, Gresten Klippen Zone, Austria). *Geologica Carpathica,* vol. 60, No. 2, pp. 151-164, ISSN 13350552

Smrečková, M. (2011). Lower turonian radiolarians from the locality Červená Skala section. Mineralia Slovaca 43, 31-38, ISSN 0369-2086

Soua, M., Zaghbib-Turki, D. & O'Dogherty, L. (2006). Les réponses biotiques des radiolaires à l'événement anoxique du cénomanien supérieur dans la marge sud téthysienne (tunisie). *Proceedings of the 10th tunisian petroleum exploration & production conference.* pp. 195-216.

Soua M. & Tribovillard, N. (2007). Modèle de sédimentation au passage Cénomanien /Turonien pour la formation Bahloul en Tunisie. *Comptes Rendus Geoscience,* vol. 339, No. 10, pp. 692-701, ISSN 1631-0713

Soua, M., Echihi, O. Herkat, M., Zaghbib-Turki, D., Smaoui, J., Fakhfakh-Ben Jemia, H. & Belghaji, H. (2009). Structural context of the paleogeography of the Cenomanian - Turonian anoxic event in the eastern Atlas basins of the Maghreb. *C. R. Geoscience,* 341, No. 12, pp. 1029–1037, ISSN 1631-0713

Talbi, R. (1991). Etude géologique et géochimique des faciès riches en matière organique d'âge Albien du bassin de Bir M'Cherga (NE de Tunisie) : déterminisme de leur genèse et intérêt pétrolier de la région. PhD Thesis, University de Tunis, 223 pp.

Thurow, J. (1988). Cretaceous Radiolarians of the North Atlantic Ocean: ODP Leg 103 (Sites 638, 640 and 641) and DSDP Legs 93 (Site 603) and 47B (Site 398). In: *Proceedings of the Ocean Drilling Program, Scientific Results,* Boillot, G., Wintere, E.L., et al. (Ed.). Proceedings of the Ocean Drilling Program, Scientific Results, 103, 379 –418, ISSN 0884-5891.

Turki, M.M. (1985). Polycinematique et contrôle sédimentaire associé sur la cicatrice Zaghouan-Nebhana. Ph.D. thesis, Univ. Tunis, 252 pp.

9

Geology and Geomorphology in Landscape Ecological Analysis for Forest Conservation and Hazard and Risk Assessment, Illustrated with Mexican Case Histories

María Concepción García-Aguirre[1], Román Álvarez[2] and Fernando Aceves[3]
[1]*Centro de Ciencias de la Complejidad(C3), Departamento de Ecología y Recursos, Naturales, Facultad de Ciencias. Universidad Nacional Autónoma de México (UNAM) Ciudad Universitaria, C.P. Coyoacán, D.F.*
[2]*Instituto de Matemáticas Aplicadas y Sistemas, Universidad Nacional Autónoma de México (UNAM)*
[3]*Instituto de Geografía, Universidad Nacional Autónoma de México (UNAM) México*

1. Introduction

The aim of landscape ecology is to understand both the effects of spatial patterns on ecological processes, and the development of those spatial patterns. It is considered holistic since it regards nature as a whole and it deals also with all human environment interactions. Application of landscape ecological principles for prioritizing rich species sites has the advantage of integrating spatial information, non-spatial information, and horizontal relationships in space and time.

Landscapes are complex systems constituted by a large number of heterogeneous components (with different geology, geomorphology, vegetation cover, ecological communities, land uses and so on) interacting in a non-linear way, that are hierarchically structured and scale-dependent (Wiens, 2009, Hall *et al.*, 2004). Landscape description and structure has traditionally been done on the basis of landscape metrics, that is, basic measures of the amount of habitat and core habitat, the number of discrete patches and the perimeter to area ratio. However many of these metrics are generally poorly tested and require of rigorous validation if they are to serve as reliable indicators of habitat loss and fragmentation (McAlpine and Eyrie, 2002). There are strategies which can help to improve the reliability of landscape pattern analysis (Shao and Gu, 2008). Since landscape pattern is spatially correlated and scale dependent, often multiscale information is required (Wu, 2004).

The study of causes, processes and consequences of land/cover change is one of the main research topics of landscape ecology. It is important to study processes and not merely spatial patterns considering cultures as a drivers of landscape change. Knowledge of temporal changes in landscape composition and structure, and their driving processes, can provide insight into regional landscape dynamics (Luke, 2000; Burgui *et al*, 2004).

Landscape information is of the out most importance to develop appropriate policies for environmental planning and nature conservation (Gulink *et al.*, 2000). An example of recognition of landscape ecology as an essential field of science for territorial planning is the project of the metropolitan region of Barcelona (Forman, 2004), in which environmental principles based on landscape ecology and sustainable use of resources and basic spatial models are applied, even when lacking quantitative regional analysis.

2. Geology and geomorphology as fundamental elements in landscape analysis

Several terrain characteristics are important for soil scientists, geologists, and geographers, because of their strong influence in the capability of the land to support various plant or animal species, or for terrain evaluation. Geologic origin and structure can be estimated by air photo interpretation and satellite image analysis. Sedimentary (sandstone, shale, limestone) or igneous rocks can be differentiated using digital analysis (remote sensing and GIS). The recognition of strike and dip attitudes, land form types, drainage patterns and the orientation of highlights and shadows, as well as susceptibility to flooding, are based on geomorphology (Sabins, 1978, Lillesand and Keiffer, 1979, Verstappen, 1988). Geomorphological analysis is greatly improved by the use of aerial photographs and satellite images, since they provide a synoptic view of terrain and a relatively rapid description of geographic distribution of major landforms and dominant land cover. Terrain classification based on landforms, lithology and genesis (historical processes) can be further specified into biogeomorphic land units on the basis of geomorphologic processes, relative age, sediment, drainage, and land cover/use (García-Aguirre *et al.*, 2010).

Ecological research provides ample evidence that topography can exert a significant influence on the processes shaping broad-scale landscape vegetation patterns. Unfortunately, the standard methods for landscape pattern analysis are not designed to include topography as a pattern shaping factor. Topography features may be derived from the digital elevation models (DEM) to obtain slope and aspect maps (Dorner *et al.*, 2002, Peiffer *et al.* 2003). A DEM and Landsat images were used to assess topographical complexity and evaluate changes in landscape composition and structure after fire (Viedma, 2008). Simultaneous analysis of maps of non-biotic elements (such as geology, geomorphology and topography) and biotic elements (land use/cover) allow to generate synthetic and systematic information of landscape in the form of biogeomorphic land unit maps (Zonneved, 1995).

3. Remote sensing and GIS in landscape analysis

Remote sensing and GIS are essential tools for generation of landscape thematic information (Gulink *et al.*, 2000) even when it is common to face problems during integration of different data sources in the GIS (Tinker *et al.*, 1998). Advances in remote sensing technologies have provided practical means for land use/cover mapping, which is particularly important for landscape ecological studies. These tools can also efficiently identify and assess areas of landscape damage at different scales and help land managers to solve specific problems. However, it is a key consideration to evaluate the remote sensing data and methods used as well the scale and information needed, for instance, to correctly define the best resolution to use (Ludwing *et al.*, 2007), as well as to find the best procedures to follow when linking data of different qualities (Falcucci, *et al.*, 2007). Land cover change information through time, combined with thematic information can be stored and managed efficiently in a GIS since it relates different layers of spatial information. In addition, it is a powerful analysis tool that allows identification of spatial relationships among different maps, through connection of spatial data with its attributes (Belda and Melia, 2000, Baysnat *et al.*, 2000). Time series remote sensing provide researchers with a valuable tool for the dynamic analysis of landscape (Staus *et al.*, 2002).

Environmental models implemented in computers have become important tools for designing management plans towards ecological and economic sustainability. Computers help to deal with the tremendous complexity reflected in the extensive temporal and spatial scales at which human and natural processes occur. Iverson and Prasad (2007) evaluated tsunami damage and built empirical vulnerability models of damage/no damage based on elevation, distance from shore, vegetation, and exposure.

4. An outline of the geology of Mexico

The geology of Mexico is the result of multiple tectonic processes that have taken place along its geologic history. Current geologic configuration of Mexico is the consequence of continental block interaction with surrounding oceanic provinces. As a result, young sedimentary and volcanic outcrops are dominant. 80% of exposed units are placed on Cenozoic and Mesozoic eras (less than 250 million years), 13% correspond to the Paleozoic and only 7 % belong to the Precambrian, belonging to the Proterozoic (up to 2500 million years ago (Ortega *et al.* 1992).

Precambrian

The oldest metamorphic Rocks in Mexico were found in Sonora State and belong to the Bamori Complex (Figure 1), it is conformed by muscovite schist, hornblende-amphibolites schist and quartzite, dated at 1755 ± 20 million years (myr) by Anderson and Silver (1981). In Chihuahua State a metamorphic complex outcrops (metagranite, metadiorite, amphibolite, gneiss, metalimestone and quartzite), these rocks have an age between 1025 $\pm$ 21 myr and 948 $\pm$ 14 myr (Blount 1983). In southeast Mexico one finds dispersed outcrops from the Proterozoic composed of metamorphic rocks (augengneiss, orthogneis, marble, amphibolite and migmatite) that belong to the Oaxaqueño Complex, with an age between 1300 to 700 myr (SGM, 2007). Along the southern coast from Zihuatanejo, Guerrero to Puerto Ángel, Oaxaca emerges a group of paragneiss, pelitic schist, boitite schist, quartzite, marble, orthogneiss, amphibolite and migmatites that has been grouped inside the Xolapa complex (between 980 and 1300 myr; SGM, 2007). Intrusive rocks of the Paleo and Mesoproterozoic, emerge in Sonora state, showing small outcrops of granite, granodiorite, and in less proportion, diorite. Their ages vary between 1440 and 1140 myr (Anderson and Silver, 1981).

Sedimentary rocks of the Proterozoic outcrop in Sonora state like small patches of dolomite, limestone and sandstone. These deposits were dated as Neoproterozoic owing to the presence of conic stromatolite fossils of the Conophyton genus (SGM, 2007).

Paleozoic

Metamorphic rocks from this period are schist, marble and quartzite. They are located in the states of Baja California, Sinaloa, Sonora and Chihuahua, their ages fluctuate between

Cambric and Carboniferous (Figure 1). Phyllites and schist with quartzite can be found at the southeast of the State of Chiapas (Late Mississippian). In Tamaulipas state metamorphic rocks appear in Huizachal-Peregrina Structure presenting mica-schist interstratified with green rocks, metaflints, serpentine, and metalimestones. In the Huizachal-Peregrina structure, near Ciudad Victoria, Tamaulipas, there are outcrops formed by mica-schist of low grade. The age for this unit was 330 myr (Stewart *et al.*, 1999). Low grade metavolcanic sedimentary rocks from the Permic (SGM, 2007) exist north of Durango city. Outcrop schists, phyllites, quartzite's and metalavas are found northeast of Puebla. These rocks are from Early Permic (280 myr; Iriondo *et al.*, 2003). There are outcrops of schists and quartzites in southeast Oaxaca. The age of these rocks ranges between 289±5 Ma and 219±6 Ma. (Grajales-Nishimura *et al.*, 1999). In Zacatecas outcrops are found of a metamorphosed sedimentary succession, from Late Paleozoic, 260.2 ± 3 Myr (Díaz-Salgado, 2004).

Fig. 1. Distribution of metamorphic rocks of Precambrian and Paleozoic

The metamorphic rocks from Mesozoic (Figure 2) in the northeast of the country are phyllites and chlorite and biotite schist, amphibolite gneiss, metatonalite and metadiorite (De Cserna *et al.*, 1962; 220 myr). A meta-volcanosedimentary succession was deposited from Late Jurassic to Early Cretaceous along the west of Mexico.

States of Michoacán, México and Guerrero present outcrops of metamorphic rocks of low grade, there are schist's, slates and quartzites. (Tejupilco Schist, Taxco Schist and Green Rock Taxco Viejo formations (De Cserna 1982) and probably belong to the Jurassic.

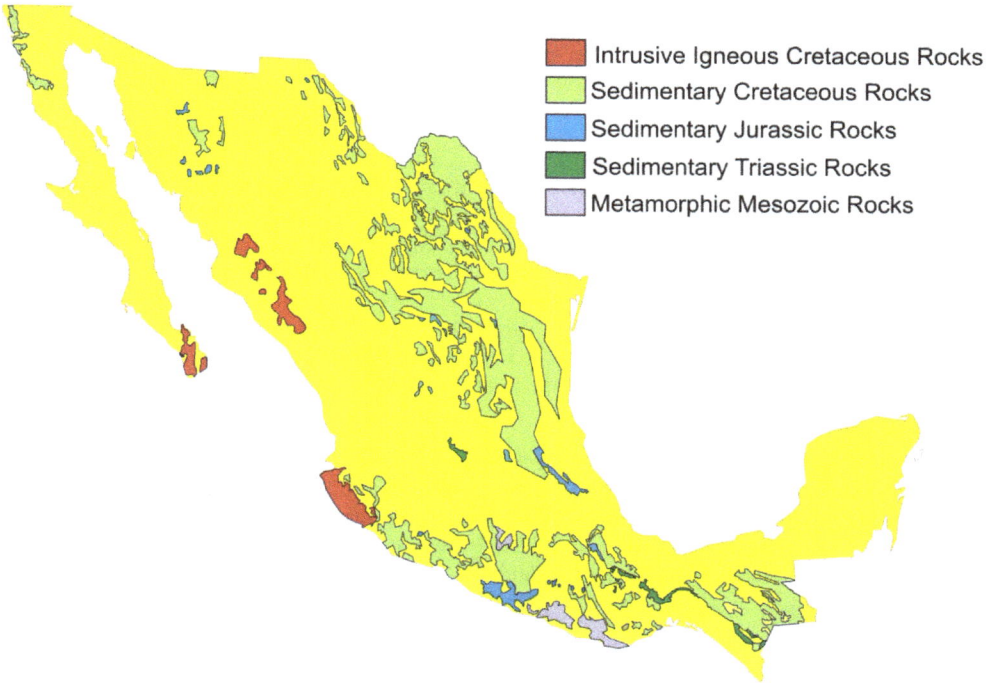

Fig. 2. Mesozoic Lithologic Units

Outcrops from Triassic are scarce. In southwest Sonora there is calcareous sandstone, alternated with limonite. This rock contains fossils of gastropods, coral, bryozoans, sponges and ammonites from this period (González-León, 1980), and the same in northeast Mexico were deposited at the Huizachal Formation conglomerate (SGM, 2007).

Along the Lower Jurassic an alternate succession of shales and limestones are deposited in the states of San Luis Potosí, Querétaro and Hidalgo. Jurassic sediments in southern Mexico (Oaxaca and Guerrero states) is composed of a conglomerate and sandstone succession with quartz clasts (Cualac Formation). The Upper Cretaceous was identified with ammonites and radiolarian fossils. In south Baja California the lower Cretaceous is represented by sandstones, limonites, shales and conglomerate successions.

To the east, in Chihuahua and Nuevo León states the great sea transgression deposited thick beds of calcareous and siliceous rocks (Formations Taraises, Cupido, La Peña, La Virgen.). At the same time calcareous anhydrite and clayish successions were deposited in Tamaulipas, San Luis Potosí, Hidalgo, Queretaro, Puebla and Veracruz states. The most important formations are Lower and Upper Tamaulipas, Otates, El Abra, Tamabra, Tamasopo and Cuesta del Cura. The last Formation is composed of limestone and chert. In south and central Mexico sedimentary rocks appear in the states of Jalisco, Michoacán, Guerrero, México, Morelos and Oaxaca, where calcareous successions settled, and clayish components are constituted by conglomerates, sandstones and limonites, with interstratified limestone, marls, and gypsum in different facies. The most important formations are Zicapa, Tepexi de Rodríguez, Xochicalco, Morelos, Cuautla y Mexcala. These formations contain a wide variety of mollusca, gasteropods, ammonites, and milliolids.

A change in sedimentation takes place in the Upper Cretaceous marked by the suffocation of the platforms with a series of terrigenous successions that show the evolution of deltas and basins. These successions form big bundles of sandstone, conglomerate, limonite, marl, and shales.

The outcrops of intrusive rocks are scarce and dispersed along the territory. The most important deposits are found in southern Baja California (granites, Peridotite, utramafic sequence, and an ophiolite sequence (Kimbrough and Moore, 2003). A sequence of granite and diorite from Middle Jurassic is located in Puebla (Macizo de Teziutlán), dated between 163 ± 13 and 134 ± 11 Myr (Manjarrez and Hernández, 1989). In Guanajuato state outcrops a sequence of granite, diorite and tonalite from the Upper Jurassic outcrops, these deposits are related to the evolution of Mesozoic insular arcs from this period (SGM, 2007). Plutonic magmatism appears in southern Mexico, probably from Late Jurassic to Early Cretaceous, with a variable composition: granite, granodiorite and diorite. Some localities in which this plutonic intrusive appears are Tumbiscatio, near Zitácuaro city.

Cenozoic

A continental sedimentation of the Early Cenozoic marks the change of sedimentary rocks to volcanic sequences (Figure 3). The Red Conglomerates with intercalation of sandstones and limonites represent them in the Balsas, Red Conglomerate of Guanajuato, and Tehuacán formations. Transgressive events are reflected by the horizons of sandstone, lutite and conglomerates. In the Eocene-Oligocene, deposits of conglomerate and sandstone, limonite,

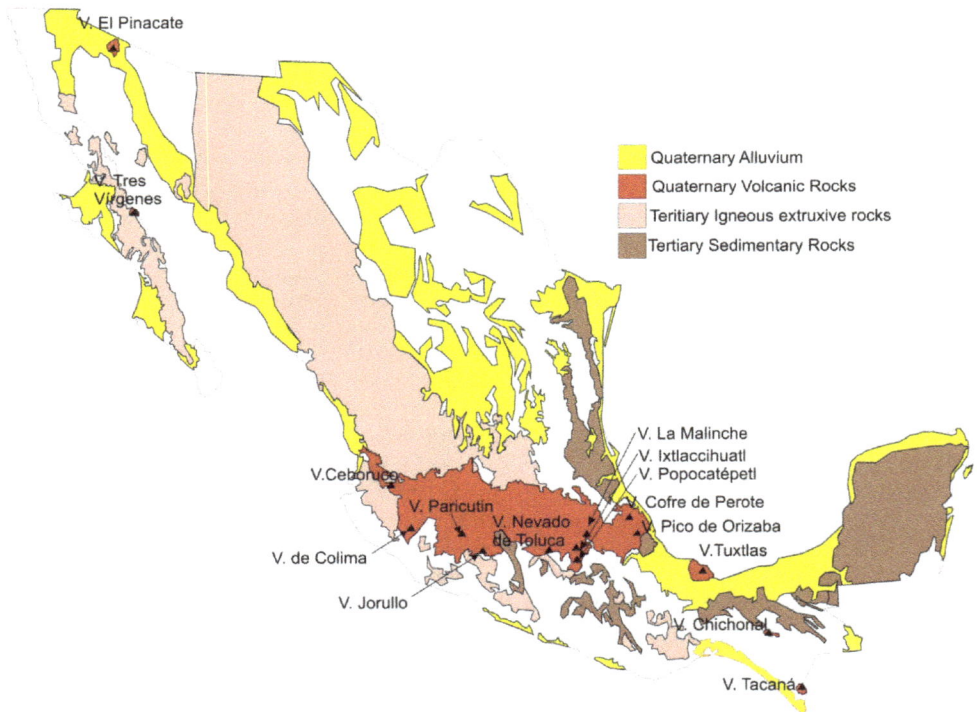

Fig. 3. Cenozoic Lithologic Units

and sandstone and limestone are distributed in grabens and synclinal valleys in the Sierra Madre Oriental and Sierra Madre Occidental. In the Sierra de Chiapas, the Miocene differed in the deposition of the clay-calcareous successions, as well as in the thin horizons of the conglomerates.

On the other hand, in the regions of the rim of the Yucatan Platform the bar reefs and lagoons keep on developing carbonate sediments of limestone and dolomites. The Holocene deposits of coastal environments of the coast of the Gulf of Mexico in the states of Tamaulipas, Veracruz, Tabasco, Campeche, Yucatan, and Quintana Roo, are still in the process of sedimentation of silts, clays and marshy sand, flat dunes of coastal sand, and continental shelf carbonate sediments.

The volcanic units of the Cenozoic are widely distributed in the Mexican territory, including the ignimbrites of the Sierra Madre Occidental and the Pliocenic-Quaternary sequences of the Transmexican Volcanic Belt (TMVB). The Sierra Madre Occidental is formed by an extensive volcanic plateau affected by grabens and normal faults. It spreads from Sonora to Guerrero states, although in the states of Jalisco, Michoacán and Guerrero it is fragmented and mixed with the Transmexican Volcanic Belt and the rocks of the Sierra Madre del Sur. The TransMexican Volcanic Belt (TMVB) extends from the Pacific Ocean up to the coast of the Gulf of Mexico along 920 km between parallels 19 and 20. It is formed by a large variety of volcanic rocks produced by a number of volcanic buildings, some of which constitute the main elevations in the country. Likewise, this activity has caused a big number of endorreic basins with the consistent development of lacustrine landscapes. The principal volcanoes of the TMVB are stratovolcanoes of variable dimensions, such as Pico de Orizaba, Popocatépetl, Iztaccíhuatl, La Malinche, Nevado de Toluca and Nevado de Colima.

5. Case studies

Landscape studies in Mexico are numerous: Ochoa (2001) undertook the integration of geologic and geomorphic units, incorporating afterwards variables of climate, hydrology, vegetation and soil in the Tehuacán-Cuicatlán, Puebla, area. Casals-Carrasco et al., (2000) performed a geomorphologic analysis in order to establish relationships among land cover, landforms and soils using interpretation of a stereoscopic pair of SPOT-PAN, and a TM false color composite image. Martínez (2002) elaborated environmental land unit maps of a sub watershed in Morelos State. García (1991) studied the influence of relief dynamics in landscape structure on the vegetation in Zapotitlán, Puebla watershed. García (1998) analized the east slope of Sierra de la Cruces, Monte Alto and Monte Bajo while, Garcia-Aguirre et al. (2007) related geology, landform and vegetation in the Ajusco volcano area in central Mexico. Aguilar (2007) performed an environmental diagnostic of the Parque Nacional Nevado de Toluca from the biogeomorphic land units of the region.

Two case studies will be described in detail. The first is related to the use of biogeomorphic land units of a region located nearby Mexico City as a basis for hydrology and vegetation analysis. The second case refers to the evaluation of risks and hazards of the fourth highest summit in México, the Nevado de Toluca.

5.1 Sierra de las Cruces

The objectives of the first study are two-fold: to identify the most degraded areas of the region through the land unit analysis, and to study the relationship between forest loss and runoff in the region (scale 1:250,000) through a conceptual and cartographic model.

Landscape analysis was performed to describe regional characteristics in an integral form; land cover features were overlaid with other landscape elements (geology, geomorphology, soils and climate) to obtain a land unit map (García-Aguirre, 2008).

Mountainous relief and flat plain may be appreciated in the shadow relief model derived from a DEM (figure 4). Chichinautzin region, towards the south of the zone, being an infiltration zone is very important from the hydrological view point. Refief of the east slope of the Sierra de las Cruces, Monte Alto and Monte Bajo is constituted by four geomorphic units: mountain, upper and lower piedmont, and hills. Notice the N-S orientation of this mountain range.

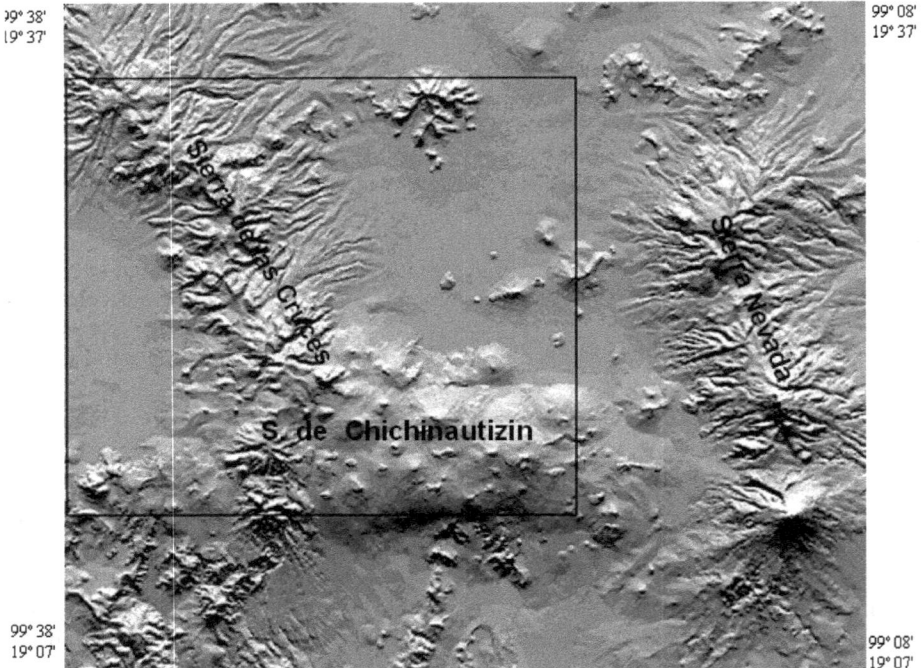

Fig. 4. Shadow relief model derived from the DEM. Study area is located inside the square. Sierra de Chichinautzin is located to the south, and Sierra Nevada toward the east

Land unit map

Remote sensing and GIS were linked to integrate geomorphological and geological information to find mayor associations among variables. Then, biogeomorphic land units were delineated on the basis of homogeneity of a dominant factor.

Figures 5 and 6 show the geology and geomorphology maps of the region. These maps were obtained by digitizing hardcopy maps of INEGI (1993) and reclassified using IDRISI (Eastman, 1997). Andesite and basalt are dominant in the area and in turn, andosols and lithosols (Figure 5). Lugo (1984) points the south of Cuenca de México as one of the zones of the country with higher concentration of young volcanoes, from the late Pleistocene and Holocene (Figure 6).

Geology and Geomorphology in Landscape Ecological Analysis for
Forest Conservation and Hazard and Risk Assessment, Illustrated with Mexican Case Histories

269

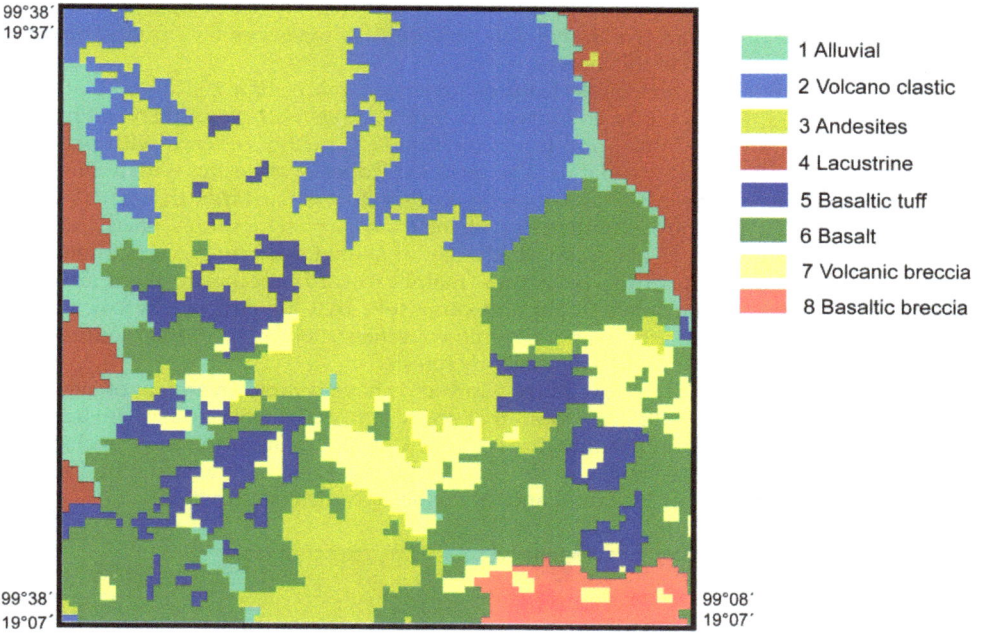

Fig. 5. Geologic map. Andesites and Basalt are dominant in this region

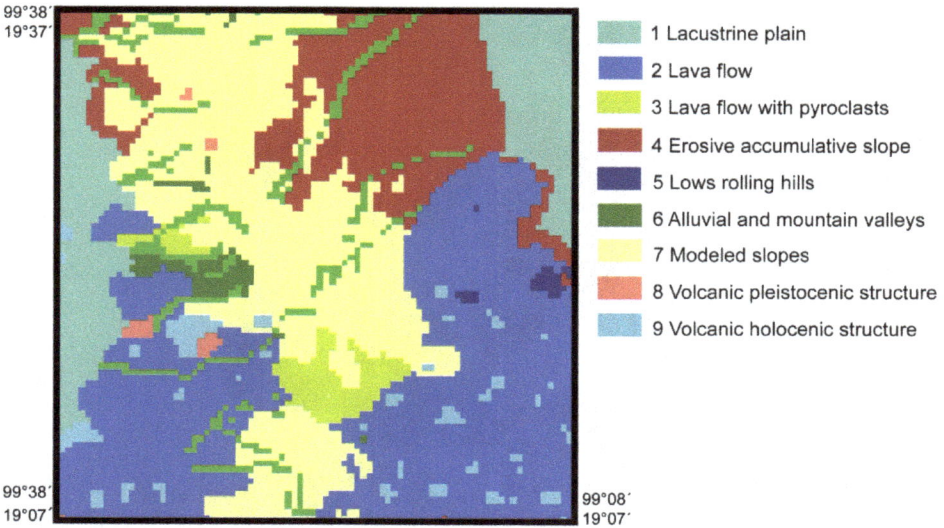

Fig. 6. Geomorphologic map. Dominant geoforms are modeled slopes and lava flows

Geomorphologic map (Figure 6) shows modeled slopes and lava flows as dominant geoforms. The region has an extense footslope, in which half the area slopes down to the piedemont and the other half is mountainous terrain. Also extensive lava flows show the active quaternary volcanism in this zone.

Combination of biotic and non biotic features generated more than 100 units, that were reclassified into 48 units on the basis of its surface (only units with more than 500 ha), for map legibility (Figure 7). Alluvial and lacustrine units are mainly covered with pheozem and hystosol, agriculture and grasslands. The modeled slopes of andesites with andosol are subdivided into those with Abies and those with Pine forest. Towards the east (BGU39, BGU40, and BGU2), there are abundant lacustrine forms with agriculture, grasslands and human settlements. Basalt is abundant to the south with forest cover, agriculture and grassland (BGU29, BGU33). In the footslopes, mainly towards the north, there are units constituted by andesites, modeled slopes and cambisols, that are covered by oak forests, crops and grasslands. There are many holocenic volcanic structures over the mountainous area, with andosol and litosol, covered mainly by forests.

The regional vision provided by this study allowed to have a rapid overview of sites that should be preserved, such as basalt zones, that are nevertheless continuously invaded by human settlements. Results indicate that Sierra de Chichinautzin is the main recharging area, but the foot slopes of the Sierra de Las Cruces are also important infiltration and recharging zones as a result of the abundance of clastic volcanic forms therein.

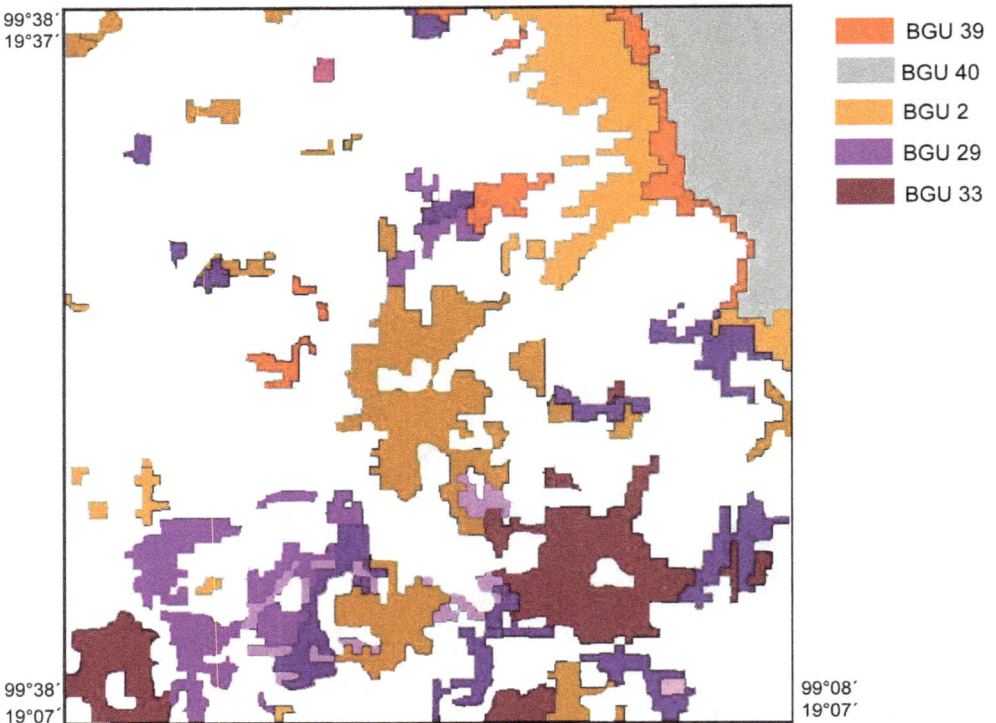

Fig. 7. Land unit map (BGU=biogeomorphic land units)

Geology and Geomorphology in Landscape Ecological Analysis for
Forest Conservation and Hazard and Risk Assessment, Illustrated with Mexican Case Histories

271

5.2 Nevado de Toluca geologic history and hazards

Nevado de Toluca Volcano (NTV), located in central Mexico, is a large stratovolcano, with an explosive history. The area is one of the most important human developing centers (>2 million people) in Mexico and in the last 30 years large population growth and urban expansion have increased the potential risk in case of a reactivation of the volcano. NTV is the fourth highest summit in Mexico (4,665 masl) and it is a potentially dangerous large stratovolcano, that lies in the southeastern part of the Toluca Basin, some 70 km east of Mexico City (Figure 8). The NTV has been characterized by very explosive eruptions with long periods of dormancy; the periods between eruptions are discussed below.

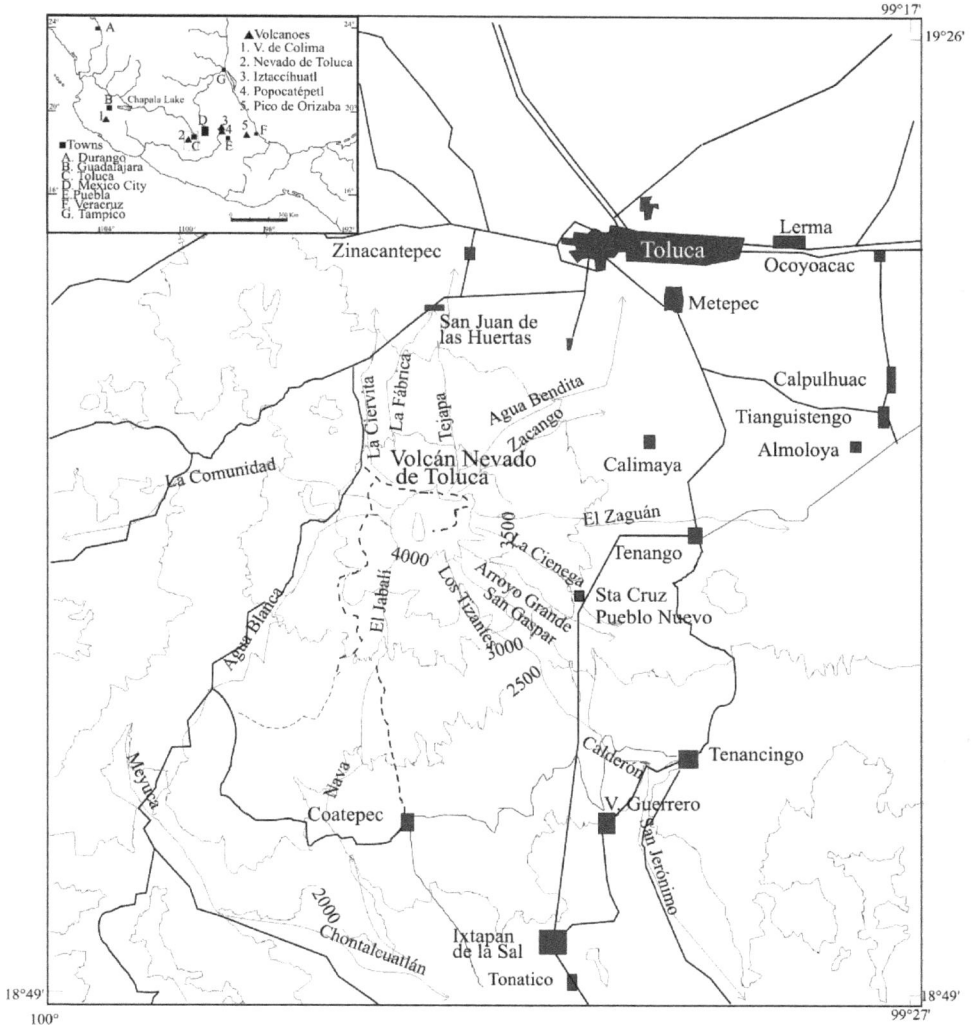

Fig. 8. Nevado de Toluca, location map

Nevado de Toluca Volcanic hazards

Pyroclastic flow Hazards: These deposits are widely spread around the volcano, filling the stream valleys where several settlements are located (Figure 8). The pyroclastic flow deposits cover a minimum area of 630 km² and assuming that they have an average thickness of 5 m, the approximate volume is 3.15 km³ (Macias *et al.* 1997). The maximum distance reached by these deposits is 32 km from the crater towards the south, in the Tizantes and Calderón stream valleys. The block and ash flows form massive units interstratified with surge horizons. The first dome collapse deposited the Zacango Block and Ash flow (37 kyr BP) composed of three massive units with associated surge horizons. The second dome collapse (28 kyr BP) deposited El Capulín Block and Ash Flow. These deposits are distributed around the volcano and cover approximately 630 km² with a volume of 2.6 km³ (Aceves *et al.*, 2007).

The pumice flows erupted by the NTV are: the Pink Pumice Flow (43 kyr BP); the White Pumice Flow (26 kyr BP). The MF2 (13.4 kyr BP) pumice flow is a gray ash flow enriched in pumice clasts (<2 cm) and charcoal (Aceves, *et al.* 2007). The Intermediate White Pumice (12.1 kyr BP) is composed of white ribbon pumice clasts and gray to reddish (altered) dacitic lithic clasts, interbedded with a surge (Cervantes 2001). The pumice flows are distributed around the volcano and cover more than 200 km², with a volume of 0.2 km³. The pumice flows are related to the Plinian eruptions, many of which result in mudflows and include charcoal and variable amounts of pumice. Some are altered to paleosoils. One of the most pumice-rich flows belongs to the Upper Toluca Pumice Formation (UTP), dated 10,500 yr BP (Macias *et al.* 1997).

Debris avalanches hazards: The debris avalanches have been located towards the south of the NTV in the Meyuca, Calderón Chontalcuatlán and San Jerónimo river valleys. Two units compose the debris avalanches. The oldest unit (DAD1) is massive, with 35% blocks up to 2.5 m in diameter, in a heavy pink partially hardened sand sized matrix. The lithological composition is heterogeneous dacitic, andesitic and schist lithics, with"Jig saw" blocks (Capra and Macias, 2000). DAD1 is around 10 m thick in Coatepec and continues towards the south to the Valley of the Chontalcuatlan River. The youngest avalanche (DAD2) is composed of two large cohesive debris flows: the Pilcaya (PDF) and the El Mogote (MDF) (Capra 2000). Thickness varies from 6 m in the proximal section to 40 m in the intermediate zone, extending out to a distance of 75 km from the crater. It covered an area of 220 km² with a volume of 2.8 km³ (Aceves *et al.*, 2007, Capra 2000).

Lahar hazards: Lahars at NTV are wide spread around the volcano filling new and old valleys. Lahars have rounded and subrounded dacite lithics (15–25 cm), small pumice fragments (<5 cm) fixed in a muddy-sand sized matrix. To the south, the lahar thickness is more than 30 m. The oldest lahars are made up of rounded and subrounded gray and red andesite blocks fixed in red clay sized matrix with scarce pumice fragments. The recent deposits contain subangular and subrounded blocks of gray and red dacite, with pumice fragments, some of which are hydrated, fixed in a siltclay sized matrix. The flow direction of these lahars was controlled by the topography, principally deep tectonic, glacial and fluvial valleys. In distal areas, the lahar deposits were transformed into fluvial mixing with the stream and river waters. To the east, the lahars contain more pumice fragments in a pale brown silt-clay matrix. There is a lahar with large hydrated pumice fragments (20–30 cm), in the Arroyo Grande channel. In this area, many secondary lahars exist, such as the one deposited in 1952 in the Ciénaga channel.

These secondary lahars are not related to volcanic activity and represent an increased hazard, because these materials are not consolidated. In torrential periods these materials can be removed by rain, triggering these secondary lahares, which can come up to the low zones affecting the populations who are settled in the mouths of the valleys as it happened to the people of Santa Cruz Pueblo Nuevo. To the north, the ravines are shallower (<30 m). The origin of the secondary lahars are uncompacted volcanic products (ash fall and, pyroclastic flows) mixed with water from the glacial melt and torrential rains (Aceves *et al.*, 2007).

Ash fall hazards: Ash and pumice fall deposits cover wide zones around the volcano. The most important deposits are the Upper and Lower Toluca Pumice. For the ash fall hazards map, the distribution and thickness of these deposits were considered as well as the present wind direction for the UTP. These isopach and isopleth maps show dispersal to the northeast. In the Toluca Basin, the maximum thickness measured for the UTP was 40 cm. In order to plot the isopach of 10 cm, map scales of 1:100,000, and 1:250,000 scale were used. The dominant wind direction was obtained from the National Meteorological Service and calculated for a height between 20 and 30 km (Fonseca 2003). The results were: east-northeast from November to March, west-northwest in April and west from May to October (Aceves et al., 2007).

In conclusion, in the last 50,000 yr, NTV had eight vulcanian, and four Plinian eruptions, three large dome collapses, and one ultraPlinian eruption. Block and ash and pumice flows are the most common deposits. The eruptive history of NTV shows cataclysmic events of Plinian and utraPlinian type, beside Vulcanian eruptions, which represent a large hazard for the Toluca Basin. The regions that would be most affected by pyroclastic flows and lahars are: (1) Toluca, Lerma, Metepec, San Mateo Atenco, Santiago Tianguistenco and Capulhuac, located to the northeast of the volcano. These areas concentrate the major population and industrial centers. (2) Calimaya, Zacango, Tenango among several others located to the east of NTV are highly populated and important agricultural areas. (3) To the south, the flower producing centers: Coatepec and Villa Guerrero, and the tourist towns of Ixtapan de la Sal and Tonatico.

Four volcanic hazards types were identified: pyroclastic flows (block and ash flows and pumice flows), lahars, debris avalanches and ash fall. The most destructive (based on energy and frequency) in the NTV are the block and ash flows and the pumice flows, both of them have reached distance of up to 35 km from the volcano summit. The principal affected areas are the northeast and south of the volcano, because these areas have major differences in altitude and present the major development of ravines. Lahars have been present in most of the eruptions. The deep and large valleys located to the east and south of the volcano are the most hazardous areas. The most active valleys are San Jerónimo, Chontalcuatlan, Grande, and El Zaguán Rivers. Debris avalanches present a hazard for areas to the east and south of the volcano, because of the active faults in the area and the instability caused by the difference in altitude, favoring the gravitational collapse of NTV. This type of event has occurred twice in the last 100,000 yrs, the deposits are located to the south of the volcano. The hazards from ash fall, arising from the dominant winds, are: from November to March, they would affect mainly the east and north-east sectors of the volcano, in April affectation would be to the north-west, and from May to October to the west. In case of small and medium eruptions (VEI = 1–3), the affected zone would be the Toluca Basin, but in case of large eruptions (VEI > 4) the affected zone would even include Mexico City.

6. Acknowledgment

Authors wishes to thank Verónica Aguilar for help in drawing of Figures 5, 6 and 7

7. References

Aceves-Quesada, F., Martin del Pozzo, A.L., López-Blanco J. 2007. Volcanic hazards zonation of the Nevado de Toluca Volcano, Center of Mexico" *Natural Hazards.* 41(1):159-180.

Aguilar, V. 2007. Diagnóstico del Parque Nacional Nevado de Toluca con base en unidades de Paisaje. Tesis de Maestría en Ciencias Biológicas (Biología Ambiental). Facultad de Ciencias. Universidad Nacional Autónoma de México. 94 p.

Anderson, T.H., Silver, L.T. 1981. An overview of precambrian rocks in Sonora: *Revista del Instituto de Geología,* 5(2): 131-139.

Basnyat, P., Teeter, L.D. , Lockaby, B.G., Flynn K.M. 2000. The use of remote sensing and GIS in watershed level analyses of non-point source pollution problems. *Forest Ecology and Management.* 128: 65-73.

Belda, F., Melia,J. 2000. Relationhips between climatic parameters and forest vegetation: application to burned area in Alicante (Spain). *Forest Ecology and Management.* 135: 195-204.

Blount, J.G., 1983, The geology of Rancho Los Filtros, Chihuahua, Mexico, in Clark, K.F., Goodel, P.C., (eds.), Geology and mineral resources of north-central Chihuahua: El Paso, Texas, El Paso Geological Society Guidebook, p. 157–164.

Burgui, M., A.M. Hersperger and N. Schneeberger. 2004. Driving forest of landscape change. Current and New directions. *Landscape Ecology,* 19:857-868.

Capra PL (2000) Colapsos de Edificios Volcánicos: transformación de Avalanchas de Escombros en flujos de escombros cohesivos. Tesis de Doctorado, UNAM, p176

Capra, L., Macías, J.L., 2000. Pleistocene cohesive debris fl ow at Nevado de Toluca volcano, central Mexico: Journal of Volcanology and Geothermal Research, 102, 149-168.

Casals-Carrasco, P., Kubo S., Madhavan, B. 2000. Application of spectral mixture analysis for terrain evaluation studies. *Int. J. Remote Sensing,* 21(16):3039-3055.

Cervante, K.E. 2001. La Pómez Blanca Intermedia: depósito producido por una erupción pliniana-subpliniana del Volcán Nevado de Toluca hace 12,100 años. Tesis de Mestría en Ciencias de la Tierra, UNAM.

De Cserna, Z., Schmitter, E., Damon, P.E., Livingston, D.E., Kulp, J.L., 1962, Edades isotópicas de rocas metamórficas del centro y sur de Guerrero y de una monzonita cuarcífera del norte de Sinaloa: *Boletín del Instituto de Geología,* 64(número), p. 71-84.

De Cserna, Z., 1982, Hoja Tejupilco 14Q-h (7), Geología de los Estados de Guerrero, México y Morelos: México, D.F., México, Universidad Nacional Autónoma de México, Instituto de Geología, Cartas Geológicas de México serie 1:100 000, 1 mapa con texto.

Díaz-Salgado, C., 2004, Caracterización Tectónica y Procedencia de la formación Taray, Región de Pico de Teyra, Estado de Zacatecas: México, D.F., México, Universidad Nacional Autónoma de México, Instituto de Geología, Tesis de Maestría, 95 p.

Dorner, B., Lertzman, K, Fall, J. 2002. Landscape in topographically complex landscapes: issues and techniques for analysis. *Landscape Ecology,* 17:729-743.

Eastman, J.R. 1997. IDRISI for windows user's guide, version 3.2, Clark laboratories for cartographic technologies and geographic analysis. Clark University, Worcester, MA.

Falcucci, A., Maiorano , L., Boitani, L. 2007. Changes in land-use/land-cover patterns in Italy and their implications for biodiversity conservation. *Landscape Ecol* 22:617–631

Fonseca R (2003) Efectos Ambientales de la Erupcio´n de 1913, del Volcán de Colima, Proyecciones a Futuro. Tesis de Maestría, Colegio de Geografía, Universidad Nacional Autónoma de México, p 90

Forman, R.T.T. 2004. Mosaico territorial para la región metropolitana de Barcelona. Barcelona regional/Editorial Gustavo Gili, S.A. Barcelona, España.

García, F. 1991. Influencia de la dinámica del paisaje en la distribución de las comunidades vegetales en la Cuenca del Río Zapotitlán, Puebla. *Investigaciones Geográficas. Boletín del Instituto de Geografía*. No. 23: 53-70.

García, A. 1998. Análisis integrado de paisajes en el occidente de la Cuenca de México (La vertiente oriental de la Sierra de las Cruces, Monte Alto y Monte Bajo) Tesis de Doctorado. Facultad de Geografía e Historia. Universidad Complutense de Madrid. 543 p.

García-Aguirre, M.C., M.A, Ortiz, J.J. Zamorano y Y. Reyes. 2007. Vegetation and landform relationships at Ajusco volcano Mexico, using a geographic information system (GIS). *Forest Ecology and Management* 239: 1-12.

García-Aguirre M.C., Álvarez, R., Dirzo, R., Ortiz, M.A., Mah Eng, M. 2010. Delineation of biogeomorphic land units across a tropical natural and humanized terrain in Los Tuxtlas, Veracruz, México. *Geomorphology*. 121:245-256.

González-León, C.M., 1980, La Formación Antimonio (Triásico Superior-Jurásico Inferior) en la Sierra del Alamo, Estado de Sonora: *Revista del Instituto de Geología*, 4(1), p. 13-18.

Grajales-Nishimura, J.M., Centeno-García, E., Keppie, J.D., Dostal, J., 1999, Geochemistry of Paleozoic basalts from the Juchatengo complex of southern Mexico: tectonic implications: *Journal of South American Earth Sciences*, 12 (6), p. 537-544.

Gulinck, H., Dufourmont, H., Coppin, P., Hermy, M.. 2000. Landscape research, landscape policy and Earth observation. *Int. J. Remote Sensing*. 21(13&14): 2541-2554.

Hall, O., Hay, G.J., Bouchard, A., Marceau, D.J. 2004. Detecting dominant landscape objects through multiple scales: An integration of object-specific methods and watershed segmentation. *Landscape Ecol* 19:59–76.

INEGI (Instituto Nacional de Estadística, Geografía e Informática). 1993. Mapas de uso del suelo y vegetación escala 1:250,000. México.

Iriondo, A., Punk, M.J., Wininck, J.A., 2003, 40Ar- 39Ar Dating studies of minerals and rocks in various areas in Mexico: USGS/CRM scientific collaboration (Part I), Open file report 03-020 on-line edition: USGS, Consejo de Recursos Minerales(COREMI), http://www.coremisgm.gob.mx/productos/novedades/estudios/USGS% 202003_Texto01.pdf.

Iverson, L. R., Prasad, A.M. 2007. Using landscape analysis to assess and model tsunami damage in Aceh province, Sumatra. *Landscape Ecol*. 22:323–331.

Kimbrough, D.L., Moore, T.E., 2003, Ophiolite and volcanic arc assemblages on the Vizcaíno Peninsula and Cedros Island, Baja California Sur, México, in Johnson, S.E., Paterson, S.R., Fletcher, J.M., Girty, G.H., Kimbrought, D.L., Martín-Barajas, A. (eds.), Mesozoic forearc lithosphere of the Cordilleran magmatic arc, in Tectonic evolution of northwestern Mexico and the southwestern: Boulder, Colorado, EE.UU., Geological Society of America, Special Paper 374, p. 43-71.

Lillesand, T.M. and R.W. Keifer, 1979. Remote sensing and Image Interpretation. Second Edition. John Wiley and Sons. 612 p.

Ludwig, J. A., Bastin, G. N., Wallace, J.F., McVicar, T.R. 2007. Assessing landscape health by scaling with remote sensing: ¿when is it not enough? *Landscape Ecol* 22:163–169.

Lugo, H. J. 1984. Geomorfología del Sur de la Cuenca de México. *Serie Varia* T.1. Núm. 8. Instituto de Geografía, Universidad Nacional Autónoma de México. 95 p.

Luque, S. 2000. The challenge to manage biological integrity of nature reserves: a landscape ecology perspective. *Int. J. Remote Sensing.* 2000, 21(13&14): 2613-2643.

Manjarrez, P.P., Hernández, R., 1989, Informe geológico final Prospecto Cardel: Petróleos Mexicanos, Superintendencia General de Exploración Geológica, IGPR-278

McAlpine, C.A. and T.J. Eyre. 2002. Testing landscape metrics as indicators of habitat loss and fragmentation in continuous *Eucaliptus* forest (Queensland, Australia). *Landscape Ecology,* 17:711-728.

Macias JL, García A, Arce JL, Siebe C, Espíndola JM, Komorowski JC, Scott K., 1997. Late Pleistocene-Holocene cataclysmic eruptions at Nevado de Toluca and Jocotitlan volcanoes, Central Mexico. BYU Geology Studies, vol 42, Part I, pp 493–528

Martínez, F. 2002. Síntesis de las unidades ambientales biofísicas de la subcuenca del río Colotepec, Edo. de Morelos, mediante la aplicación del enfoque geomorfológico y un GIS. Tesis de Maestría en Ciencias (Ecología y Ciencias Ambientales). Fac. de Ciencias, División de Estudios de Posgrado. UNAM. 91p.

Ochoa, V. 2001. Geomorfología, clima y vegetación del valle de Tehuacan-Cuicatlán, Pue-Oax. México. Tesis de licenciatura en Biología. Facultad de Ciencias, UNAM. 80 p.

Ortega-Gutiérrez, F., Mitre-Salazar L.M., Roldán-Quintana J., Aranda-Gómez J.J., Morán-Zenteno D., Alaniz-Álvarez S.A., Nieto-Samaniego A.F. 1992. Texto Explicativo de la Quinta Edición de la Carta Geológica de la Republica Mexica a Escala 1:2,000,000. Instiuto de Geología, Universidad Nacional Autónoma de México, Consejo de Recursos Minerales. 74 p.1 Mapa.

Peiffer, K., Pebesma, E.J., Burrough, P.A. 2003. Mapping alpine vegetation using vegetation observations and topographic attributes. *Landscape Ecology,* 18:759-776.

Servicio Geológico Mexicano (SGM). 2007, Texto Explicativo de la Sexta Edición de la Carta Geológica de la República Mexicana Escala 1:2,000,000. Servicio Geológico Mexicano. Secretaria de Economía. 30 p. 1 Mapa.

Shao, G., Gu, J. 2008. On the accuracy of landscape pattern analysis using remote sensing data. *Landscape Ecology,* 23:505-511.

Sabins, F.F 1978. Remote Sensing. Principles and Interpretation. Second Edition. W.H. Freeman and Company. 449 p.

Staus,N.L., Striitholt, J.R., Delta, D.A., Robinson, R. 2002. Rate and pattern of forest disturbance in the Klamath Siskiyoa ecoregion, USA, between 1972 and 1992. *Landscape Ecology,* 17:455-470.

Stewart, J.H., Blodgett, R.B., Boicot, A.J., Carter, J.L., López, R., 1999, Exotic Paleozoic strata of Gondwanan provenance near Ciudad Victoria, Tamaulipas, México, in Ramos, V.A., Keppie, J.D. (eds.), Laurentia Gondwanan connections befote Pangea: Geological Society of America, Special Paper 336, p. 227-252.

Tinker, D.B., C.A.C. Resor, G. P. Beauvais, K.F. Kipfmueller, Ch. I. Fernandes and W.L. Baker. 1998. Watershed analysis of forest fragmention by clearcuts and roads in a Wyoming forest. *Landscape Ecology* 13:149-165.

Verstappen, H.Th. 1988. Remote Sensing in geomorphology. Elsevier Scientific Publishing Company. Amsterdam. 214 p.

Viedma, O., 2008. The influence of topography and fire in controlling landscape composition and structure in Sierra de Gredos (Central Spain). *Landscape Ecology,* 23: 657-672.

Wiens, J.A. 2009. Landscape ecology as a foundation for sustainable conservation. *Landscape Ecology,* 24:1053-1065.

Wu, J. 2004. Effects of changing scale on landscape pattern analysis: scaling relation. *Landscape Ecology,* 19:125-138.

Zonneveld, I.S. 1995. Land Ecology. SPB Academic Publishing, Amsterdam. 90 p.

10

Miogypsinid Foraminiferal Biostratigraphy from the Oligocene to Miocene Sedimentary Rocks in the Tethys Region

Kuniteru Matsumaru
Innovation Research Organization, Saitama University, Saitama
Japan

1. Introduction

Tan Sin Hok (1936, 1937) has done the anatomical and morphometrical analysis of the family Miogypsinidae at the first time. This is regarded as the important contribution for the Micropaleontology on foraminiferal studies. His studies have developed from the phylogenetic history of the genus *Cycloclypeus* and their relative species of the family Nummulitidae de Blainville, 1827 (Tan Sin Hok, 1932). The basic materials of these two families have been gathered based on detailed geological fieldworks of many geologists for a long time from the East Indies (Indonesia and its surroundings) as eastern Tethys region or Indo – Pacific region. Therefore the research results of Miogypsinid foraminiferal Biostratigraphy through Tan Sin Hok's morphogenetic method could compare easily with the results of Miogypsinid foraminiferal biostratigraphic research from many areas (Drooger, 1993).

The purpose of this study is to describe the introduction of the Miogypsinid foraminiferal Biostratigraphy and its evolutional lineage based on the author's research and other colleagues results, and research of materials from three areas (Maraş, Palu, and Muş) of Menderes – Taurus Platform, Turkey, respectively.

2. Method of study

All microscopical studies were conducted by examination of all sectioned foraminiferal specimens from sample materials collected from the biostratigraphial columnar sections or spot samplings in order to reinforce the space and time distribution of species. Concerning to the observation of outer and inner structure of the foraminiferal test, the microscope used had the lens combination from x 20 to x 200. The biometrical measurement of the equatorial sectioned nucleoconch and peri-nucleoconchal chambers were made by means of a curvimeter and/or scale protractor from a drawing or direct thin section of the nucleoconch and peri-nucleoconchal chambers at magnification of x 200. The present study is based on the sectioned specimens and free specimens, which were collected from various localities and/or drill core in Japan, Taiwan, and Turkey.

In the present paper, the morphological terms used are given in the glossary and the important criteria for detailed measurements (Figure 1). The measurements were taken from the equatorial and axial (= vertical) sections of megalospheric specimens which exhibited considerable variations in measurements. Also those of microspheric specimens are used supplementary for the measurements and for observation of structure. Since all characters are not measurable, the statistical analysis and consideration from measurements are not perfectly alternative to traditional description, but merely supplementary to it, but provides a more objective basis for comparison between the measurable characters of important morphology and/or structure. The measured parameters are explained for the following terminology as defined by many authors (Figure 1).

On nucleoconch (= embryonic chambers), as showing the development of embryonic chambers is explained as below.

DI: diameter of protoconch, the first chamber of embryonic chambers, consisting protoconch and deuteroconch (Figure 1a). Generally inner protoconch is measured at right angle for the center line of both protoconch and deuteroconch. DI is Drooger's (1952) symbol or definition. The next two is also his symbol and definition.

DII: diameter of duteroconch, the second chamber of embryonic chambers (Figure 1a).

DII/DI: ratio of diameter between protoconch and deuteroconch.

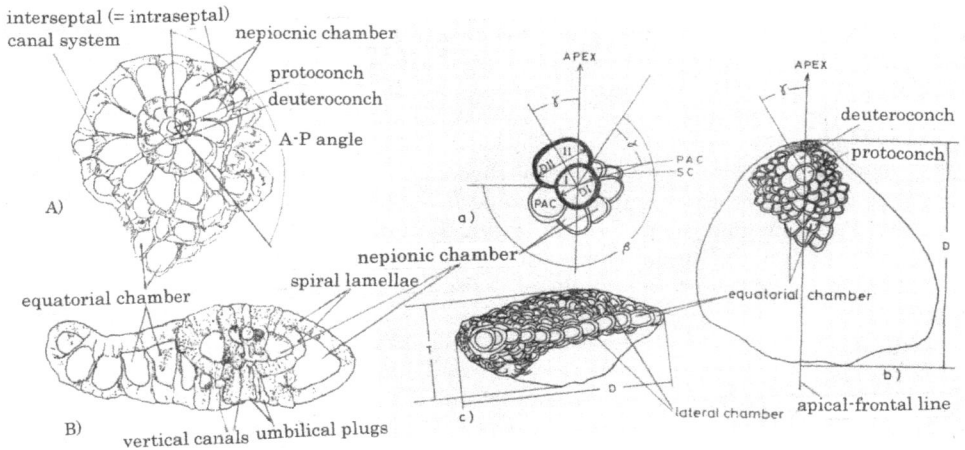

Fig. 1. Terminology of the Miogypsinid foraminifera. A. Equatorial section of *Miogypsinella boninensis* Matsumaru, and B. Axial (= Vertical) section of *Miogypsinella boninensis* Matsumaru. a. Enlargement of the apical portion of equatorial section of *Miogypsina globulina* (Michelotti). b. Equatorial section of *Miogypsina globulina* (Micgelotti). c. Axial (=Vertical) section of *Miogypsina globulina* (Michelotti). I = protoconch. II = deuteroconch. PAC = principal auxiliary chamber. SC = symmetrical (= closing) nepionic chamber. DI = diameter of protoconch. DII = diameter of deuteroconch. α = angle between a line joining the center of the protoconch and the junction of both walls of the protoconch and deuteroconch, and another line connecting the said center and a mid-point of the posterior wall of the symmetrical chamber. ß = angle representing the whole development area of both nepionic spirals of the protoconch surround with both spirals starting from both PAC. V = 200 α/ß = indicate from the absence of the second PAC representing by the value 0 to the protoconchal spirals of equal length by the value 100. γ = angle between the apical – frontal line and line joining the center of the protoconch and deuteroconch.

On nepionic chambers, as base character of the principle of nepionic acceleration (including nepionic retardation, but with additional development of new nepionic spirals) by Tan Sin Hok's (1936, 1937) theory is explained as below.

Parameter X: number of spiral nepionic chambers developed in peri-nucleoconch (= peri-embryonic chambers). This is Tan Sin Hok's (1936) "*Rotalia*-Anfang und mit intraseptalen Spalten". Drooger (1952) counts and used as one of important characters as symbol or definition, parameter X. The value X is progressed until the new nepionic spirals of parameter α as stated below. In the genus *Miogypsinella* with trochoid spirals (Figure 2, top left) and genus *Miogypsinoides* with planispiral (Figure 2, top right), the nepionic chambers are counted in number as parameter X in a spiral arrangement. This is continued until the presence of *Miogypsina primitiva* (Tan Sin Hok) (Figure 2, second left from top), which is considered to be the genus *Miogypsinopsis* Hanzawa, 1940, and *Miogypsina borneensis* Tan Sin Hok, 1936. The number of parameter X is getting decrease from *Miogypsinella boninensis* Matsumaru, 1996 (Figure 1A, X = 25) to *Miogypsina primitiva* (Figure 2, X = 10) and *Miogypsina borneensis* (Plate 2, figures 1-3, X = 6), and it is regarded as nepionic retardation. The next step is beginning at the development of secondary nepionic spirals from the second prinicipal auxiliary chamber and situated in the opposite side of the primary principal auxiliary chamber (Figure 1a; parameter α). Therefore this evolutionary lineage is generally regarded as Tan Sin Hok's (1936, 1937) nepionic acceleration from the genus *Miogypsinella* to genus *Miogypsina*. The following four parameters are based on Drooger (1952, 1963).

Parameter α: small nepionic spiral developed from the second principal auxiliary chambers (Figure 1a). This is the secondary or short nepionic spirals, situated under the outer wall of protoconch. This spiral is arc length and measured by the angle between the line connecting the center of protoconch and rough inscribed line touchrd between deuteroconch and second principal auxiliary chamber, and the line connecting the center of protoconch and center line of closing chamber, which is situated between large and small nepionic spirals developed from opposite direction of two principal auxiliary chambers (Figure 1a). Parameter α in Figure 1a = 40°.

Parameter ß: total nepionic spirals including a closing chamber (= symmetrical chamber, Figure 1a, sc) under the outer wall of protoconch, developed from two primary and secondary principal auxiliary chambers. This spiral is also arc length and measured by the angle between two rough inscribed lines from both two principal auxiliary chambers and deuteroconch, connecting with the center of protoconch (Figure 1a). Parameter ß in Figure 1a = 240°. Generally the primary or long nepionic spirals from the primary principal auxiliary chamber is larger arc than the secondary or short nepionic spirals.

Parameter V (= 200 α/ß): ratio of small or short nepionic spirals (α) for total short and long nepionic spirals (ß), and parameter α will be stopped at the midpoint of parameter ß. Then the ratio times 200 are expressed as a continuous scale with units from 0 to 100. When a closing or symmetry chamber is situated at the midpoint of both protoconchal nepionic spirals, V value is indicated as 100. When a short nepionic spiral or a closing chamber isn't present, V value is indicated as 0. Parameter V in Figure 1a = 40°/240° x 200 = 33.332. When measuring these parameters in numerous specimens in a sample, there is sometimes exceed over 100 in V value, but it is few case.

Parameter γ: angle between the apical-frontal line of test through the center of protoconch and the line connecting centers of embryonic chambers (Figure 1b). If the primary principal auxiliary chamber is situated below the line connecting of centers of embryonic chambers, parameter γ is positive, and reverse is negative (Plate 2, figure 2). Parameter γ in Figure 1a = positive 30°.

Genus *Miogypsinella* Hanzawa, 1940

Genus *Miogypsinoides* Yabe and Hanzawa, 1928

trochoid spirals

planispiral

Miogypsina primitiva (Tan Sin Hok, 1936), included in Genus *Miogypsinopsis* Hanzawa, 1940, and is regarded as the genus *Miogypsina*

Genus *Miogypsina* Sacco, 1893

Genus *Miolepidocyclina* A. Silvestri, 1907

Tania inokoshiensis Matsumaru, 1990 type species of Genus *Tania* Matsumaru, 1990.

Genus *Lepidosemicyclina* Rutten, 1911

Miogypsina mexicana Nuttall, 1933, type species of Genus *Miogypsinita* Drooger, 1952, and is regarded as intermediate form between the *Miogypsina* and *Miolepidocyclina*

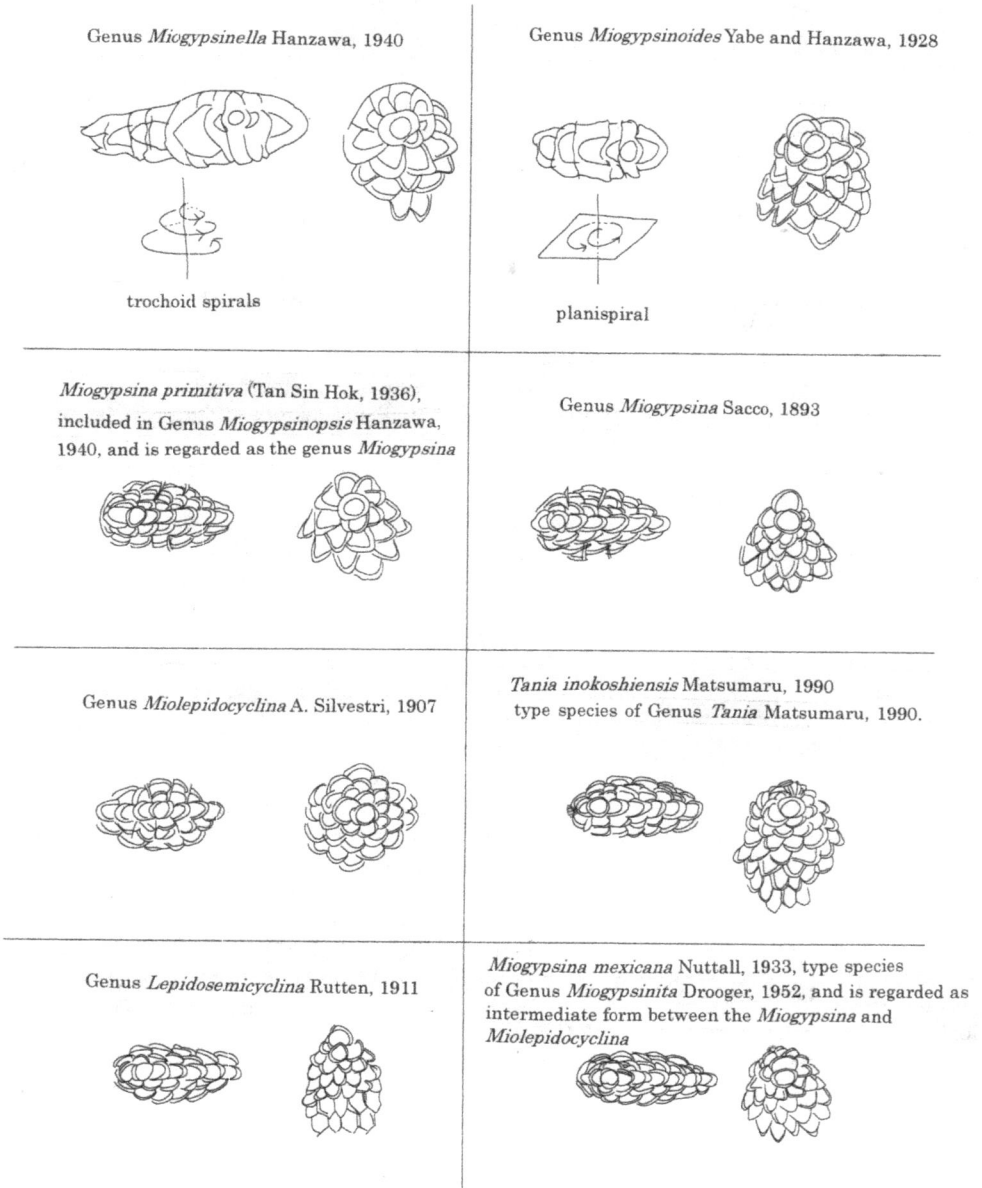

Fig. 2. Sketch of several genera and species of the family Miogypsinidae Vaughan, 1928 in the Tethys region. Two nepionic spirals are shown as trochoid spirals and planispirals.

Parameter A-P angle: arc length of nepionic spirals starting from embryonic chambers, and ending to the apical point of test (Hanzawa, 1957, p. 91; table 6), and A-P angle in Figure 1A , AP = 360° + 145° = 505°.

Designation of the species: In the previous investigation by many authors, the species units of miogypsinid foraminifera have mainly been established by applying from the mean X value to mean V value (mean X – mean V value scale), in addition to traditional main observation, i.e. presence and/or absence of lateral chambers, shape and arrangement of equatorial chambers, arrangement of stolons and canal system in chamber walls, and development of pillars and/or sometimes spines. Species of Figure 1A, B is *Miogypsinella boninensis* Matsumaru, carrying Parameter X (X = 25) and A-P angle (AP = 505°), and that of Figure 1 a-c is *Miogypsina globulina* (Michelotti), carrying Parameter V (V = 33.3) and Parameter γ (γ = + 30°). Parameter X exists until 5, and doesn't exist 4.

3. Stratigraphy, faunal succession, and correlation

In this chapter, the author describes the introduction of the fundamental Miogypsinid foraminiferal Biostratigraphy, faunal succession and phylogenetic lineage, and correlation.

1.   Ogasawara Islands, Japan

The basal Oligocene carbonate sedimentary rocks in Japan has been known in Chichi-Jima (island) and Minami-Jima, Ogasawara Islands, Japan (Matsumaru, 1996) (Figures 4-7). In there, six stratigraphic sections in the Minamizaki cape, SW of Chichi-Jima, and two stratigraphic sections and several spot samplings in the Minami-Jima, Ogasawara Islands were examined for the larger foraminiferal biostratigraphy of the Minamizaki Limestone (Formation) with maximum 244 m thick, overlying the basement volcanic rocks (boninite, andesite, dacite, and others) (Figures 5-6). Two larger foraminiferal assemblages (Assemblage IV and Assemblage V) during Oligocene age were recognized in the respective sections of biostratigraphic sequence, based on the stratigraphic range of larger and smaller foraminifera in association with planktonic foraminifera. The Assemblage IV is the *Eulepidina dilatata* (Michelotti) - *E. ephippioides* (Jones and Chapman) - *Heterostegina borneensis* van der Vlerk Assemblage and the Assemblage V is the *Miogypsinella boninensis* - *Spiroclypeus margaritatus* (Schlumberger) - *Austrotrillina howchini* (Schlumberger) Assemblage. Both assemblages were correlative with Tertiary c and/or Tertiary d, and Tertiary e1-2 to Tertiary e4 of the East Indies Letter Stages (Leupold and van dr Vlerk, 1931), respectively, and were also correlative with Zone P 18?-21 or *Globigerina sellii* (Borsetti) Zone – *Globorotalia opima opima* Bolli Zone, and Zone P 21? or P 22 of planktonic foraminiferal zonations. The Assemblage IV is correlated with the fauna of the Tertiary beds of 1629 to 2687 feet, in Eniwetok Atoll Drill Holes (Cole, 1957), and 1723.5 to 2359.5 feet, in Bikini Atoll Drill Holes (Cole, 1954), respectively, because of the coexistence and range of *Eulepidina ephippioides* (Jones and Chapman), *Heterostegina borneensis* van der Vlerk, *H. duplicamera* Cole, and *Halkyardia minima* (Liebus) (Figure 4). The Assemblage V is also correlated with Tertiary e limestones in bore-holes at Eniwetok Atoll Drill Holes at depth from 1210 to 1599 feet, and at Bikini Atoll Drill Holes at depth from 1597.5 to 1671 feet, respectively, where *Miogypsinella grandipustula* (Cole) and *Miogypsinella ubaghsi* (Tan Sin Hok) were reported by Cole (1954, 1957) (Figure 4).

Two assemblages (IV and V), Ogasawara Islands are referable in the geological age to Early to late Early Oligocene, and early Late Oligocene, respectively (Figures 6-7). According to Kaneoka et al. (1970) and Tsunakawa (1983), K-Ar radiometric ages on boninite, andesite, dacite and quartz dacite of basal volcanic rocks of Chichi-Jima, Ogasawara Islands is regarded as 43.0 to 29.4 Ma, and the most young age of volcanics is 26.7 Ma. The Minamizaki Limestone is regarded as submerged karst topography, with summits sticking up from the sea as peninsulas of Minamizaki Cape, Chichi-Jima and Minami-Jima (island) and a lot of islets. The largely submerged Minamizaki Limestone is estimated to be more than 244 m thick and overlies the basement volcanic rocks of lavas and pyroclastics of boninite and other rocks as stated above.

The Assemblage IV is at least regarded as Tertiary d, in this study, due to occurrence of *Heterostegina borneensis* van der Vlerk, *H. duplicamera* Cole, *Eulepidina dilatata* (Michelotti), *E. ephippioides* (Jones and Chapman), *Pararotalia mecatepecensis* (Nuttall), *Paleomiogypsina boninensis* Matsumaru, *Borelis pygmaeus* (Hanzawa) and *Nephrolepidina marginata* (Michelotti), with associated planktonic foraminifera of Zone P 21(Blow, 1969) such as *Globorotalia opima nana* Bolli, *G.* cf. *opima opima* Bolli, and *G.* gr. *opima* Bolli. They are correlated with the Late Eocene to Neogene time scale (official website of ICS, 2004; Berggren et al., 1995) (Figures 6-7, 9). The Assemblage V is assigned to Tertiary e1-2 to Tertiary e3, in this study, due to occurrence of *Miogypsinella boninensis* Matsumaru, *Spiroclypeus margaritatus* (Schlumberger), *Cycloclypeus eidae* Tan Sin Hok, which is junior synonym of *C. koolhoveni* Tan Sin Hok and/or *C. oppenoorthi* Tan Sin Hok, *Paleomiogypsina boninensis*, *Boninella boninensis* Matsumaru, *Flosculinella reicheli* Mohler, which is a synonym of *Flosculinella globulosa* Rutten, and *Austrotrillina howchini* (Schlumberger). This fauna didn't associate with diagnostic planktonic foraminifera, but it should be assumed to be Zone P 22 from the biostratigraphical occurrence (Figure 6). Although the basal part of the Minamizaki Limestone is obscure due to subsidence under the sea, *Pararotalia mecatepecensis* may evolve into *Paleomiogypsina boninensis* due to nepionic acceleration and well-developed subsidiary chambers during early Oligocene (Rupelian) and/or latest Eocene (Priabonian?) due to basal volcanic radiometric age (Figures 3-4, 7). Also *Paleomiogypsina boninensis* evolved into *Miogypsinella boninensis* due to biostratigraphical occurrence, Tan Sin Hok's nepionic acceleration, and development of equatorial chambers (Figures 3-4, 7). *Miogypsinella boninensis* has the character of number of nepionic chambers (mean X = 27) and A-P angle (mean AP = 578°) (Figure 4).

2.   Komahashi-Daini Seamount, Japan

The larger foraminiferal assemblage has been discovered from limestone blocks dredged at two sites on the Komahashi-Daini Seamount of the Kyushu-Palau Ridge, Japan (sample DG-04-01; 30°02.98′N. lat., 133°19.88′E. long.; sample DG-05-02; 29°53.98′N. lat., 133°22.66′E. long.; Mohiuddin et al., 2000) (Figure 5). The assemblage is dominated by the occurrence of *Miogypsinella ubaghsi* (Tan Sin Hok), *Spiroclypeus margaritatus*, *Heterostegina borneensis*, *Eulepidina dilatata*, *E. ephippioides*, *Nephrolepudina marginata*, and *Austrotrillina howchini*, and was correlated with the top part of the Minamizaki Limestone of Ogasawara Islands. In this study, the Komahashi-Daini larger foraminiferal fauna may be regarded as the fauna from the covering limestone of the Minamizaki Limestone, Ogasawara Islands, because *Miogypsinella ubaghsi* didn't occur from the top member of the Minamizaki Limestone (Figures 4, 6-7). *Miogypsinella ubaghsi* has the character such as number of nepionic chambers (X = 21) and A-P angle (AP = 395°) (Mouhiddin et al, 2000, fig.8-3). Judging from the stratigraphic correlation and nepionic acceleration, *Miogypsinella boninensis* evolved into *Miogypsinella ubaghsi* as the author's consideration (Matsumaru, 1996, p. 39, fig. 24) (Figure 4).

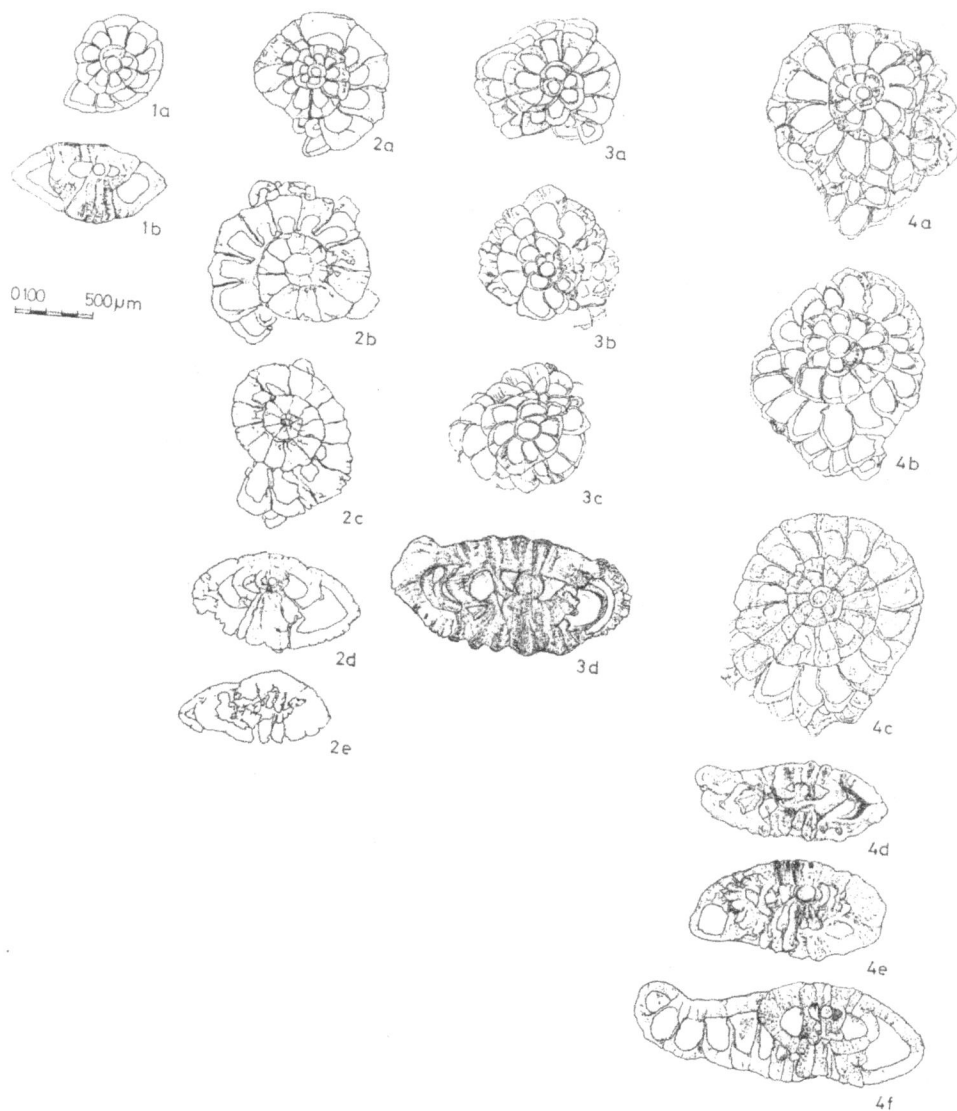

Fig. 3. Drawings of the embryonic, nepionic, and neanic stages in the equatorial and axial sections of species of: 1. *Pararotalia mecatepecensis* (Nuttall), 2. *Paleomiogypsina boninensis* Matsumaru, 3. *Boninella boninensis* Matsumaru, 4. *Miogypsinella boninensis* Matsumaru (Matsumaru, 1996, fig. 23).

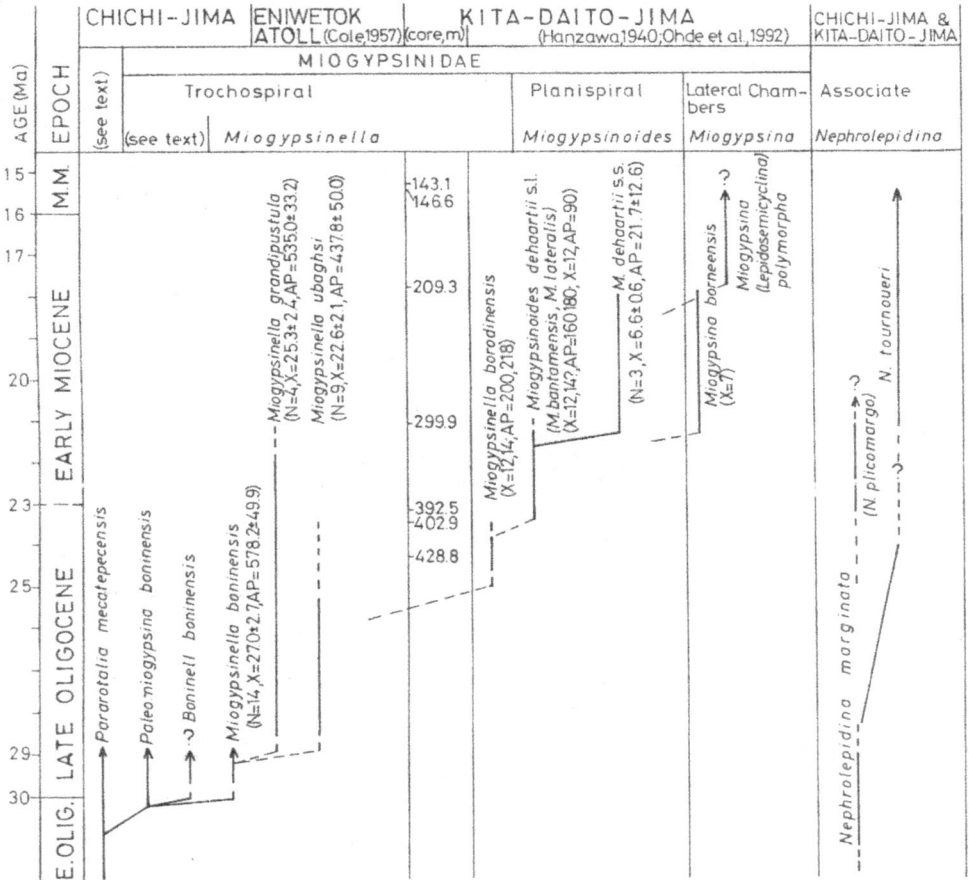

Fig. 4. Evolution of the western Pacific Miogypsinids from Chichi-Jima, Eniwetok Atoll, and Kita-Daito-Jima, and the stratigraphic position of associated *Nephrolepidina* species from Chichi-Jima and Kita-Daito-Jima (Matsumaru, 1996, fig. 24).

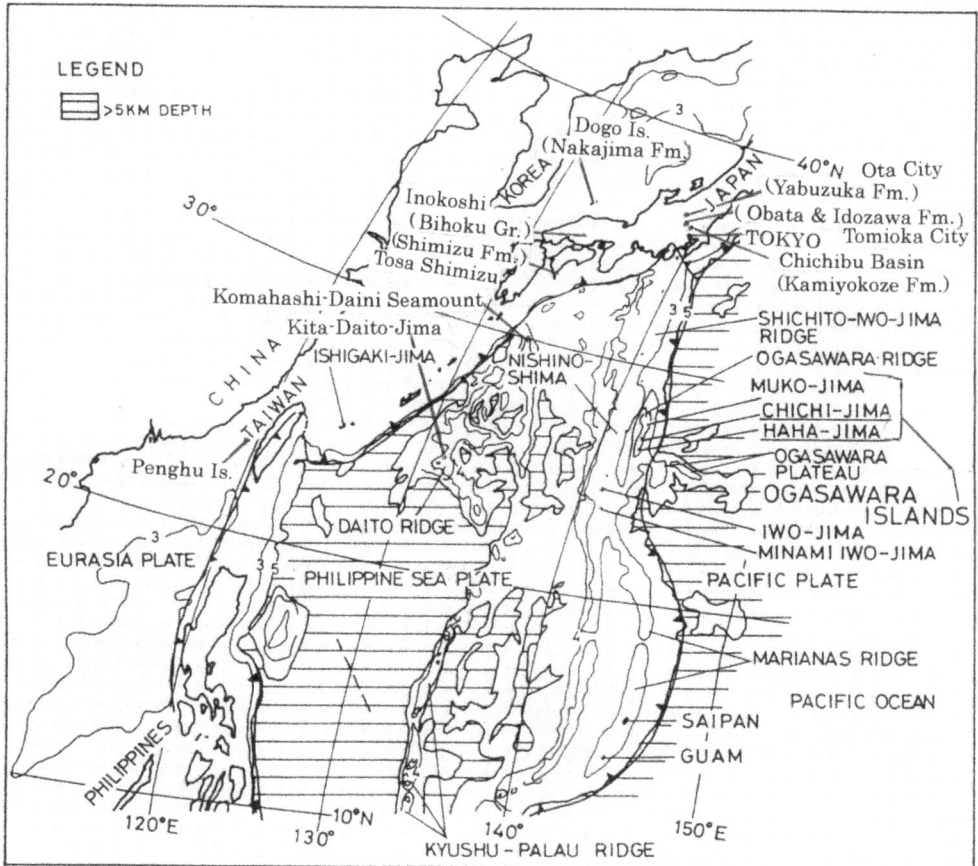

Fig. 5. Geographical locations from Japan and Taiwan treated in this study (Retouch to Matsumaru, 1996, fig. 1).

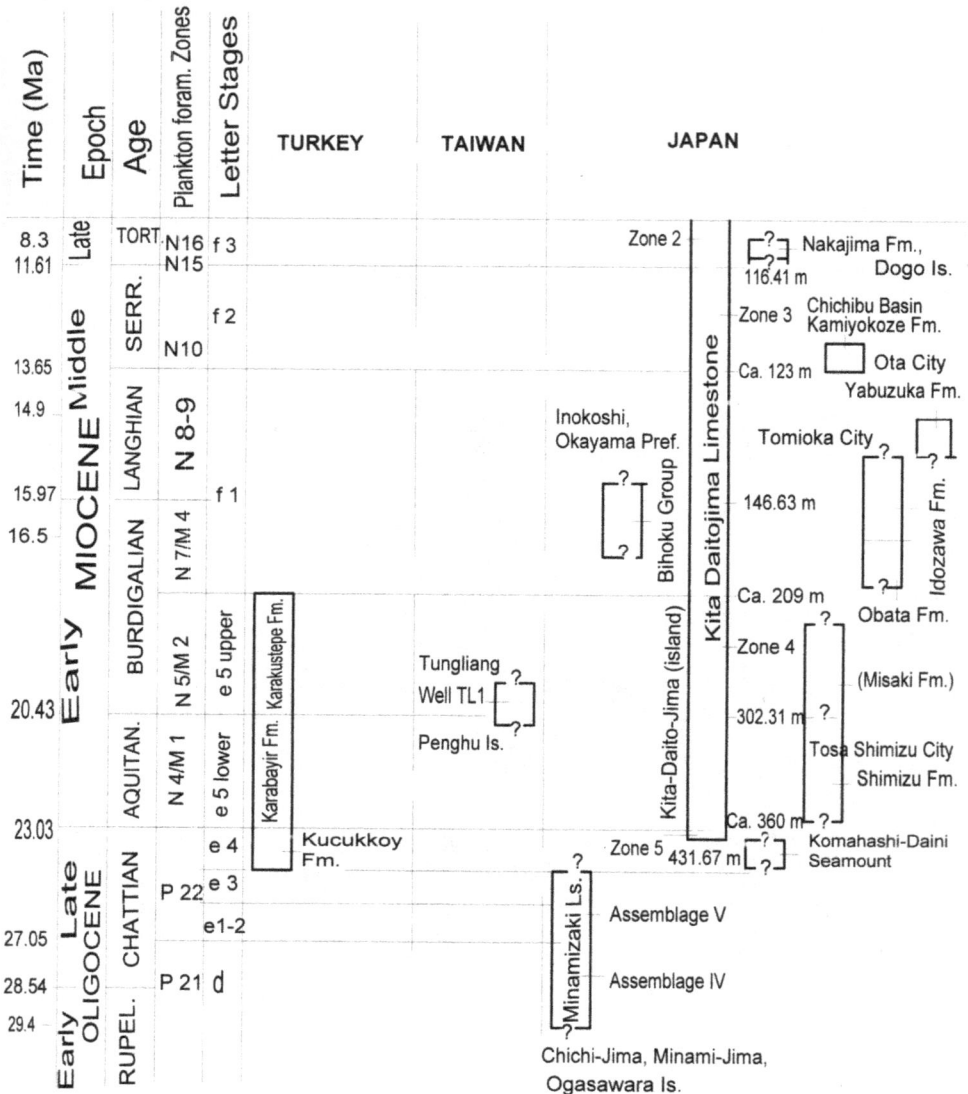

Fig. 6. Correlation chart between the stratigraphic columnar sections treated in Turkey (Matsumaru et al., 2010); Taiwan (Matsumaru, 1968); and Japan (Hanzawa, 1940; Matsumaru, 1967, 1971, 1972, 1977, 1980, 1982, 1996; Matsumaru et al., 1993; Mohiuddin et al., 2000; Nomura et al., 2003).

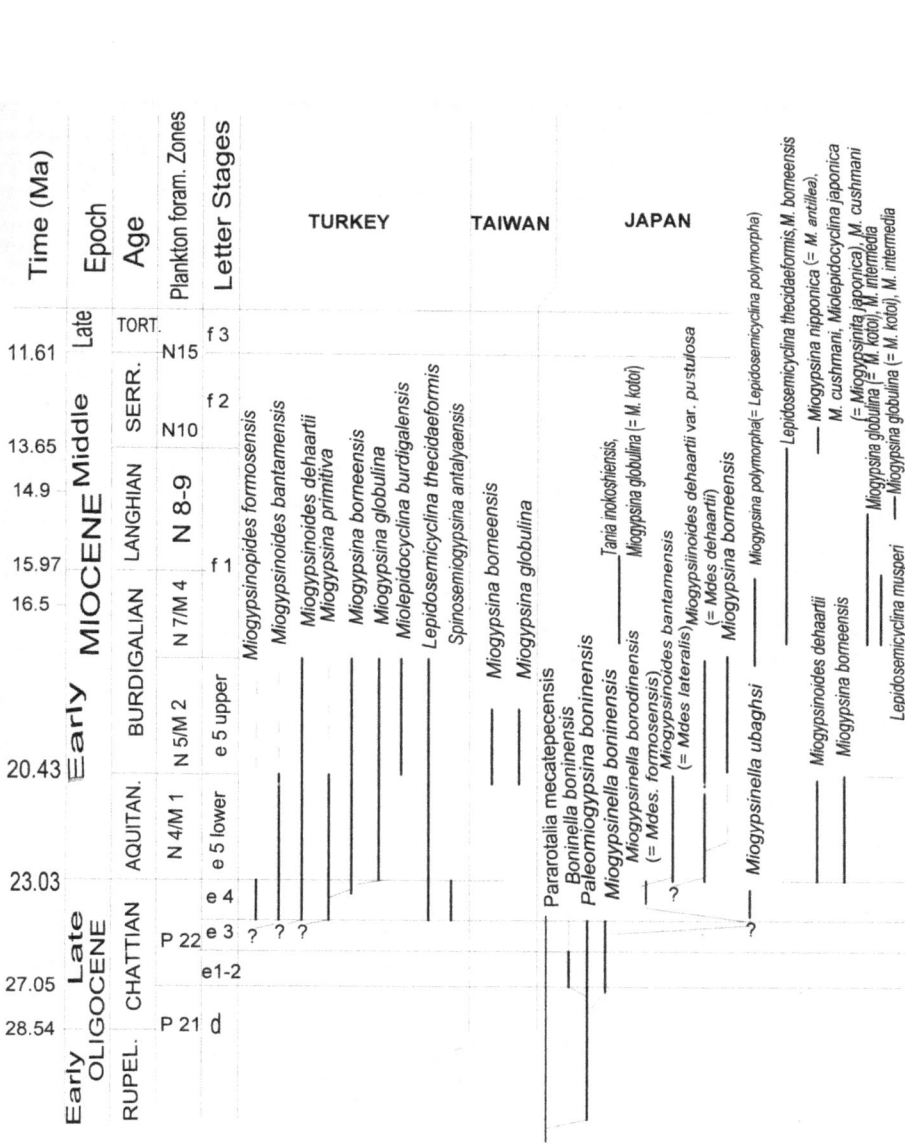

Fig. 7. Biostratigraphic occurrence of Miogypsinid foraminifera from Turkey, Taiwan and Japan.

3.　Kita-Daito-Jima, Okinawa Prefecture, Japan

Five foraminiferal fauna have been established into depth zones of the drill cores (431.67-2.68 m thick) of the Kita Daitojima Limestone, at Kita-Daito-Jima (North Borodino Island, 25°56′47″N. lat., 131°17′30″E. long.), Okinawa Prefecture, Japan (Hanzawa, 1940) (Figures 4-7). Hanzawa's Zone 5 (431.67-394.98 m) is characterized by the occurrence of *Miogypsinella borodinensis* Hanzawa, which is later assigned to *Miogypsinoides borodinensis* (Hanzawa) without lateral chambers or with incipient lateral chambers by Hanzawa (1964). This species evolved from *Miogypsinoides formosensis* Yabe and Hanzawa (Hanzawa, 1964, p. 309, 311) due to decrease of number of nepionic chambers. In this study, *Miogypsinella borodinensis* of probable holotype specimen has the character of both number of nepionic chambers (X = 13) and A-P angle (AP = 220°) (Hanzawa, 1940, pl. 39, fig.6), and *Miogypsinoides formosensis* of probable holotype specimen has the character of both number of nepionic chambers (X = 16) and A-P angle (AP = 240°) (Yabe and Hanzawa, 1928, fig. 1a). Hanzawa's consideration is right, but there is unknown on species variation of both species. Both forms could fortunately be found from the Küçükkoy Formation in Korkuteli area, Bey Dağlari Autochton, Menderes-Taurus Platform, SW Turkey (Matsumartu et sl., 2010) (Figures 6-9). The author in Matsumaru et al. (2010) described *Miogypsinoides formosensis* (Yabe and Hanzawa) and regarded their all specimens of schizont (A1 form) and gamont (A2 form) of sexual reproduction, rather planispiral, and carrying rudimentary lateral chambers (Matsumaru et al., 2010, pl. 2, fig. 1) from the Küçükkoy Formation. A specimen (Matsumaru et al., 2010, pl. 1, fig. 8) has the character of number of nepionic chambers (X = 13) and A-P angle (AP = 210°), while a specimen (Matsumaru et al., 2010, pl. 1, fig. 9) has the character of number of nepionic chambers (X = 16) and A-P angle (AP = 250°). Another specimen (Matsumaru et al., 2010, pl. 1, fig. 10) has the character of number of nepionic chambers (X = 13) and A-P angle (AP = 260°). As such the Küçükkyoy Formation carrying *Miogypsinoides formosensis* was correlated with the Zone 5 drill cores (431.67-ca. 360 m, as stated below) of the Kita Daitojima Limestone due to occurrence of *Miogypsinella borodinensis* (= *Miogypsinoides formosensis*) (Matsumaru et al., 2010).

Sr isotope age of Hanzawa's Zone 5 is regarded as 24.3 to 23.5 Ma (Ohde and Elderfield, 1992), and then Hanzawa's Zone 5 is applied for drill cores from 431.67 to ca. 360 m from their age assignment (Figures 6-7). Judging from the Tan Sin Hok's nepionic acceleration, *Miogypsinella ubaghsi* occurred from the limestone of Komahashi-Daini Seamount evolved into *Miogypsinella borodinensis* (= *Miogypsinopides formosensis*) occurred from Zone 5 drill cores of Kita Daitojima Limestone due to reduction of number of nepionic chambers and low value of A-P angle (Figure 7).

4.　Tosa-Shimizu City, Shikoku, Japan

The Shimizu Formation, Ashizuri Cape, Tosa Shimizu City, Shikoku, Japan occupies the southernmost part of the Shimanto Belt, one of Japanese Tectonic Zones, and consists of calcareous sandstone and volcanic conglomerate into the coherent rock facies and chaotic rock facies (Figures 5-7). The calcareous sandstone of the Shimizu Formation occurred *Miogypsinoides dehaartii* (van der Vlerk) (Plate 2, figure 12), *Miogypsina* sp, which is assigned to *M. borneensis*, *Nephrolepidina praejaponica* Matsumaru, *Spiroclypeus margaritatus* (Schlumberger) and *Victoriella conoidea* (Rutten) in the location of Ashizuri Cape, Tosa Shimizu City, Kochi Prefecture, Japan (32°47′15.4 ″N. lat., 132°57′34.6″E. long., Matsumaru et al., 1993). The Misaki Formation of Tosa Shimizu City is composed on alternation of sandstone and mudstone, and crops out in 4 km NW of Ashizuri Cape and there is no contact with the Shimizu Formation. The lower member of the Misaki Formation occurs *Nephrolepidina praejaponica* Matsumaru, *Amphistegina radiata* (Fichtel and Moll), *Sphaerogypsina globulus* (Reuss) and *Rotalia* spp. and also occurrs planktonic foraminifera such as *Catapsydrax stainforthi* Bolli, *Globigerina altiapertura* Bolli, *G. immaturus* Leroy, *G. subquadratus* Brönnimann, *Globorotalia zealandica* Hornibrook, *Globorotaloides suteri* Bolli and *Praeorbulina sicana* (de Stefani) (Matsumaru and Kimura, 1989). At least *Nephrolepidina praejaponica*-bearing Misaki Formation is assumed to be correlated with Zone N5 to lower Zone N7 of planktonic foraminiferal Zonations (Blow, 1969). Therefore the *Miogypsinoides dehaartii* and *Miogypsina borneensis* bearing Shimizu Formation is underlain the Misaki Formation, and its age is regarded as Zone N4 or Zone N 5? (Blow, 1969) (Figures 6-7). Both embryonic and nepionic stages of *Miogypsinoides dehaartii* and *Miogypsina borneensis* from the Shimizu Formation are insufficient in oblique and vertical sections. The Shimizu Formation carrying *Miogypsinoides dehaartii* and *Miogypsina borneensis* may mostly be correlated with the upper Zone 4 drill cores (302.31 to ca, 209 m thick) of the Kita Daitojima Limestone due to the upper occurrence of *Miogypsinoides dehaartii* var. *pustulosa* (= *M. dehaartii*, s. l.) and *Miogypsina borneensis*, and without *Miogypsinoides bantamensis* (Hanzawa, 1940) (Figures 4-7). *Miogypsinoides dehaartii* s. l. of the Kita Daitojima Limestone has the character of number of nepionic chambers (X = 7, 7 and 7) and A-P angle (AP = 20°, 10° and 20°) in three specimens (Hanzawa, 1940, pl. 40, figs. 29, 27 and 26), and *Miogypsina borneensis* of the Kita Daitojima Limestone has the character of number of nepionic chambers (X = 7 and 7) and A-P angle (AP = 10° and 30°) in two specimens (Hanzawa, 1940, pl. 41, figs. 19-20). However *Miogypsinoides bantamensis* (Tan Sin Hok) in the lower Zone 4 drill cores (ca. 360 to 302.31 m thick) has the character of number of nepionic chambers (X = 12, 12 and 13) in three specimens and A-P angle (AP = 165°, 180° and 195°) (Hanzawa, 1040, pl. 39, figs. 16-17, 19). Then *Miogypsinoides bantamensis* evolved into *Miogypsinoides dehaartii* s. l. due to biostratigraphic occurrence and Tan Sin Hok's nepionic acceleration, with reduction of number of nepionic chambers as explained above (Figure 7). Moreover *Miogypsinoides dehaartii* without lateral chambers evolved into *Miogypsina borneensis* with lateral chambers (Figure 7).

5.   Tungliang Well TL1, Paisa Island, Penghu Islands, Taiwan

Miogypsina globulina (B form, but not microspheric form; Matsumaru, 1968, pl. 36, figs. 1-6) with nepionic chambers arranged single type (= *Miogypsina borneensis* Tan Sin Hok) and *Miogypsina globulina* (A form; Matsumaru, 1968, pl. 35, figs. 1-6) with two unequal protoconchal nepionic spirals (= *Miogypsina globulina* (Michelotti)) have been found in *Miogypsina* bearing calcareous sandstone at about 500 m depth in the Tungliang Well TL1, located at about 800 m NE of Tungliang Village, Paisa Island, Penghu Islands, Taiwan (Figure 5). *Miogypsina borneensis* has the character of number of nepionic chambers (X = 6, 5, 6, 6, 7, and 6; mean X = 6) in 6 specimens (n = 6) and A-P angle (AP = 10°, 15°, 6°, 6°, 15° and 25°) (Matsumaru, 1968, pl. 36, figs. 1-6). *Miogypsina globulina* has the characters of ratio of two nepionic spirals (V = 32.56, 18.18, 15.20 and 28.36; mean V = 23.58) in 4 specimens (n = 4) (Matsumaru, 1968, pl. 35, figs. 1-4). Therefore the *Miogypsina* bearing sandstone of Well TL1, carrying *Miogypsina borneensis* and *M. globulina*, but not *Miogypsinoides dehaartii*, is stratigraphically younger than the Shimizu Formation and upper Zone 4 of the Kita Daitojima Limestone, which carry *Miogypsinoides dehaartii* and *Miogypsina borneensis* (Figures 6-7). As such *Miogypsina borneensis* carrying number of nepionic chambers (mean X = 6) from the *Miogypsina* sandstone, Well TL1, Paisa Island, Penghu Islands is necessarily fewer number of nepionic chambers than *M. borneensis* carrying number of nepionic chambers (mean X = 7) from the upper Zone 4 drill cores of Kita Daitojima Limestone, Kita-Daito-Jima.

6. Early to Middle Miocene *Miogypsina* from Honshu, Japan
The Obata and Idozawa Formations, Tomioka Group, Honshu, Central Japan are known as representative sedimentary rocks of late Early to early Middle Miocene age in Japan (Matsumaru, 1967, 1977) (Figures 5-7). Japanese *Miogypsina* has been known to occur from the Lower/Middle Miocene sedimentary rocks in Honshu, Japan as *Miogypsina kotoi* Hanzawa, 1931, *Miolepidocyclina* (= *Miogypsinita*) *japonica* Matsumaru, 1972, *Miogypsina japonica* Ujiie, 1973 (= *M. globulina* (Michelotti)), *M. nipponica* Matsumaru, 1980 (= *M. antillea* (Cushman) and *M. cushmani* Vaughan steps of nepionic acceleration), and *Tania inokoshiensis* Matsumaru, 1990, in addition to *Miogypsina borneensis* Tan Sin Hok, *Lepidosemicyclina thecidaeformis* (Rutten), and *Lepidosemicyclina musperi* (Tan Sin Hok) (Plates 1-2). According to Matsumaru and Takahashi (2004), Japanese *Miogypsina* is discussed as the followings: The measurement data of topotype specimen of *Miogypsina kotoi* Hanzawa is as follows: V = 20, DI = 120 x 70 micron, DII/DI = 1.0, and γ = 35°, and *Miogypsina kotoi* Hanzawa is junior synonym of *Miogypsina globulina* (Michelotti). *Miogypsina japonica* Ujiie from type locality and other three stations has the following data: V = 40.9 and DII/DI = 1.34, V = 39.2 and DII/DI = 1.28, V = 36.1 and DII/DI = 1.32, and V = 47.2 and DII/DI = 1.24. Then *Miogypsina japonica* Ujiie doesn't represent *Miogypsina cushmani* Vaughan, 1924 of V scale of Drooger (1963), but represent *M. globulina* due to having of V value (mean V = 40.85) (n = 4). The *Miogypsina* population at Nogami locality found from the Obata Formation, Tomioka Group, Tomioka City, Gunma Prefecture is known as V value (mean V = 43.93 in 20 specimens), which is assigned to *Miogypsina globulina* (Matsumaru, 1967, 1977) (Figures 5-6). However, critical viewing Miogypsinid population, *Lepidosemicyclina musperi* (Rutten) and *Miogypsina cushmani* Vaughan with V value (V = 77) can be found from the Obata Formation (Figure 7). The Obata Formation is conformably overlain by the basal tuff (T6 Tuff or Wagoubashi Tuff, Matsumaru, 1967) of the Idozawa Formation carrying *Miogypsina globulina,* and the fission truck age of the T6 Tuff is 16. 5 ± 1.9 Ma by Nomura and Ohira (1998). The Idozawa Formation is conformably overlain by the basal tuff (T5 Tuff, Matsumaru, 1967) of the Haratajino Formation, which yields *Orbulina suturalis* Brönnimann, *O. universa* d'Orbigny, *Globorotalia birnagea* Blow, *Globigerinoides sicanus* de Stefani and others (Matsumaru, 1977). The fision truck age of T5 Tuff is 15.2 ± 0.5 Ma (Nomura and Ohira, 2002). Then the geological age of the Idozawa Formation is roughly Langhian of early Middle Miocene based on Zone N8 (Blow, 1969; Berggren et al., 1995), and the age of T5 Tuff is regarded to be the age of the *Orbulina* datum-plane. The *Miogypsina* population at Kanayama locality found from the Yabuzuka Formation at Ota City, Gunma Prefecture is known as V value (mean V = 47.38 in 24 specimens), which is assigned to *Miogypsina intermedia* of Drooger's mean V scale (Figures 5-7). The Kanayama *Miogypsina* population is found from the medium sandstone below the pumice tuff of the Yabuzuka Formation, and fission truck age of this tuff is 14.9 ± 0.5 Ma (Nomura et al., 2003). As such Miogypsinid foraminifera from the Obata, Idozawa and Yabuzuka Formations are known as *Miogypsina globulina* (Michelotti) and *Miogypsina intermedia* Drooger Assemblage. Raju (1974) and Mishra (1996) regarded *Miogypsine globulina* as population with mean V value between zero and 45 and positive γ, but *Miogypsina intermedia* couldn't find from the study of Indian *Miogypsina*. These criteria are arbitrary, and why they cannot find *Miogypsina intermedia* between *Miogypsina globulina* and *Miogypsina cushmani* or *M. antillea* in a series of mean V value scale? (Figure 7).

Miogypsina nipponica Matsumaru is found from the middle member of the Kamiyokoze Formation, at sample location UN-1 (35°58′31″ N. lat., 139°5′42″E. long.), Chichibu Basin, Saitama Prefecture (Matsumaru, 1980) (Figures 5-6). Also the planktonic foraminifera such as *Globigerinoides immaturus* (Leroy), *Globigerinoides subquadratus* Brönnimann, *Globigerina praebulloides* Blow, *Globorotalia* (*Turborotalia*) *peripheroacuta* Blow and Banner, and *Globorotalia* (*Turborotalia*) *birnagea* Blow are found from the upper member of the Kamiyokoze Formation at sample location NG-1 (35°58′25″ N. lat., 139°6′20″ E. long.), Chichibu Basin, and these fauna is shown in the lower part of Zone N10 of Blow (1969) (Matsumaru, 1980). *Miogypsina nipponica* has the character of V value in 23 specimens (mean V = 88.38 ±5.20) and this mean V value is regarded as *Miogypsina antillea* (Cushman) step of Drooger (1952). V value of *Miogypsina nipponica* varies from 78 to 100, and more than 42 % of specimens having V value larger than 90. Then *Miogypsina nipponica* Matsumaru has both V value of *Miogypsina cushmani* (Cushman) step and *M. antillea* step of Drooger (1952). Raju (1974) regarded *Miogypsina cushmani* as miogypsinid population with mean V value between 70 and 100, usually less than 90, and more than 50 % carrying less than 90. Also *Miogypsina antillea* has mean V value between 70 and 100, and more than 50 % carrying greater than 90.

Miogypsina nipponica resembles the topotype of *Miogypsina antillea* according to Cole (1957, pl. 29, fig. 1), but *M. nipponica* frequently possess small nepionic chambers on the deuteroconch (Matsumaru, 1980, pl. 25, figs. 4-5) (Plate 1, figure 10). The ogival to lozengic equatorial chambers and very short hexagonal equatorial chambers are sometimes distributed in *Miogypsina nipponica*, but the elongate hexagonal equatorial chambers are distributed in the frontal margins in *Miogypsina antillea* by Raju (1974, pl. 2, figs. 25-26). *Miogypsina nipponica* is distinguished from *Miogypsina antillea* by Frost and Langenheim (1974) in view of the description and illustration of Mexican specimens. In this study, *Miogypsina nipponica* is regarded as *Miogypsina antillea* step of V value, and associated with *Miogypsina cushmani* Vaughan and *Miolepidocyclina* (= *Miogypsinita*) *japonica* Matsumaru (Figure 7).

Miogypsina kotoi, which is junior synonym of *Miogypsina globulina*, is found from the Nakajima Formation, Dogo Island, Oki Islands (Matsumaru, 1982) (Figures 5-6). The sample location is Kumi at left side of the Kumi River about 1.1 km NW of Kumi Tunnel, Goka-Mura (village), Oki-Gun, Shimane Prefecture, Japan (36°18′11″ N. lat., 133°15′5″ E. long.). *Miogypsina kotoi* has the character of V value of 34 specimens (mean V = 31.29 ± 11.0), and is in association with planktonic foraminifera such as *Globorotalia acostaensis acostaensis* Blow, *G. continuosa* Blow, *G. quinifalcata* Saito and Maiya, *G. scitula* (Brady), *Globigerinoides quadrilobaturus* Leroy and *Globigerina* sp. Then the geological age of the Nakajima Formation carrying *Miogypsina kotoi* is early Late Miocene based on Zone N16 (Blow, 1969). Therefore it is inferred from the planktonic foraminiferal zones that *Miogypsina kotoi* with low mean V value from the Nakajima Formation were reworked from the pre-Nakajima Formation. Moreover *Miogypsina kotoi* (= *M. globulina*) from the Nakajima Formation is associated with *Miogypsina borneensis* Tan Sin Hok, which is known to occur from the Lower/Middle Miocene sedimentary rocks in Japan in the Yatsuo Formation, Toyama Prefecture; and Hirashio Formation, Ibaraki Prefecture (Matsumaru and Takahashi, 2004).

Tania inokoshiensis Matsumaru is found from the sandstone of the Lower Formation ("Yamaga" Formation) of the Bihoku Group, Okayama prefecture (Matsumaru, 1990) (Figures 2, 5-7). The sample location is the same place as Hanzawa's (1935) and Tan Sin Hok's (1937) Inokoshi, Koyamaichi Village, Kawakami-Gun, Okayama Prefecture (34°45′ N. lat., 133°24′ E. long.) (Figures 5-6). *Tania inokoshiensis* Matsumaru is characterized by having

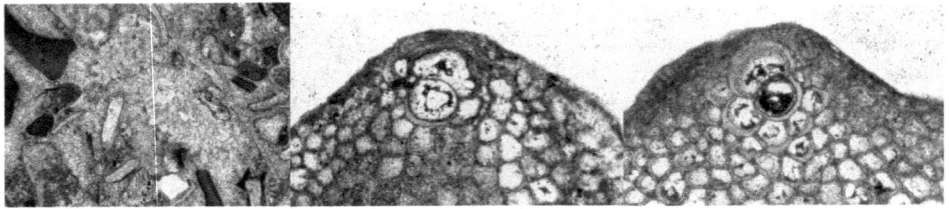

1   2   3

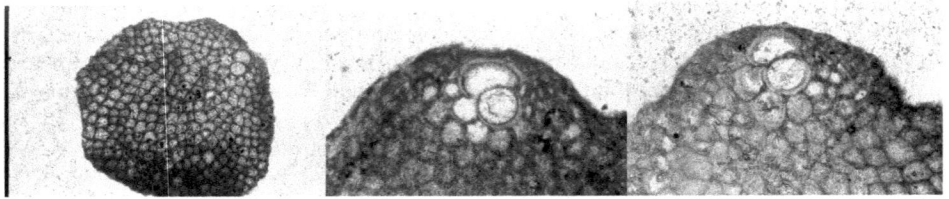

4   5   6

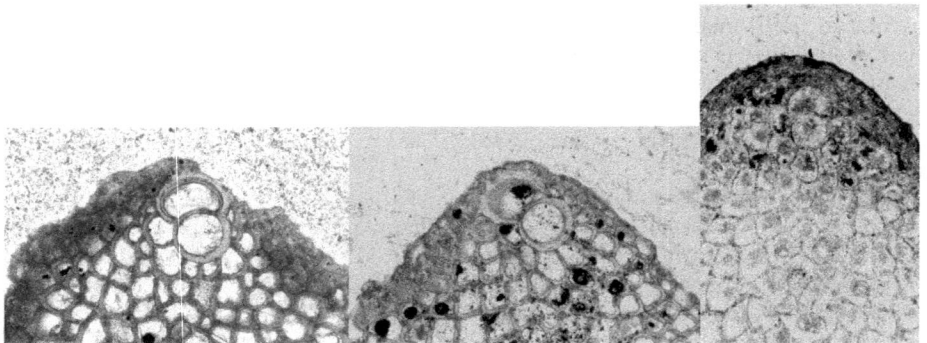

7   8   9

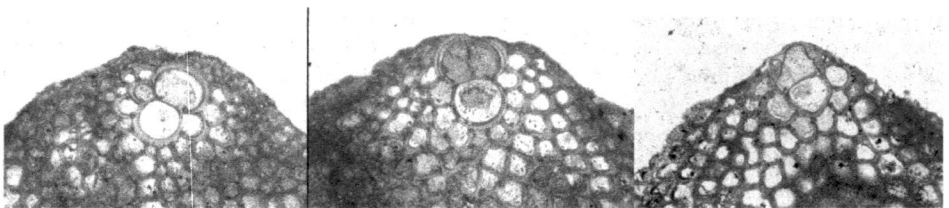

10   11   12

Plate 1.
Figures 1-5. *Miogypsina globulina* (Michelotti)
1. Centered oblique section. Topotype specimen of *Miogypsina kotoi* Hanzawa, 1931.

Otsuki Limestone at Otsuki locality, Yamanashi Prefecture, Japan. x 19. 2-5. Equatorial sections. 2-3, 5. Miyato Formation, correlated with Obata Formation, at Komori locality, Saitama Prefecture, Japan, x 53. 2: V = 25, γ = 20°. 3; V = 29, γ = 25°; 5: V = 40, γ = 20°; 4. Shiomizaki Formation at Todoroki locality, Aomori Prefecture, Japan. x 19. V = 35, γ = 35°. Figures 6-8. *Miogypsina intermedia* Drooger. Equatorial sections. 6-7. Ichinokawa Formation, correlated with Idozawa Formation, at Kawabata locality, Saitama Prefecture, Japan. x 53. 6: V = 46, γ = 25°; 7: V = 48, γ = 20°. 8. Nakahara Formation at Hota locality, Chiba Prefecture, Japan. x 53. V = 62, γ = 40°. Figure 9. *Miogypsina cushmani* Vaughan Equatorial section. Yabuzuka Formation, correlated with Idozawa and/or Haratajino Formations, at Kanayama locality, Gunma Prefecture, Japan. x 53. V = 70, γ = 10°. Figures 10-12. *Miogypsina nipponica* Matsumaru Equatorial sections. 10-12. Kamiyokoze Formation, at Une locality, Saitama Prefecture, Japan. x 53. 10. V = 84, γ = 30° (This specimen possess *Miogypsina cushmani* step of V value, and small nepionic spirals on the outer wall of deuteroconch). 11. Holotype, Saitama Univ. coll. no. 800301, V = 93, γ = 5°; 12. V = 90, γ = 20°.

the peculiar structure of two unequal sets of spiral nepionic chambers, situated along the outer wall of deuteroconch, but not outer wall of protoconch, and having lozengic and short hexagonal shaped equatorial chambers and rectangular shaped lateral chambers (Plate 2, figures 8-9). Then *Tania inokoshiensis* represents more primitive arrangement of embryonic chambers and more advanced hexagonal equatorial chambers than each one of *Miogypsina globulina*, and is associated with *Miogypsina globulina*, which carry the character of V value (mean V = 44.31 in 37 specimens) at Inokoshi (Matsumaru and Takahashi, 2004). *Tania inokoshiensis* is similar to *Lepidosemicyclina thecidaeformis* due to having short hexagonal shaped equatorial chambers. But *Tania inokoshiensis* is different from *Lepidosemicyclina thecidaeformis* in having characteristic structure of two sets of spiral nepionic chambers developed along the outer wall of deuteroconch, but not along the outer wall of protoconch. *Tania inokoshiensis* is similar to *Miogypsina primitiva* Tan Sin Hok due to having deuteroconch situated on the frontal side of test and/or situated beside protoconch along the outer wall of protoconch. However *Tania inokoshiensis* is different from *Miogypsona primitiva* in having two sets of nepionic spirals along the outer wall of deuteroconch. The *Miogypsina* Sandstone of the Lower Formation, Bihoku Group is correlated with the Obata and Idozawa Formations, Tomioka Group due to similar mean value of parameter V of *Miogypsina globulina*. Moreover the *Miogypsina* Sandstone of the Lower Formation, Bihoku Group occurs *Miolepidocyclina japonica* Matsumaru, and is at lest correlated with the lower Zone 3 drill cores (ca. 209-146.63 m) of the Kita Daitojima Limestone due to occurrence of *Miogypsina* (= *Lepidosemicyclina*) *polymorpha* (Rutten) with hexagonal equatorial chambers (Hanzawa, 1940) (Figures 6-7). *Miolepidocyclina japonica* is known to occur from the Lower/Middle Miocene Yatsuo Formation, Toyama Prefecture; Shiomizaki Formation, Aomori Prefecture; Saigo Formation, Shizuoka Prefecture; Naeshiroda Formation, Ibaraki Prefecture; Gassanzawa Sandstone, Yamagata Prefecture; Saginosu Formation, Saitama Prefecture; and Nakahara Formation, Chiba Prefecture, respectively. Their locations are shown in Honshu, Japan (Matsumaru and Takahashi, 2000, fig. 1).

7.   Korkuteli Area, Bey Dağlari Autochton, Menderes - Taurus Platform, Turkey
The Oligocene - Miocene succession of the Küçükkoy, Karabayir and Karakuştepe Formations is known to occur in the Bey Dağlari Autochton in the Menderes - Taurus Platform of main

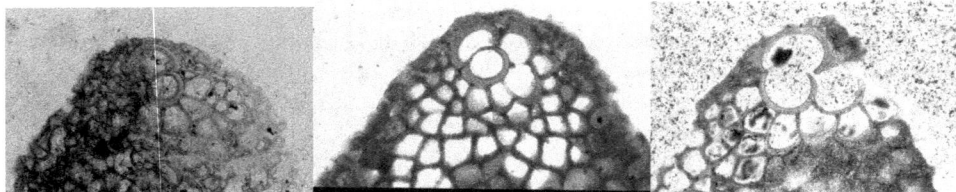

1   2   3

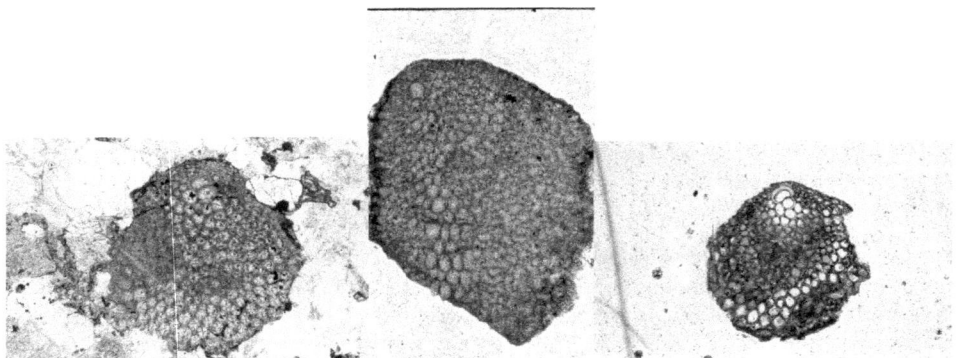

4   5   6

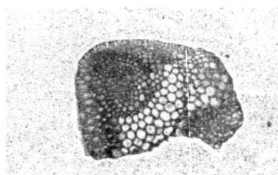

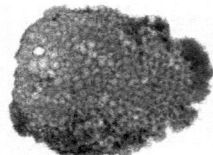

7   8   9

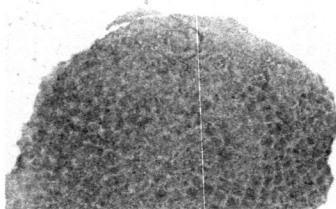

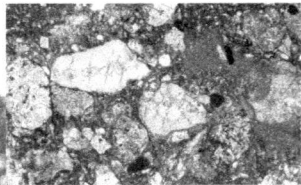

10   11   12

Plate 2.
Figures 1-3. *Miogypsina borneensis* Tan Sin Hok

Equatorial sections. 1. Naeshiroda Formation at Tsukiori-Toge locality, Ibaraki Prefecture, Japan. x = 6, γ = 25°; 2. Nakajima Formation at Dogo locality, Shimane prefecture, Japan. x = 6, γ = – 30°; 3. Hirashio Formation at Tanagura locality, Ibaraki Prefecture, Japan. x = 6, γ = 20°. x 53.

Figure 4. *Lepidosemicyclina musperi* (Tan Sin Hok)
Equatorial section. Obata Formation at Nogami locality, Gunma Prefecture, Japan. x 19.

Figures 5-7. *Lepidosemicyclina thecidaeformis* Rutten
Equatorial sections. 5-6. Megalospheric specimens, 5. Koguchi Formation at Kushimoto, Wakayama Prefecture, Japan, 6. Nakahara Formation at Hota locality, Chiba Prefecture, Japan. 7. Microspheric specimen. Nakahara Formation at Hota locality, Chiba prefecture, Japan. x 19.

Figures 8-9. *Tania inokoshiensis* Matsumaru
Equatorial sections. 8-9. Lower ("Yamaga") Formation, Bihoku Group, at Inokoshi, Okayama Prefecture, Japan. 8. Holotype, Saitama Univ. coll. no. 8803. x 19.

Figures 10-11. *Miolepidocyclina japonica* Matsumaru
Equatorial sections. 10. Gassanzawa Sandstone at Gassanzawa, Yamagata Prefecture, Japan. Holotype, Saitama Univ. coll. no. 720301. 11. Saigo Formation at Shinzaike locality, Shizuoka Prefecture, Japan. 10. x 19, 11. x 53.

Figure 12. *Miogypsinoides dehaartii* (van der Vlerk)
12 left. Axial section. 12 right. Oblique section. Shimizu Formation at Ashizuri Cape, Kochi Prefecture, Japan. x 19.

tectonic units, 40 km NW Antalya City, Turkey (Figure 8). 13 columnar sections from Korkuteli to Karabayir Villages in Korkuteri area, Bey Dağlari Autochton are examined for Miogypsinid biostratigraphy (Matsumaru et al., 2010, figs. 1-3). As results, three larger foraminiferal assemblages are established as the following: *Miogypsinoides formosensis - Miogypsinoides bantamensis - Miogypsinoides dehaartii - Miogypsina primitiva - Spiroclypeus margaritatus* Assemblage (Assemblage 1), *Miogypsinoides bantamensis - Miogypsinoides dehaartii - Miogypsina primitiva - Miogypsina borneensis – Miogypsina globulina – Spiroclypeus margaritatus* Assemblage (Assemblage 2), and *Miogypsinoides dehaartii – Miogypsina borneensis – Miogypsina globulina – Miolepidocyclina burdigalensis* Assemblage (Assemblage 3). The Assemblage 1 is known from the Küçükkoy Formation, and 7 species of *Miogypsinoides formosensis, Miogypsinoides bantamensis, Miogypsinoides dehaartii, Miogypsina primitiva Miogypsina borneensis, Lepidosemicyclina thecidaeformis*, and *Spinosemiogypsina antalyaensis* Matsumaru, Özer and Sari are found in this Assemblage 1 (Figure 7). However two species of *Paleomiogypsina boninensis* Matsumaru and *Miogypsinella complanata* (Schlumberger) in the Assemblage 1 are considered to be reworked. The Assemblage 1 is a younger assemblage than *Miogypsinella boninensis – Spiroclypeus margaritatus – Austrotrillina howchini* Assemblage (Assemblage V) from the upper Minamizaki Limestone, Ogasawara Islands, Japan (Matsumaru, 1996). Because the Ogasawara assemblage (V) has the occurrence of *Miogypsinella boninensis* carrying primitive nepionic spirals and probable planktonic foraminifera belonging to Zone P22 than Zone P21 of Blow (1969) (Matsumaru, 1996) (Figures 6-7). Moreover the Assemblage 1 of Turkey is correlated with Zone 5 drill cores (431.67-ca.360 m) of the Kita Daitojima Limestone (Hanzawa, 1940) due to occurrence of *Miogypsinella borodinensis* (= *Miogypsinoides formosensis*) (Matsumaru et al., 2010). The measurement data of Miogypsinid foraminifera in the Assemblage 1 is as follows: A schizont specimen (A1 form; DI = 88 x 92 micron, DII = 96 x 40 micron)) of *Miogypsinoides formosensis* (Matsumaru et al., 2010, pl. 1, fig. 8) from locality 97-95 in Section 7 has the character of number of nepionic chambers (X = 13) and A-P angle (AP = 210°), while a schizont specimen (DI = 88 x 96 micron, DII = 96 x 48 micron) of *Miogypsinoides formosensis* (Matsumaru et al., 2010, pl. 1, fig. 9) from locality 97-96 in Section 7 has the character of number of nepionic chambers (X = 16) and A-P angle (AP = 250°). A gamont specimen (A2 form; DI = 160 x 160 micron, DII = 128 x 48 micron) of *Miogypsinoides formosensis* (Matsumaru e al., 2010, pl. 1, fig. 10) from locality 96-136 in Section 11 has the character of nepionic chambers (X = 13) and A-P angle (AP = 260°), and also a schizont specimen (DI = 104 x 112 micron, DII = 84 x 44 micron) of *Miogypsinoides bantamensis* (Matsumaru et al., 2010, pl. 2, fig. 3) from locality 97-95 in Section 7, in associated with *Miogypsinoides formosensis*, has the character of number of nepionic chambers (X = 13) and A-P angle (AP = 180°). A gamont specimen (DI = 184 x 168 micron, DII = 192 x 136 micron) of *Miogypsinoides bantamensis* (Matsumaru et al., 2010, pl. 2, fig. 4) from locality 96-137 in Section 11, associated with *Miogypsinoides formosensis*, has the character of number of nepionic chambers (X = 11) and A-P angle (AP = 150°). A gamont specimen (DI = 136 x 120 micron, DII = 128 x 80 micron) of *Miogypsinoides bantamensis* (Matsumaru et al., 2010, pl. 2, fig. 5) from locality 96-137 in Section 11 has the character of number of nepionic chambers (X = 10) and A-P angle (AP = 150°). Moreover on *Miogypsinoides dehaartii*, associated with *Miogypsinoides formosensis* and *Miogypsinoides bantamensis*, a gamont specimen (DI = 152 x 116 micron, DII = 168 x 88 micron; Matsumaru et al., 2010, pl. 2, fig. 7) from locality 97-95 in Section 7 has the character of number of nepionic chambers (X = 9) and A-P angle (AP = 60°), while a gamont specimen (DI = 176 x 168 micron, DII = 208 x 160 micron; Matsumaru et al., 2010, pl. 2, fig. 8) from locality 96-119 in Section5 has the character of nepionic chambers (X = 7) and A-P angle (AP = 50°). A gamont specimen (DI = 224 x 200 micron, DII = 244 x 112 micron; Matsumaru et al., 2010, pl. 3, fig. 6) from locality 97-95 in Section 7 has the character of number of nepionic chambers (X = 10) and A-P angle (AP = 60°).

On *Miogypsina primitiva*, associated with *Miogypsinoides formosensis, Miogypsinoides bantamensis* and *Miogypsinoides dehaartii*, a schizont specimen (DI = 84 x 88 micron, DII = 72 x 56 micron) from locality 97-95 in Section 7 has the character of nepionic chambers (X = 12) and A-P angle (AP = obscure due to twist) (Matsumaru et al., 2010, pl. 3, fig. 4), while a gamont specimen (DI = 200 x 176 micron, DII = 176 x 128 micron) from locality 97-95 in Section 7 has the character of nepionic chambers (X = 10) and A-P angle (AP = 110°) (Matsumaru et al., 2010, pl. 3, fig. 5).

The Assemblage 2 is known from the Karabayir Formation, and 6 species of *Miogypsinoides bantamensis, Miogypsinoides dehaartii, Miogypsina primitiva, Miogypsina borneensis, Miogypsina globulina* and *Lepidosemicyclina thecidaeformis* are found from the Assemblage 2 (Figur 7). The Assemblage 2 is correlated with lower Zone 4 drill cores (ca. 360-302.31 m) of the Kita Daitojima Limestone (Hanzawa, 1940) due to occurrence of *Miogypsinoides bantamensis, Miogypsinoides lateralis*, and *Miogypsinoides dehaartii* var. *pustulosa* (= *M. dehaartii*) (Matsumaru et al., 2010). The measurement data of Miogypsinid foraminifera of the Assemblage 2 is described as follows: a schizont specimen (DI = 112 x 96 micron, DII = 120 x 72 micron) of *Miogypsinoides bantamensis* (Matsumaru et al., 2010, pl. 2, fig. 2) from locality 96-121 in Section 5 has the character of number of nepionic chambers (X = 12) and A-P angle (AP = 180°). On *Miogypsina primitiva*, two specimens are measured: a schizont specimen (DI = 96 x 72 micron, DII = 88 x 40 micron) from locality 97-90 in Section 8 has the character of nepinic chambers (X = 11) and A-P angle (AP = 145°) (Matsumaru, 2010, pl. 3, fig. 2), and a schizont specimen (DI = 96 x 90 micron, DII = 88 x 64 micron) from locality 97-153 in Section 4 has the character of nepionic chambers (X = 9) and A-P angle (AP = 110°) (Matsumaru et al., 2010, pl. 3, fig. 3). On *Miogypsiona borneensis*, three specimens are measured: a gamont specimen (DI = 120 x 112 micron, DII = 128 x 84 micron) from locality 97-90 in Section 8 has the character of nepionic chambers (X = 8) and A-P angle (AP = 50°) (Matsumaru et al., 2010, pl. 3, fig. 9). A gamont specimen (DI = 120 x 116, DII = 140 x 83 micron) from locality 97-152 in Section 4 has the character of nepionic chambers (X = 7) and A-P angle (AP = 25°) (Matsumaru et al., 2010, pl. 3, fig. 10), while a schizont specimen (DI = 96 x 72, DII = 88 x 40 micron) from locality 97-90 in Section 8 has the character of nepionic chambers (X = 7) and A-P angle (AP = 20°) (Matsumaru et al., 2010, pl. 4, fig. 1). Also a schizont specimen (DI = 128 x 104, DII = 144 x 80 micron) of *Miogypsina globulina* from locality 97-90 in Section 8 shows the character of V value (V = 30) and γ value (γ = 10°) (Matsumaru et al., 2010, pl. 4, fig. 7).

The Assemblage 3 is known from the Karakuştepe Formation, and 5 species of *Miogypsinoides dehaartii, Miogypsinopides borneensis, Miogypsina globulina, Miolepidocyclina burdigalensis* and *Lepidosemicyclina thecidaeformis* are found from the Assemblage 3 (Figure 7). The Assemblage 3 is correlated with upper Zone 4 drill cores (302.31 – ca. 209 m) of the Kita Daitojima Limestone (Hanzawa, 1940) due to occurrence of *Miogypsinoides dehaartii* var. *pustulosa* (= *M. dehaartii*) and *Miogypsina borneenisis* (Matsumaru et al., 2010). In the Assemblage 3, the following Miogypsinid foraminifera are measured: On *Miogypsina globulina*, two specimens are measured; a gamont specimen (DI = 200 x 136, DII = 216 x 120 micron) from locality 97-142 in Section 4 has the character of V value (V = 35) and γ value (γ = 40°) (Matsumaru et al., 2010, pl.4, fig. 5), and a gamont specimen (DI = 176 x 152, DII = 248 x 152 micron) from locality 97-125 in Section 4 has the character of V value (V = 25) and γ value (γ = 34°) (Matsumaru et al., 2010, pl. 4, fig. 6).

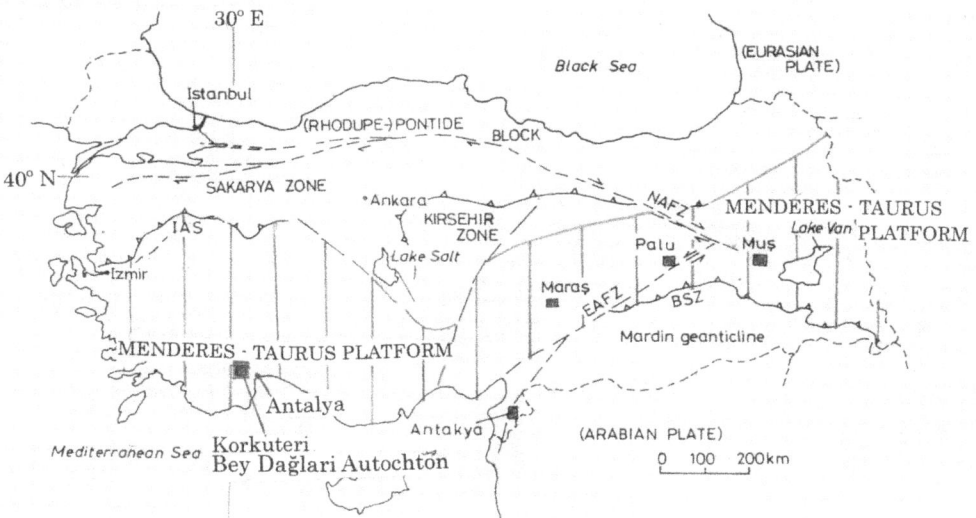

Fig. 8. Locations of research areas (Maraş Palu and Muş) in Turkey treated in this study, in addition to Korkuteri area (Matsumaru et al., 2010). Antakya area without Miogypsinid samples is described in the text.

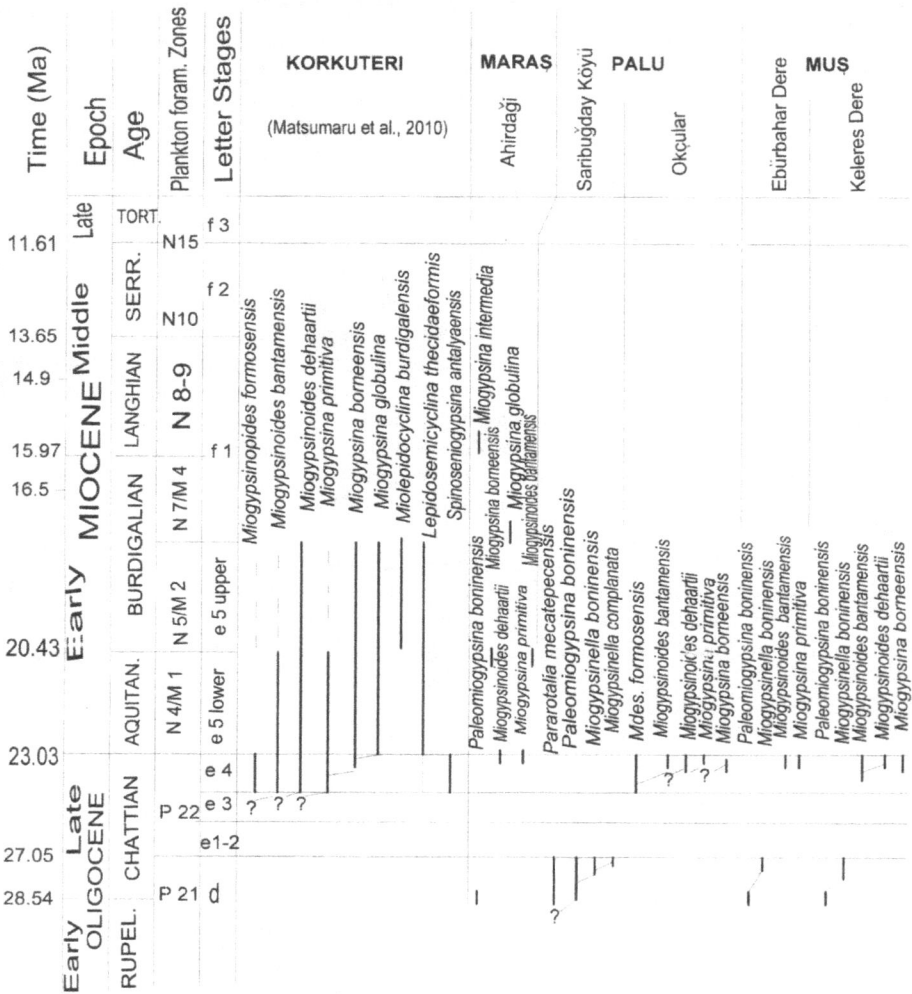

Fig. 9. Biostratigraphic occurrence of Miogypsinid foraminifera from Korkuteri, Maraş, Palu, and Muş areas in the Menderes – Taurus Platform, Turkey.

4. Biostratigraphic occurrence of Miogypsinid foraminifera from Maraş, Palu and Muş areas in the Menderes - Taurus Platform, Turkey

While the author has visited the General Directorate of Mineral Resaerch and Exploration (MTA; T.C. Maden Tetkik Ve Arma Genel Müdürlüğü), Ankara as the Japan Society for the Promotion of Science (JSPS) fellowship researcher for a half year in 1992, he has investigated the foraminifers from the upper Cretaceous to middle Miocene sedimentary rocks in Turkey. This study is to give a note on Miogypsinid foraminiferal Biostratigraphy of Maraş, Palu, and Muş Areas in the eastern Turkey after Uysal et al. (1985), except Antakya Area due to lack of

Miogypsinid foraminifera-bearing samples, where *Nummulites fabianii* (Prever), *N. perforatus* (Montfort), *Pellatispira orbitoidea* (Provale) and others are found from samples A-1 to A-8. Here the author describes useful scientific contribution for Miogypsinid foraminifera (Figures 8-9).

1.   Ahirdaği Section, Maraş Area

According to Uysal et al (1985), there are four columnar sections in Maraş Area. In Ahirdaği section in this study, there are five Miogypsinid horizons. Basal samples 77/59 to 77/57 yield *Paleomiogypsina boninensis* Matsumaru, *Lepidocyclina boetonensis* van der Vlerk, carrying the character (DI = 300 x 188 micron, DII = 390 x 223 micron, and DI = 305 x 200 micron, DII = 335 x 215 micron, and thickness of embryonic wall (T = 12 micron)) in two specimens, *Nephrolepidina marginata* (Michelotti), *Heterostegina borneensis* van der Vlerk, carrying the character (1 or 2? operculine chamber(s) and 19 to 20 nepionic septa), and *Cycloclypeus koolhoveni* Tan Sin Hok, carrying the character (more than 23 and 25 heterostegine septa, and DI = 118 x 118 micron), and *Eulepidina dilatata* (Michelotti). They are regarded as the age of late Early to early Late Oligocene. Also they are assigned to probable Tertiary d of the Letter Satges (Leupold and van der Vlerk, 1931; Matsumaru, 1996), because of the occurrence of *Paleomiogypsina boninensis, Heterostegina borneensis,* and *Eulepidina ephippioides,* which are dominated in the Minamizaki Limestone, Ogasawara Islands, Japan (Matsumaru, 1996) (Figures 6-7). Moreover *Paleomiogypsina boninensis* in the Assemblage 1 was found in samples 97-486, 97-502, and 97-503 from the Küçükkoy Formation in the Korkuteri Area, Bey Dağlari Autochton, but *Paleomiogypsina boninensis* was regarded as the reworked species in those samples as stated before (Matsumaru et al., 2010, pl. 1, figs. 1-4) (Figure 9). Sample 77/60, 10 m below from Sample 77/59, yields *Eulepidina dilatata, Lepidocyclina boetonensis* van der Vlerk, carrying the character (DI = 320 x 220 micron, DII = 325 x 125 micron, and 6 nepionic spirals), *Nephrolepidina marginata,* and *Heterostegina borneensis*, carrying the character (1 operculine chambers, 8 chambers in 1 whorl, and 21 chambers in 2 whorls), although there is no Miogypsinid foraminifera, but it is worth to describe the fauna.

More than 40 m above from sample 77/57, sample E479 in Ahirdaği Section yields *Miogypsinoides dehaartii* (van der Vlerk), carrying number of nepionic chambers (X = 7), *Miogypsina primitiva* Tan Sin Hok, carrying the character (X = 9), *Elphidium* spp., and *Planorbulinella larvata* (Parker and Jones). This horizon is probable assigned to the boundary between the Oligocene and Miocene, due to occurrence of *Miogypsinoides dehaartii* and *Miogypsina primitiva* based on foraminiferal biostratigraphic occurrences between the Küçükkoy and Karabayir Formations, Korkuteri Area (Matsumaru et al., 2010) (Figures 6-7, 9). Above 90 m from sample E479, Sample E477 occurs *Miogypsina borneensis* Tan Sin Hok, carrying the character (X = 6 to 8), and *Miogypsinoides bantamensis* (Tan Sin Hok), carrying the character (X = 14), and *Elphidium* spp. This horizon is correlated with the Karabayir Formation due to occurrence of *Miogypsinoides bantamensis* and *Miogypsina borneensis*, and then the age of the horizon is assigned to the Early Miocene (Aquitanian) (Figure 9).

Sample E475, at about 50 m above from sample E477, yields *Miogypsina globulina* and *Operculina complanata* (Defrance). *Miogypsina globulina* with 11 specimens has been measured, and they are the character of the followings: DI = 108 x 108 micron, DII = 150 x 125 micron, DII/DI = 1.39, V = 40.7, γ = 40°; DI = 200 x 163 micron, DII = 238 x 175 micron, DII/DI = 1.29, V = 37.0, γ = 10°; DI = 165 x 140 micron, DII = 173 x 103 micron, DII/DI = 1.05, V = 43.5, γ = 15°; DI = 148 x 105 micron, DII = 198 x 118 micron, DII/DI = 1.34, V = 29.6, γ = 5°; DI = 165 x 135 micron, DII = 207 x 116 micron, DII/DI = 1.25, V = 38.5, γ = 30°; DI = 140 x 130 micron, DII = 210 x 125 micron, DII/DI = 1.50, V = 44.4, γ = 15°; DI = 125 x 123 micron, DII = 163 x 125 micron, DII/DI = 1.30, V = 40.0, γ = 25°; DI = 168 x 166 micron, DII = 235 x

123 micron, DII/DI = 1.39, V = 37.0, γ = 20°; DI = 175 x 140 micron, DII = 240 x 150 micron, DII/DI = 1.37, V = 34.8, γ = 20°; DI = 118 x 108 micron, DII = 125 x 116 micron, DII/DI = 1.06, V = 33.3, γ = 15°, and DI = 145 x 165 micron, DII = 213 x 116 micron, DII/DI = 1.47, V = 30.4, γ = 25°. *Miogypsina globulina* is generally characterized by the data of mean V of 37.2 and mean γ of 20° (n = 11), and is regarded as the form of the stratigraphic position between *M. globulina* of Tungliang Well TL1, Taiwan (mean V = 23.58, mean γ = 15°, n = 4) and *M. globulina* of Nogami Area, Obata Formation, Tomioka Group, Japan (mean V = 43.93, mean γ = 40°, n = 20)(Figures 6-7). Therefore the age of *Miogypsina globulina* from sample E475 is probably situated in Early Miocene (Burdigalian) (Figure 9).

Sample E471, about 20 m above from sample E475, yields *Miogypsina intermedia* Drooger, which has the character such as DI = 173 x 148, DII = 175 x 75 micron, DII/DI = 1.01, V = 49, and γ = 30°. As such *Miogypsina intermedia* from sample E475 is correlated with *M. intermedia*, associated with *M. globulina* from the Obata and Idozawa Formations, Tomioka Group, and other Japanese Miocene sedimentary rocks, i.e. Yabuzuka Formation in Ota City (Matsumaru and Takahashi, 2000) (Figures 6-7). The age of *M. intermedia* bearing sample E471 horizon is assigned to the Middle Miocene (Langhian) (Figure 9).

2.   Saribuğday Köyü Section, Palu Area

Sample M230 in Saribuğday Köyü Section, Palu Area (Uysal et al., 1985) yields *Pararotalia mecatepecensis* (Nuttall), *Paleomiogypsina boninensis* Matsumaru, *Nummulites fichteli* Michelotti, *Lepidocyclina isolepidinoides* van der Vlerk, *Eulepidina dilatata* (Michelotti), *Borelis pygmaeus* (Hanzawa), and *Austrotrillina* spp. As such this fauna is correlated with the fauna of samples 77/59 to 77/57 in Ahirdaği Section, Maraş Area, as stated above, due to occurrence of *Paleomiogypsina boninensis* (Figure 9). As the author has described the evolution from *Pararotalia mecatepecensis* (Nuttall) to *Paleomiogypsina boninensis* Matsumaru, both species could be found in sample M230 (Matsumaru, 1996, p. 56, fig. 24) (Figure 4). The basal Sample M224, about 740 m below from Sample M230, yields *Lepidocyclina isolepidinoides* van der Vlerk, carrying the character (DI = 163 x 110 micron, DII = 175 x 105 micron, and 6 nepionic spirals), *Eulepidina dilatata, Nummulites fichteli, N. vascus* Joly and Leymerie, and *Borelis pygmaeus* (Hanzawa), and is regarded as the basal Tertiary d of the Letter Stages, although there is no miogypsinid foraminifera. However it is worth to describe the basal Oligocene in this area.

Sample M232, about 40 m above from sample M230, yields *Miogypsinella boninensis* carrying the character (X = 23, DI = 110 x 90 micron, and DII = 90 x 60 micron), *Nummulites fichteli, Nephrolepidina marginata, Eulepidina dilatata, Borelis pygmaeus, Heterostegina* spp., *Halkyardia minima* (Liebus), and *Operculina* spp. Top sample M233, about 130 m above from sample M232, yields *Miogypsinella boninensis,* carrying the character (X = 26, DI = 110 x 95 micron, and DII = 85 x 40 micron), *Miogypsinella complanata* (Schlumberger), carrying the character (X = more than 18, DI = 110 x 103 micron, and DII = 88 x 30? micron), *Paleomiogypsina boninensis, Pararotalia mecatepecensis,* and rarely *Nummulites fichteli.* Therefore these fauna from three samples (M230, M232, and M233) are correlated with the Assemblage IV from the Minamizaki Limestone due to occurrence of *Paleomiogypsina boninensis, Miogypsinella boninensis, Eulepidina dilatata, Nephrolepidina marginata, Borelis pygmaeus,* and *Halkyardia minima* (Matsumaru, 1996). Then these fauna from Saribuğday Köyü is assigned to Tertiary d stage of the Letter Stages. *Nummulites fichteli* is known in the fauna from Saribuğday Köyü, but isn't known in the Assemblage IV from Ogasawara Islands. Also *Miogypsinella complanata* carrying the number of nepionic chambers (X = more than 18), is known to occur from the fauna of Saribuğday Köyü, but this species isn't known in association with *Nummulites fichteli* in the Tethys region as far as the author knows. Then *Nummulites fichteli* in sample M233 is considered to be reworked, and the fauna of sample M233 may be partly correlated with the Assemblage V from the uppermost Minamizaki Limestone, Ogasawara Islands, Japan (Matsumaru, 1996).

3.   Okçülar Section, Palu Area

Sample 77/27B below the basalt layer in the Okçülar Section, Palu Area (Uysal et al., 1985) yield *Miogypsinoides formosensis* (Yabe and Hanzawa), carrying the character (X = 15, DI = 116 x 110 micron, DII = 95 x 63 micron, and AP = 210°), *Eulepidina dilatata*, *Nephrolepidina marginata* (Michelotti), *Cycloclypeus* spp. and *Operculina complanata* (Defrance). The fauna of sample 77/27B is correlated with the Assembalge 1 of the Küçükkoy Formation in the Korkuteri Area due to occurrence of *Miogypsinoides formosensis* (Matsumaru et al., 2010). In sample 77/27B, *Miogypsinella complanata*, carrying the character (X = more than 18, DI = 100 x 88 micron, and DII = 75 x 50 micron) is, however, found in association with *Miogypsinoides formosensis*, but *Miogypsinella complanata* is considered to be reworked.

Sample 77/26 above the same basalt layer yield *Miogypsinoides formosensis*, *Spiroclypeus* spp., *Heterostegina* spp., and *Operculina complanata*, in addition to *Nummulites vascus* Joly and Leymerie, which is characterized by having the character (DI = 100 x 98 to 263 x 193 micron, DII = 83 x 40 to 208 x 108 micron, distance and number of chambers in 1/2 whorl = 360 to 400 micron and 3, those in 1 whorl =825 to 875 micron and 8, those of 1 1/2 whorl = 1125 to 1200 and 13 to 17, and those in 2 whorl = 1405 to 1475 micron and 20 to 24), *Cycloclypeus koolhoveni* Tan Sin Hok, carrying the character (DI = 125 x 125 to 120 x 118 micron, DII = 168 x 73 to 175 x 58 micron, number of opeculine chamber = 3, and number of nepionic septa = more than 16), and *Miogypsinella complanata*, carrying the character (DI = 88 x 78 micron, DII = 72 x 38 micron, and X = 19), respectively. The latter three species of *Nummulites vascus, Cycloclypeus koolhoveni*, and *Miogypsinella complanata* are considered to be reworked due to non coexistence.

Sample 77/20, about 50m above from Sample 77/26, yield *Miogypsinoides dehaartii*, carrying the character (X = 6, DI = 145 micron, DII/DI = 1.14; X = 6, DI = 153 x 125 micron, DII = 150 x 123 micron, DII/DI = 0.98; and X = 7, DI = 175 micron, DII/DI = 0.74) in three specimens, *Miogypsina borneensis*, carrying the character (X = 6, DI = 145 x 125 micron, γ = 40°; and X = 7, DI = 175 x 175 micron, DII = 130 x 63 micron, γ = 40°) in two specimens, *Eulepidina dilatata*, *Nephrolepidina marginata*, and *Operculina complanata*. Sample 77/19, 20 m above from Sample 77/20 yields *Miogypsinoides dehaartii*, *Operculina complanata*, and *Cycloclypeus* spp., in addition to *Paleomiogypsina boninensis*, and *Miogypsinella ubaghsi* (Tan Sin Hok), carrying the character (X = 24, DI = 63 x 58 micron, DII = 50 x 43 micron, AP = 425°, and diameter of spiral chambers = 650 x 763 micron). The latter two species are considered to be reworked, but the discovery of *Miogypsinella ubaghsi* is important to consider the evolutional lineage from *Miogypsinella boninensis* to *Miogypsinella ubaghsi* based on Tan Sin Hok's nepionic acceleration (Figure 4). This lineage has been considered as the evolution from *Miogypsinella boninensis* in the Assemblage V of the Minamizaki Limestone, Ogasawara Islands to *Miogypsinella ubaghsi* in the dredge limestones of the Komahashi-Daini Seamount, Kyushu – Palau Ridge as stated before (Figure 6-7).

Sample 77/18, about 20 m above from Sample 77/19 yields *Miogypsinoides formosensis*, carrying the character (X = 16, DI = 65 x 45 micron, DII = 112 x 80 micron, and AP = 270°), *Miogypsinoides bantamensis* (Axial section), *Miogypsina primitiva*, carrying the character (X = 11, DI = 125 x 116 micron, DII = 138 x 78 micron, and AP = 160°), *Heterostegina* spp., carrying the character (DI = 230 x 200 micron, and number of operculine chambers =3), and *Planorbulinella larvata* (Parker and Jones). Moreover Sample 77/16, about 40 m above from Sample 77/18, yields probable *Miogypsinoides formosensis*, *Heterostegina* spp., *Cycloclypeus* spp, and *Operculina complanata*, and is overlain by the basalt. In this section, all samples treated in the study belong to upper Oligocene and can be correlated with the Küçükkoy Formation in Korkuteri Area (Figure 9).

4.   Ebürbahar Dere Section, Muş Area

Sample M27 in the Ebürbahar Dere, Muş Area (Uysal et al., 1985) yields *Paleomiogypsina boninensis*, carrying the character (DI = 125 x 125 micron, DII = 130 x 83 micron, 17 to 20 spiral chambers in 2 whorls, and diameter of nepionic spirals = 900 to 1050 micron), *Nephrolepidina marginata* (Michelotti), *Borelis pygmaeus*, *Operculina* spp., and *Peneroplis* spp. This sample is a horizon about 800 m above from the boundary between the Eocene and Oligocene sedimentary rocks, and is correlated with Samples 77/59 to 77/ 57 in the Ahirdaği Section, Maraş Area, and Sample M230 in the Saribuğday Köyü Section, Palu Area, due to occurrence of *Paleomiogypsina boninensis* (Figure 9). Sample M5, about 722 m above from Sample M27 yields *Miogypsinella boninensis*, carrying the character (DI = 70 x 50 micron, and diameter of nepionic spirals =745 micron), *Eulepidina dilatata*, *Nephrolepidina* spp., and *Spiroclypeus* spp. This horizon is correlated with Samples M 232 and M233 in Saribuğday Köyü Section, Palu Area, due to occurrence of *Miogypsinella boninensis* (Figure 9). Moreover Sample M1, about 450 m above from Sample M5 yields *Miogypsinoides bantamensis*, carrying the character (dimension of protoconch (diam. x height.) of 223 x 208 micron, and dimension of deuteroconch (diam. x height.) of 175 x 118 micron in axial section, and form ratio of diameter/thickness (F. R. = 1.5 mm/ 0.58 mm = 2.61)), *Miogypsina primitiva*, carrying the character (X = more than 10, DI = 183 x 175 micron, and DII = 173 x 123 micron), *Eulepidina dilatata*, *Heterostegina* spp., carrying the character (DI = 318 x 283 micron, DII = 365 x 188 micron, and 7 nepionic chambers in 1 whorl), and *Spiroclypeus* spp. This fauna from Sample M1 is correlated with the fauna of Sample 77/18 in the Okçülar Section, Palu Area, due to occurrence of *Miogypsinoides bantamensis* and *Miogypsina primitiva* (Figure 9).

5.   Keleres Dere Section, Muş Area

Sample O155 in the Keleres Dere Section, Mus Area (Uysal et al., 1985) yields *Paleomiogypsina boninensis*, carrying the character (X = more than 20, DI = 110 x 108 micron, and DII = 112 x 58 micron), *Heterostegina* spp., *Borelis pygmaeus*, *Peneroplis* spp. and *Austrotrillina* spp. This fauna is correlated with the fauna of Sample M27 in the Ebürbahar Dere Section, Muş Area; Sample M230 in the Saribuğday Köyü Section, Palu Area; and Samples 77/59 to 77/57 in the Ahirdaği Section in Maraş Area, due to occurrence of *Paleomiogypsina boninensis*, respectively. Sample O148, placed about 310 m thick above from Sample O155 yields *Miogypsinella boninensis*, *Heterostegina* spp. and *Planorbulinella larvata*. Sample 142, placed more than 1000m above from Sample O148, yields *Miogypsenella boninensis*, carrying the character (X = more than 23, DI = 110 x 105 micron, DII = 112 x 73 micron, and diameter of nepionic spirals = 865 micron), and *Operculina complanata*, carrying the character (DI = 250 x 208 micron, DII = 238 x 135 micron, and distance and number of chambers in 1/2 whorl = 700 micron and 3, those in 1 whorl = 1125 micron and 8, those in 1 1/2 whorl = 1400 micron and 18, and those in 2 whorl = 3700 micron and 28). The fauna from Samples O148 to M142 is correlated with the fauna of Samples M232 to M233 in the Saribuğday Köyü Section, Palu Area, due to occurrence of *Miogypsinella boninensis*, respectively.

Sample M139, placed about 650 m above from Sample M142, yields *Miogypsinoides bantamensis*, carrying the character (X = 12, DI = 118 x 120 micron, DII = 100 x 60 micron, and diameter of nepionic spirals = 600 micron), *Spiroclypeus* spp., *Eulepidina dilatata*, and *Lepidocyclina boetonensis* van der Vlerk. Sample M133, placed about 300 m above from Sample M139, yields *Miogypsina borneensis*, carrying the character (X = 7, DI = 171 x 170 micron, DII = 190 x 100 micron, γ = 30°, and diameter of nepionic spirals = 625 micron),

Heterostegina spp., and *Operculina complanata*. Sample M131, placed obscure rightly, but about 60 m thick above from Sample M133, yields *Miogypsinoides bantamensis*, carrying the character (X = 11, DI = 113 x 105 micron, DII = 113 x 75 micron, γ = 20°, and diameter of nepionic spirals = 525 micron, and AP = 170°), *Miogypsinoides dehaartii*, carrying the character (X = 6, DI = 158 x 113 micron, DII = 158 x 105 micron, and γ = 20°), and *Miogypsina borneensis*, carrying the character (X = 7, DI = 135 x 120 micron, DII = 150 x 85 micron, and γ = 20°). As such the fauna of Samples M139 to M131 is correlated with the fauna of Sample M1 in the Ebürbahar Dere Section, Muş Area; and Samples 77/20 to 77/18 in the Okçülar Section, Palu Area, due to occurrence of *Miogypsinoides bantamensis, Miogypsinoides dehaartii, Miogypsina primitiva*, and/or *Miogypsina borneensis*, respectively. Moreover these fauna are correlated with the fauna of the Assemblage 1 in the Küçükkoy Formation in Korkuteri Area (Figure 9).

5. Conclusion

The Miogypsinid foraminifera (Order Foraminiferida) in the Tethys Region are known to occur from the Early Oligocene (Rupelian) to Middle Miocene (Serravallian) age. Characteristic faunal assemblages from the Miogypsinid foraminiferal Biostratigraphy in Japan, Taiwan and Turkey have been known and correlated each other, respectively (Figures 6-7, 9). Judging from the correlation and analysis of faunal assemblages, the following evolution is established: *Paleomiogypsina boninensis* was proved to be a diagnostic species for the basal assemblage of the Early Oligocene (Rupelian), and *Paleomiogypsina boninensis* evolved from *Pararotalia mecatepecensis* due to having co-existence, and trochoid nepionic spirals in the Minamizaki Limestone, Ogasawara Islands, Japan (Matsumaru, 1996) (Figures 6-7). *Miogypsinella boninensis* evolved from *Paleomiogypsina boninensis*, and evolved into *Miogypsinella ubaghsi* during Late Oligocene (Chattian), based on the biostratigraphic relationship between the Minamizaki Limestone and limestones of Komahashi-Daini Seamount, Kyushu – Palau Ridge, Japan (Figures 6-7). *Miogypsinella ubaghsi* may evolve into *Miogypsinella complanata* due to nepionic acceleration, but there is no discovery on direct evidences in the field. However there is evidence of the evolution from *Miogypsinella ubaghsi* to *Miogypsinella borodinensis* (= *Miogypsinoides formosensis*) during the Late Oligocene (Chattian), based on the biostratigraphic relationship between limestones of Komahashi-Daini Seamount and basal Zone 5 drill cores of the Kita-Daitojima Limestone, Kita-Daito-Jima, Okinawa Prefecture, Japan (Figures 6-7). However *Miogypsinella complanata* is missing in both limestones as stated above, but probably has been existed as co-existence. *Miogypsinella complanata* and *Miogypsinoides formosensis* are found together, but both species are associated with *Paleomiogypsina boninensis* and *Miogypsinoides bantamensis* in Sample 97-486 in Section 4 and Sample 97-502 in Section 6 in the Küçükkoy Formation, Bey Dağlari Autochton, Menderes–Taurus Platform, Turkey (Matsumaru et al., 2010). Then *Paleomiogypsina boninensis* and *Miogypsinella complanata* are regarded as the reworking. During Late Oligocene (Chattian), *Miogypsinoides formosensis* evolved into *Miogypsinoides dehaartii* through *Miogypsinoides bantamensis* due to the nepionic acceleration, and *Miogypsinoides dehaartii* evolved into *Miogypsina primitiva* due to having the lateral chambers during the depositional age of the Küçükkoy Formation (Figures 6-7, 9). Moreover, *Miogypsina primitiva* evolved into *Miogypsina borneensis* due to the nepionic acceleration in the Küçükkoy Formation during Late Oligocene (Chattian) (Figures 6-7, 9). In the Early Miocene (Aquitanian), *Miogypsina borneensis* from the Küçükkkyoy Formation evolved into *Miogypsina globulina* from the Karabayir Formation, Bey Dağlari Autochton (Figures 6-7, 9). Further *Miogypsina globulina* evolved into *Miogypsina intermedia* due to occurrence and nepionic acceleration (Drooger, 1952, 1963). *Miogypsina intermedia* evolved into *Miogypsina cushmani* due to the nepionic acceleration during Early Miocene (Burdigalian)/ Middle Miocene (Langhian) age, and these are shown in Indian and Japanese *Miogypsina* (Raju, 1974; Matsumaru, 1967, 1977; Matsumaru and Takahashi, 2004). Also their evolution is shown in the biostratigraphical correlation between the Karakuştepe Formation carrying *Miogypsina globulina* in Korkuteri Area, Bey Dağlari Autochton, Menders - Taurus Platform, and Sample E471 beds carrying *Miogypsina intermedia* in Ahirdaği Section in Maraş Area, Menderes - Taurus Platform, Turkey (Figure 9). *Miogypsina nipponica* (= *M. antillea* and *M. cushmani* steps of nepionic acceleration) was found from the Kamiyokoze Formation of the Middle Miocene (Serravallian) age, and this species evolved from *Miogypsinid cushmani* of nepionic acceleration by Indian and Japanese Miogupsinid researches (Raju, 1974; Matsumaru, 1980; Matsumaru and Takahashi, 2004). The ancestor of *Miolepidocyclina burdigalensis* (Gümbel), *Lepidosemicyclina thecidaeformis* (Rutten), *Tania inokoshiensis* Matsumaru, *Boninella boninensis* Matsumaru and *Spinosemiogypsina antalyaensis* Matsumaru, Özer and Sari isn't known, although some are considered, and it will be solved from further Miogypsinid foraminiferal Biostratigraphy. Some new genera by the author's research have been known from Miogypsinid foraminifera from the Philippines Archipelago, eastern Tethys region, and they will contact the unknown lineage soon.

6. Acknowledgments

The author thanks the Japan Society for the Promotion of Science (JSPS) for his fellowship researcher in 1992; and the General Directorate of Mineral Research and Exploration (MTA), Ankara, Turkey and their colleagues (Drs. and Messrs. E. Yazgan, E. Sirel, S. Acar, G. Tunay, S. Örcen, and K. Erdoğan) for their kind facility and comment. The author thanks Mr. M. Matsuo, Emeritus Professor of Saitama University, for his kind facility, and Mr. V. Grebro, and Ms. D. Duric, for their kind managing.

7. References

Berggren, W. A., Kent, D. V., Swisher III, C. C., and Aubry, M. P., 1995: A revised Cenozoic Geochronology and Chronostratigraphy. *In*, Berggren, W. A., Kent, D. V., Aubry, M. P. and Hardenbol, J. eds., *Geochronology, Time Scale and Global Stratigraphic Correlation*. SEPM Special Publication, Tulsa, 54: 129-212.

Blainville, H. M. Ducrotay de, 1827: *Manuel de malacologie et de conchyliologie* (1825). Paris: F. G. Levrault.

Blow, W. H., 1969: Late Middle Eocene to Recent planktonic foraminiferal biostratigraphy. *In*, Brönnimann, P. and Renz, H. H., eds, *Proceedings of the First International Conference on Planktonic Microfossils*. 1. Geneva: E. Brill, 199-422.

Cole, W. S., 1954: Larger foraminifera and smaller diagnostic foraminifera from Bikini Drill Holes. *U. S. Geological Survey Professional Paper* 260-0: 569-608.

_____, 1957: Larger foraminifera from Eniwetok Atoll. *U. S. Geological Survey Professional Paper* 260-V: 743-784.

Cushman, J. A., 1918: The larger fossil Foraminifera of the Panama Canal Zone. *Bulletin of the U. S. National Museum*, 103: 89-102.

Drooger, C. W., 1952: *Study of American Miogypsinidae*. Utrecht University, Thesis, 80 pp.

_____, 1963: Evolutionary trends in the Miogypsinidae. *In*, Von Koenigswald, G.H. R., Emeis, J. D., Buning, W. L. and Wagner, C. W. eds., *Evolutionary Trends in Foraminifera*. Elsevier, Amsterdam, 315-349.

_____, 1993: *Radial Foraminifera, Morphometrics and Evolution*. Verhandelingen der Koninklijke Nederlandse Akademie van Wetenschappen, Afd. Natuurkunde, Eerste Reeks 41:242 pp.

Ehrenberg, C. G., 1839: Uber die Bildung dr Kreidefelsen und des Kreidemergels durch unsichtbare Organismen. *Physikalische Abhandlungen der Koniglichen Akademie der Wissenschaften zu Berlin*, 1838: 59-147.

Frost, S. H., and Langenheim, R. L., 1974: Tertiary larger foraminifera and scleractinian corals from Mexico. *In, Cenozoic reef biofacies*. 388 pp. Northern Ilinois University press, Dekalb.

Hanzawa, S., 1940: Micropaleontological studies of drill cores from a deep well in Kita-Daito-Zima (North Borodino Island). *Jubilee Publication of Professor H. Yabe's 60th birthday*, 755-802.

_____, 1957: Cenozoic foraminifera of Micronesia. *Geological Society of America, Memoir*, 66: 163 pp.

_____, 1964: The phylomorphogeneses of the Tertiary foraminiferal families, Lepidocyclinidae and Miogypsinidae. *Science Reports of the Tohoku University, second ser. (Geology)*, 35: 295-313.

Kaneoka, I., Issiki, N., and Zashu, S., 1970: K-Ar ages of the Izu-Bonin Islands. *Geochemical Journal*, 4: 53-60.

Le Calvez, Y., 1949: Révision des foraminifères Lutetiens du Bassin de Paris. II. Rotaliidae et familles affines. *Mémoires du Service de la Carte Géologique Détaillée de la France*, 1-54.

Leupold, W., and Vlerk. I. M. van der, 1931: The Tertiary. *Leidsche Geolofische Mededeelingen, Leiden*, 5: 611-648.

Matsumaru, K., 1967: Geology of the Tomioka area, Gunma Prefecture with a note on "*Lepidocyclina*" from the Abuta Limestone Member. *Science Reports of the Tohoku University, second ser. (Geology)*, 39: 113-147.

_____, 1968: Miogypsinid population from the Tungliang Well TL-1 of the Penghu Islands, China. *Transactions and Proceedings of the Paleontological Society of Japan, New Ser.* 72: 340-344.

_____, 1971: Studies on the genus *Nephrolepidina* in Japan. *Science Reports of the Tohoku University, second ser. (Geology)*, 42: 97-185.

_____, 1972: The genus *Miolepidocyclina* from Japan. *In, Prof. Jun-Ichi Iwai Memorial Volume*, 679-681. Sasaki Pub. Co., Sendai.

_____, 1977: Neogene stratigraphy of the northern to northeastern marginal areas of the Kanto Mountainland, Central Japan. *Journal of the Geological Society of Japan*, 83: 213-225. (in Japanese with English abstract)

_____, 1980: Note on a new species of *Miogypsina* from Japan. *In, Professor Saburo Kanno Memorial Volume*, 213-219. Sasaki Pub. Co., Sendai.

_____, 1982. On *Miogypsina* (*Miogypsina*) *kotoi* Hanzawa from Zone N. 16 on Dogo Island, Oki Islands,Japan. *Proceedings of the Japan Academy*, 58, ser. B: 52-55.

_____, 1990: A new genus of the Miogypsinid foraminifera from Southwest Japan. *Transactions and Proceedings of the Paleontological Society of Japan, New Ser.* 158: 535-539.

_____, 1996: Tertiary larger foraminifera (Foraminiferida) from the Ogasawara Islands, Japan. *Paleontological Society of Japan, Special Papers*, 36: 239 pp.

_____, Myint Thein, and Ogawa, Y., 1993: Early Miocene (Aquitanian) larger foraminifera from the Shimizu Formation, Ashizuri Cape, Kochi Prefecture, Shikoku, Japan. *Transactions and Proceedings of the Paleontological Society of Japan, New Ser.* 169: 1-14.

_____, and Takahashi, M., 2004: Studies on the genus *Miogypsina* (Foraminiferida) in Japan. *Journal of Saitama University, Faculty of Education (Mathmatics and Natural Sciences)*, 53: 17-39.

_____, Sari, B., and Özer, S., 2010: Larger foraminiferal biostratigraphy of the middle Tertiary of Bey Dağlari Autochton, Menderes-Taurus Platform, Turkey. *Micropaleontology*, 56: 439-463.

Mishra, P. K., 1996: Study of Miogypsinidae and associated planktonics from Cauvery, Krishna-Godavari and Andaman Basins of India. *Geoscience Journal*, 17: 123-251.

Mohiuddin, M. M., Ogawa, Y., and Matsumaru, K., 2000: Late Oligocene larger foraminifera from the Komahashi-Daini Seamount, Kyushu-Palau Ridge and their tectonic significance. *Paleontological Research*, 4: 191-204.

Nomura, M., and Ohira, Y., 1998: Fission track age of the Miocene tuff in the Tomioka area, Gunma Prefecture. *Journal of Gunma Museum of Natural History*, 2: 35-42. (in Japanese)

_____, and _____, 2002: Fission track age of the Miocene Haratajino and Shiohatado tuffs in the Tomioka Group, Gunma Prefecture, Central Japan. *Journal of Gunma Museum of Natural History*, 6: 75-80. (in Japanese)

_____, Ishikawa, H., Kaneko, M., and Matsumaru, K., 2003; The discovery of *Miogypsina* from the Miocene Series of Kanayama Hill, Ota City, Gunma Prefecture, Central Japan. *Journal of the Geological Society of Japan*, 109: 611-614. (in Japanese)

Ohde, S., and Elderfield, H., 1992: Strontium isotope stratigraphy of Kita-Daito-Jima Atoll, North Philippine Sea: implications for Neogene sea-level change and tectonic history. *Earth and Planetary Science Letters*, 113: 473-486.

Raju, D. S. N., 1974: Study of Indian Miogypsinidae. *Utrecht Micropaleontological Bulletins*, 9: 148 pp.

Rutten, L. M. R., 1911: Over Orbitoiden uit de omgeving or Balik Papan-baai (Oostkust van Borneo). *Verhandlingen van het Geologisch Mijnbouwkundig Genootschap voor Nederlandsch Kolonien, Amsterdam, 25 Febr., 1911*. 1/1122-17/1139.

Sacco, F., 1893: Sur quelques Tinoporinae du Miocène de Turin. *Bulletin de la Sociéte Belge de Géologie, de Paléontologie, et d'Hydrologie* (1893-1894) 7: 204-207.

Tan Sin Hok, 1932: Over *Cycloclypeus* voorloopige resultaten einer biostratigrafische studies. *Mijngenieur Bandoeng*, 11: 233-242.

_____, 1936: Zur Kenntniss der Miogypsiniden. *Ingenieur in nederlandsch-Indie, IV, Mijnbouw en Geologie*, 3 (3): 45-61; 3 (5); 84-98; 3 (7): 109-123.

_____, 1937: Weitere Untersuchungen uber die Miogypsiniden I,und II. *Ingenieur in nederlandsch-Indie, Mijnbouw en Geologie*, 4 (3): 35-45; 4 (6): 87-113.

Tsunakawa, H., 1983: K-Ar dating on volcanic rocks in the Bonin Islands and its tectonic implication. *Tectonophysics*, 95: 221-232.

Ujiie, H., 1973: Distribution of the Japanese *Miogypsina*, with description of a new species. Restudy of the Japanese Miogypsinids, Part 3. *Bulletin of the National Science Museum, Tokyo*, 16: 99-114.

Uysal, S., Sirel, E., and Gündüz, H., 1985: Guneydogu anadolu boyunca (Muş-Palu-Maraş-Hatay) bazi Tersiyer Kesitleri. *Maden Tetkik Ve Arama Genel Müdürlüğü (Jeoloji Etudleri Dairesi) Jeol. Elid. Ar.*, No. 236 (in Turkish)

Vaughan, T. W., 1928: Subfamily Miogypsininae Vaughan. *In*, Cushman, J. A., 1928, *Foraminifera their classification and Economic Use, Special Publications Cushman Laboratory for Foraminiferal Research*, 1: 401 pp.

Yabe, H., and Hanzawa, S., 1928: Tertiary foraminiferous of Taiwan. *Proceedings of Imperial Academy of Tokyo*, 4: 222-225.

Permissions

The contributors of this book come from diverse backgrounds, making this book a truly international effort. This book will bring forth new frontiers with its revolutionizing research information and detailed analysis of the nascent developments around the world.

We would like to thank Imran Ahmad Dar, for lending his expertise to make the book truly unique. He has played a crucial role in the development of this book. Without his invaluable contribution this book wouldn't have been possible. He has made vital efforts to compile up to date information on the varied aspects of this subject to make this book a valuable addition to the collection of many professionals and students.

This book was conceptualized with the vision of imparting up-to-date information and advanced data in this field. To ensure the same, a matchless editorial board was set up. Every individual on the board went through rigorous rounds of assessment to prove their worth. After which they invested a large part of their time researching and compiling the most relevant data for our readers. Conferences and sessions were held from time to time between the editorial board and the contributing authors to present the data in the most comprehensible form. The editorial team has worked tirelessly to provide valuable and valid information to help people across the globe.

Every chapter published in this book has been scrutinized by our experts. Their significance has been extensively debated. The topics covered herein carry significant findings which will fuel the growth of the discipline. They may even be implemented as practical applications or may be referred to as a beginning point for another development. Chapters in this book were first published by InTech; hereby published with permission under the Creative Commons Attribution License or equivalent.

The editorial board has been involved in producing this book since its inception. They have spent rigorous hours researching and exploring the diverse topics which have resulted in the successful publishing of this book. They have passed on their knowledge of decades through this book. To expedite this challenging task, the publisher supported the team at every step. A small team of assistant editors was also appointed to further simplify the editing procedure and attain best results for the readers.

Our editorial team has been hand-picked from every corner of the world. Their multi-ethnicity adds dynamic inputs to the discussions which result in innovative outcomes. These outcomes are then further discussed with the researchers and contributors who give their valuable feedback and opinion regarding the same. The feedback is then collaborated with the researches and they are edited in a comprehensive manner to aid the understanding of the subject.

Apart from the editorial board, the designing team has also invested a significant amount of their time in understanding the subject and creating the most relevant covers. They scrutinized every image to scout for the most suitable representation of the subject and create an appropriate cover for the book.

The publishing team has been involved in this book since its early stages. They were actively engaged in every process, be it collecting the data, connecting with the contributors or procuring relevant information. The team has been an ardent support to the editorial, designing and production team. Their endless efforts to recruit the best for this project, has resulted in the accomplishment of this book. They are a veteran in the field of academics and their pool of knowledge is as vast as their experience in printing. Their expertise and guidance has proved useful at every step. Their uncompromising quality standards have made this book an exceptional effort. Their encouragement from time to time has been an inspiration for everyone.

The publisher and the editorial board hope that this book will prove to be a valuable piece of knowledge for researchers, students, practitioners and scholars across the globe.

List of Contributors

Kaan Erarslan
Dumlupinar University, Mining Engineering Department, Kutahya, Turkey

Bambang P. Istadi and Nurrochmat Sawolo
Energi Mega Persada, Indonesia

Handoko T. Wibowo
Independent geologist, Indonesia

Edy Sunardi
Universitas Padjajaran, Indonesia

Soffian Hadi
Sidoarjo Mudflow Mitigation Agency, Indonesia

Zhang Yuangao, Chen Shumin, Feng Zhiqiang, Jiang Chuanjin Zhang Erhua, Xin Zhaokun and Dai Shili
Daqing Oilfield Company Ltd., Heilongjiang Daqing, China

J. R. Harris, E. Schetselaar and P. Behnia
Geological Survey of Canada, Ottawa, Canada

Tim Webster
Applied Geomatics Research Group, Nova Scotia Community College, Middleton, Canada

Gil Oudijk
Triassic Technology, Inc., USA

Imran Ahmad Dar, K. Sankar and Mithas Ahmad Dar
Department of Industries and Earth Sciences, Tamil University- Thanjavur, India

Dimitris Alexakis
Centre for the Assessment of Natural Hazards and Proactive Planning, Laboratory of Reclamation Works and Water Resources Management, National Technical University of Athens, Athens, Greece

Ben Fadhel Moez, Zouaghi Taher, Amri Ahlem and Ben Youssef Mohamed
CERTE, Technopole de Borj Cédria, Tunisia

Soua Mohamed
Entreprise Tunisienne d'Activités Pétrolières, ETAP-CRDP 4 Rue des Entrepreneurs, 2035 la Charguia II, Tunisia

Layeb Mohsen
ISMP, Tunis, Tunisia

María Concepción García-Aguirre
Centro de Ciencias de la Complejidad(C3), Departamento de Ecología y Recursos, Naturales, Facultad de Ciencias, Universidad Nacional Autónoma de México (UNAM), Ciudad Universitaria, C.P. Coyoacán, D.F., Mexico

Román Álvarez
Instituto de Matemáticas Aplicadas y Sistemas, Universidad Nacional Autónoma de México (UNAM), Mexico

Fernando Aceves
Instituto de Geografía, Universidad Nacional Autónoma de México (UNAM), México

Kuniteru Matsumaru
Innovation Research Organization, Saitama University, Saitama, Japan

www.ingramcontent.com/pod-product-compliance
Lightning Source LLC
Chambersburg PA
CBHW070733190326
41458CB00004B/1144